Modern Research Topics in Aerospace Propulsion

G. Angelino L. De Luca W.A. Sirignano
Editors

Modern Research Topics in Aerospace Propulsion

In Honor of Corrado Casci

With 201 Illustrations

Springer Science+Business Media, LLC

G. Angelino
Politecnico di Milano
Dipartimento di Energetica
Italy

L. De Luca
Politecnico di Milano
Dipartimento di Energetica
Italy

W.A. Sirignano
Dean, School of Engineering
University of California
Irvine, CA 92717
USA

Library of Congress Cataloging-in-Publication Data
Modern research topics in aerospace propulsion : in honor of Corrado Casci / G. Angelino, L. De Luca, W.A. Sirignano, editors.
p. cm.
Includes bibliographical references.
ISBN 978-1-4612-6956-4 ISBN 978-1-4612-0945-4 (eBook)
DOI 10.1007/978-1-4612-0945-4
1. Jet propulsion. I. Casci, Corrado. II. Angelino, G. (Gianfranco) III. De Luca, L. IV. Sirignano, W.A.
TL709.M63 1991
629.132'38—dc20

90-19778
CIP

Printed on acid-free paper.

Typeset by Asco Trade Typesetting, Ltd., Hong Kong.

9 8 7 6 5 4 3 2 1

ISBN 978-1-4612-6956-4

Preface

This volume, published in honor of Professor Corrado Casci, celebrates the life of a very distinguished international figure devoted to scientific study, research, teaching, and leadership.

The numerous contributions of Corrado Casci are widely admired by scientists and engineers around the globe. He has been an impressive model and outstanding colleague to many researchers. Unfortunately, only a few of them could be invited to contribute to this honorific volume. Every one of the invited contributors responded with enthusiasm.

Corrado Casci

Contents

Contributors

Andrews, M.J., Department of Mechanical and Aerospace Engineering, Princeton University, Princeton, NJ 08544, USA.

Angelino, G., Dipartimento di Energetica, Politecnico di Milano, Piazza Leonardo Da Vinci 32, 20133 Milano, Italy.

Arts, T., von Karman Institute for Fluid Dynamics, Chaussée de Waterloo, 72 B-1640 Rhode Saint Genèse, Belgium.

Bassi, F., Istituto di Macchine, Universitá di Catania, Italy.

Borrelli, S., Centro Italiano Ricerche Aerospaziali, Capua, Italy.

Bracco, F.V., Department of Mechanical and Aerospace Engineering, Princeton University, Princeton, NJ 08544, USA.

Breugelmans, F.A.E., Von Karman Institute for Fluid Dynamics, Chaussée de Waterloo, 72 B-1640 Rhode Saint Genèse, Belgium.

Brezinsky, K., Department of Mechanical and Aerospace Engineering, Princeton University, Princeton, NJ 08544, USA.

Continillo, G., Istituto di Ricerche sulla Combustione CNR, P. le Tecchio 80, 80125 Napoli, Italy.

Culick, F.E.C., Karmen Laboratory of Fluid Mechanics and Jet Propulsion, California Institute of Technology, Pasadena, CA 91125, USA.

Davis, K.A., Department of Mechanical and Aerospace Engineering, Princeton University, Princeton, NJ 08544, USA.

De Palma P., Istituto di Macchine ed Energetica, Universitá di Bari, Via De David 200, 70125 Bari, Italy.

De Luca, L., Dipartimento di Energetica, Politecnico di Milano and CNPM/CNR Laboratories, 32 Piazza Leonardo Da Vinci, 20133 Milano, Italy.

Galfetti, L., Dipartimento di Energetica, Politecnico di Milano and CNPM/CNR Laboratories, 32 Piazza Leonardo Da Vinci, 20133 Milano, Italy.

Glaskova, A.P., Soviet Academy of Sciences, Institute of Chemical Physics, Moscow 117334, USSR.

Glassman, I., Department of Mechanical and Aerospace Engineering, Princeton University, Princeton, NJ 08544, USA.

Grasso, F., Dipartimento di Meccanica e Aeronautica, Universitá di Roma "La Sapienza," 00184 Roma, Italy.

Invernizzi, C., Dipartimento di Ingegneria Meccanica, Universitá di Brescia, via Diogene Valotti 9, 25060 Brescia, Italy.

Macchi, E., Dipartimento di Energetica, Politecnico di Milano, Piazza Leonardo Da Vinci, 32, 20133 Milano, Italy.

Magi, V., Istituto di Macchine ed Energetica, Universitá di Bari, 70125 Bari, Italy.

Miele, A., Aero-Astronautics Group, Rice University, Houston, Texas, 77252 USA.

Napolitano, M., Istituto di Macchine ed Energetica, Universitá di Bari, Via De David 200, 70125 Bari, Italy.

Oppenheim, A.K., Mechanical Engineering Department, University of California, Berkeley, CA 94720, USA.

Osnaghi, C., Dipartimento di Energetica, Politecnico di Milano, Piazza Leonardo Da Vinci, 20133 Milano, Italy.

Pandolfi, M., Dipartimento di Ingegneria Aeronautica e Spaziale, Politecnico di Torino, Torino, Italy.

Penner, S.S., Center for Energy and Combustion Research, Department of Applied Mechanics and Engineering Sciences, University of California, San Diego, La Jolla, CA 92093, USA.

Perdichizzi, A., Dipartimento di Meccanica, Universitá di Brescia, Via Diogene Valotti 9, 25060 Brescia, Italy.

Richards, M.B., Center for Energy and Combustion Research, Department of Applied Mechanics and Engineering Sciences, University of California, San Diego, La Jolla, CA 92093, USA.

Sirignano, W.A., Department of Mechanical Engineering, University of California, Irvine, CA 92717, USA.

Tomboulides, A., Department of Mechanical and Aerospace Engineering, Princeton University, Princeton, NJ 08544, USA.

Wang, T., Aero-Astronautics Group, Rice University, Houston, Texas, 77252 USA.

Winterfeld, G., DFVLR, Institut für Antriebstechnik, 5000 Köln 90, FRG.

Wu, G.D., Aero-Astronautics Group, Rice University, Houston, Texas, 77252 USA.

Zarko, V.E., Soviet Academy of Sciences, Siberian Branch, Institute of Chemical Kinetics and Combustion, Novosibirsk 630090, USSR.

Curriculum Vitae

Corrado Casci was born in Pavia during World War I, on 19 February 1917, of Tuscan parents. His father came from Casentino and was a draughtsman in the Civil Engineers. Casci's childhood was spent in Pavia, which he left to go to Arezzo on the 22d of December 1922. This city and its surroundings played an important role in developing the moral, cultural, and religious principles already instilled by his parents. When he speaks, in that voice with its light Tuscan undertones, one hears echoes of those teachers who gave him his deep philosophical, historical, literary, and scientific knowledge: "Casentino and Pavia are to me the places to which my soul belongs, the reserves of smells and tastes of my childhood and youth, the age when fundamental and critical choices are made."

In 1936, he graduated brilliantly from high school and won a place at the Collegio Ghislieri in Pavia for Lombardy university students, a college renowned for its high scholastic standards where students "pay" for the period spent there with success in their chosen studies. He enrolled in the introductory two-year course of studies at the Faculty of Engineering even though he had, until the very last moment, been considering the Faculty of Medicine.

In 1938, having completed this two-year course in Pavia, he enrolled in the third year of Applied Engineering at the University of Pisa, graduating at the end of June 1941 with honors in Mechanical Engineering. Some results of his thesis, displaying great original thought, were used in several publications.

But in that summer of 1941, Europe was in flames. Starting in the ranks as a private in the artillery, he went on to become an officer in the air force. Having won a ministerial bursary, he attended the School of Aeronautical Engineering at the Politecnico in Turin, directed by Prof. Modesto Panetti, where Casci obtained his second degree, in aeronautical engineering, again graduating with honors. He was then given the rank—as were Luigi Crocco, Antonio Ferri, and Luigi Broglio—of lieutenant in the Aeronautic Engineers and was appointed assistant and researcher of propulsion problems at the same school.

It was in this period that he began dealing with space propulsion, using as guidelines the works of H. Obert and E. Sänger. During this time he became a member of a group of academics and researchers engaged in the construction of a particular arms system, a radio-controlled long-range torpedo. On 8 September 1943, when Italy asked for an armistice and the group split up. Wishing to distance himself from the Fascist regime, Corrado Casci became a member of the Resistance. He was seized by the S.S. in a raid and sentenced to death but managed to escape and cross the border.

The war had left Casci and his family homeless and in distressed financial conditions, which compelled him to devote himself to civil engineering work designing iron, reinforced concrete, brick, and wooden bridges of which Italy was badly in need.

The adventures of this period further strengthened the character and mind of the young Casci—he was only 27—and as soon as he could he returned to the Politecnico in Turin to continue his studies and research activities under the guidance of Professors M. Panetti, A. Capetti, C. Ferrari, and P. Cicala, frequently meeting with L. Crocco and G. Gabrielli, who were among the first in Italy to design jet-propelled aeroplanes.

But as space and air propulsion at that time commanded little financial support, Corrado Casci had to devote part of his research, theoretical as well as experimental, to the study of internal combustion engines. He collaborated with foreign industries—in particular Dutch "Shell"—conducting various experiments on reciprocating engines fed by petrol-methanol or petrol-ethanol mixtures and demonstrating the beneficial influence of air humidity on combustion. During a long period spent in the Shell engine laboratories and at the Technical University of Delft, he worked with J.J. Broese, invariably on problems relative to combustion.

In the period between 1946 and 1952 he was an assistant at the Politecnico in Turin, first of air propulsion and subsequently of heat engines. In 1951 he obtained a Ph.D. in Heat Engines and in 1954 a Ph.D. in Propulsion Systems. Since 1947 he has been in charge of the course on engine construction for aircraft at the same Politecnico. In 1951, he was put in charge of aircraft engines at the Politecnico in Milan.

The years spent in Turin were marked by in-depth studies of thermodynamics, combustion, and fluid dynamics, leading to numerous scientific papers born of the many ideas that have always crowded the mind of Corrado Casci. The miniaturization of rotating exchangers for application to gas turbines, car traction, and air propulsion were among his ideas. During this period he also designed an experimental compressor (whose purpose it was to consolidate the studies of C. Ferrari on rotating blades) whose blades were adjustable in order to study, also experimentally, optimum incidence. Since that time Corrado Casci has always dealt with internal fluid dynamic problems. Numerous studies, not only his own but also those originated by his students, came about from the experience acquired from Casci in this sector.

His new teaching and scientific commitments did not distract him from his studies on propulsion or astronautical navigation. This led him, during this same period, to the study of missile motion; he then began tests on liquid propellant rockets, often putting his life at risk by the use of rudimentary experimental equipment in the open countryside. Once on a Sunday, while engaged in these tests in the fields on the outskirts of Turin, a combustion chamber exploded with a loud bang. In a nearby church, a mass was being held and all the people attending the service rushed out into the churchyard. Chased by policemen, Casci fled.

In 1951, he was given the chair of Propulsion at the Airforce Academy in Naples but he continued with his course at the Politecnico in Milan. In this period, he held a seminar on an Earth-Mars mission, using known propellants that demonstrated the need to make the vehicle start its flight from Earth and transferring, in the process, to the orbit of Deimos before landing on Mars.

Now a full professor, he returned to the Politecnico in Milan and set himself two aims: to renew the course on machines and to prepare the groundwork for his own laboratory. The teaching structure given by Casci to this course was based on the fundamental laws of thermodynamics applied to all types of machines, both with compressible and incompressible fluid, while his scheme for his laboratories—later implemented—was the organization of several sections with common centralized coordinating services.

In 1959, two events played a key role in the orientation and development of Corrado Casci's research work. He was called by T. von Karman (whom Casci had met in Turin when he was assistant to M. Panetti) and by G. Gabrielli to participate in the Combustion and Propulsion Panel (CPP), now the Propulsion and Energetics Panel (PEP), of the Advisory Group for Aerospace Research and Development (AGARD). In addition, he was made a member of the first Italian space committee, under the chairmanship of L. Broglio, whose members also included such eminent scientists as Amaldi, Occhialini, Margaria, and Righini.

The first CPP meeting in which Casci participated, "The Chemistry of Propellants" held in Paris, was the fifth. The 73rd meeting of the PEP, "Aircraft Fire Safety" held in Lisbon in 1989, marked 30 years of working in AGARD. At the end of the meeting, the PEP chairman asked all of the delegates to show their appreciation for everything Casci had done on behalf of PEP, in terms of improving studies, research, seminars, and meetings. Prolonged applause greeted the chairman's words.

In 1959, he had an idea for a seminar on astronautical propulsion which he shared with L. Crocco, A. Ferri, and S.S. Penner, all AGARD members. This idea became a reality in the shape of a seminar held the following year at the Villa Monastero in Varenna (Como). It was sponsored by the Istituto Lombardo (Academy of Science and Letters) and chaired by Casci. This seminar gathered together the cream of scientists and others responsible for western space programs. Prof. T. von Karman opened the seminar and it was closed by H. Dryden, Deputy Administrator of NASA, with a speech on the technologies needed to reach the moon. He described takeoff, transfer, and moon landing. Ten years later, on 20 July 1969, these events happened exactly as he had described them. Since this time, Casci had been continually engaged in this type of research work which has brought him into touch with the most famous international scientists and technicians. This work led him to the study of the use of hydrazine as a monopropellant, perfecting a catalyst, aimed at determining the decomposition of hydrazine for small impulses of short duration, intended to control the trim of artificial satellites.

During the academic year 1961–62, Corrado Casci left his course on aircraft engines to take up his position as a full professor of machines. From this point, beginning with the foundation of the Institute of Machines, he went on to create, what is today, the Department of Energy. In those years he began to train young researchers carefully selected from different theoretical and experimental fields, bringing into being a school of energy and propulsion which has become one of the most efficient and highly regarded both in Italy and abroad. Owing to his vast acquaintanceship and personal friendships in the international scientific academic field, he has been able to send some of his students abroad to perfect their knowledge in different universities and research laboratories such as Princeton, Stanford, University of California at San Diego, University of California Berkeley, New York University, University of Pennsylvania, UCLA, Imperial College, NASA, ONERA, and the laboratories of the USSR Academy of Sciences in Novosibirsk. In the 1960s he began the construction of research laboratories, which were enlarged as time went by, and which, after occupying several locations, are now situated in the Linate Airport lot by an arrangement, with the Italian Air Force.

In 1963, he founded CNPM (Centro di Studio sulla Propulsione and sul-Energetica) under the sponsorship of the National Research Council and a consortium of industries, including Montecatini, BPD, Breda, Edison, and OTO-Melara.

Despite dedicating himself to research and organization of this institute, he never neglected his institutional teaching duties in preparing young engineering undergraduates. He enlarged the number of courses, increased his own commitment, and trained his colleagues so that they could take over his duties. When Casci finally became head of the Institute of Machines, there were only two courses, Machines and Aircraft Engines. Thanks to his endeavors many other courses were added: Bioengineering, Machine Complements, Power Plants (subsequently Machines II), Energetics, Mechanical Power Plants, Rocket Engines, Aerospace Propulsors, Theory and Techniques of Missile Control, and Biomedical Thermokinetics.

Casci also enlarged post-degree education for young engineers, and in collaboration with insurance companies promoted a specialization course (Master in Insurance Engineering). In 1978, he created a school of energetics, of a theoretical-experimental type, with a three-year course and an international teaching staff.

In 1981, he proposed the constitution of the first Energetics Department in Italy at the Politecnico of Milan, combining within it the institutes of Technical Physics and Machines, taking over its direction in the first, certainly difficult, years of its life in an attempt to ensure harmonious collaboration between the academics and researchers of the institute.

From 1969 to 1980, with increasing funds, researchers, and technical personnel, he was finally able to implement many of his ideas. To this period belongs an arms system project, discussed with Tony Ferri, for an airbreathing missile flying at a speed of Mach 3.5, with a range of approximately 3000 km, presented to the Ministry of Aeronautical Defense; research on the use of energy recovery in stream plants for naval propulsion; and the design of a hybrid inverted propellant rocket (solid fuel for the exterior and liquid oxidizer for the interior). Casci tested the rocket, using oxidizers of an ablative type in order to cut down on material and production costs.

The reader of these notes on Corrado Casci might believe that he has a 'bellicose spirit.' Proof that this is not so is the fact that he was the first in Italy to introduce bioengineering courses (biomachines) as well as promoting studies and experiments in blood circulation and the behavior of heart valves in the human body. The results obtained in this sector were discussed and used, following a meeting in Houston, by De Bakey and Cooley.

At the end of the 1970s, he was asked to take part in the National Committees for Finalized Energy Projects. In Energetics I, he acted as coordinator of the research of alternative fuels (methanol, ethanol, and hydrogen) in internal combustion engines and in Energetics II, he was coordinator of the Machine and Turbomachine sector.

From 1975 on, he dedicated himself to the application of electronics to engine automation, to the carburetor to decrease consumption and pollution, and to intake and outlet valves in order to free these from the rigidity of mechanical transmissions so as to increase engine performance even under partial loads.

In the 1980s, in conjunction with a qualified team, he studied combined gas-steam cycles with steam injection in the gas cycle.

In the beginning of the 1980s, he anticipated the use of a shock tube to inject pellets in tokamaks for nuclear fusion. In the laboratories of the Department of Energetics, shock tubes aimed at achieving an injection speed of deuterium particles, represented by pellets of 4000 m/s (a worldwide breakthrough), were designed and implemented.

Since 1983, he has been the head of a research group on combustion in joint collaboration with the National Research Council (Italy) and the Academy of Sciences (USSR).

On 17 December 1987, Corrado Casci, as a full professor at the Politecnico of Milan, retired from teaching. He gave the last lecture of his academic career under the title of "... Towards the future." The lecture was held in the presence of university authorities, colleagues, and former students—most of whom have attained very high positions in the industrial and political life of Italy. It was held in the Great Hall of the Politecnico under whose 19th century eaves he had begun his teaching career in 1951 with nothing but a desk and a drawing table. This lecture was an excursus of his life, a weaving together of events and reflections on many research themes, with the "lecturer" revealed in every sentence, the story and a summation of a life in which the roots of the past penetrate the future, a crescendo of historic events covering the central span of our century.

The considerations of the 25-year-old Casci in 1942 on the Mustang engine system (in which propulsion given by the propellor received an increased thrust as a reaction given by the heat dispersed by the ducted radiator) are at the root of his love for the traditions of jet propulsion studied at that period to evaluate possible fall-out on air propulsion.

This youthful passion for propulsion problems never left him and, even now, his eyes sparkle when he speaks about them. Over the years he has written and spoken about propulsion, fluid dynamics, and combustion with Obert and Sänger, Sthulinger, von Braun, von Zobrowski, Crocco, Ferri, Zeldovich, Summerfield, and Spalding.

While a professor at the Politecnico of Milan, he put in a request for the new research laboratories being built to be named with the initials CNPM —"Centro nazionale di ricerche sulle tecnologie della propulsione e dei relativi materiali." The word propulsion is a constant in the life of Corrado Casci.

In his last lecture, Corrado Casci also explained how the basics of his teaching of machines using deductive methods, from the basic laws of thermodynamics and mechanics, energetics, and dynamics, which characterize the functioning of the typological classes of machines in their diverse aspects, are deduced. But in some cases, provocatively, one can start from the effects in order to reach the causes and then postulate the theory explaining the phenomenology per se and in the context of the field of potential application.

Teaching in the field of energetics at the Politecnico of Milan has been remarkably widened by Professor Casci, who understood the importance of engineering in a country that, after a ruinous war, first rebuilt itself and then assumed a key role in the world economy. This explains why Casci also held, both in Italy and abroad, many seminars of an international nature such as Supersonic Turbomachinery, Remote Handling, Design and Digital Control Systems, Cardiovascular Flow Dynamics and Measurements (in Houston) Solid Rocket Motor Technology (in Ankara, Turkey), Engineering Problems in Propulsion Systems (in Ankara, Turkey), Combustion Problems in Solid Propellants (at the Science Academy, Novosibirsk, USSR), Advances in Propulsion Air-Breathing (at the Academy of Sciences, Moscow, USSR), and Advances of Propulsion Systems (at the University of Alma Ata, Kazakistan, USSR).

When talking about Corrado Casci one should not forget Wanda, his wife and the mother of their children, who has shared his life for 45 years. Without a backward look she gave up a comfortable life to follow her husband in his wanderings and, perhaps most importantly, she relieved him of family and everyday problems by looking after the children, Simonetta and Frederico, providing them with a sound moral basis rooted in her Catholic faith, and directing them in their further educational studies.

When rereading what we have written, we realize that these lines express only inadequately what Corrado Casci's contribution to engineering has meant. We only hope that we may have given some small idea of his works, his publications, and his teachings to his many students who all bear witness to the work of the man. Although the life of a man of science and culture cannot be summarized, we nevertheless trust that this tribute may help towards some understanding of the life, the work, and the personality of Corrado Casci. They are certainly proof of his commitment to and love for study, teaching, and research. Corrado Casci is a legend in his lifetime.

Honors

International Academy of Astronautics, Paris (Member)
Accademia delle Scienze, Torino (Member)
Istituto Lombardo, Accademia di Scienze e Lettere, Milano (Member)
Accademia Petcerca di Lettere Arti e Scienze, Arezzo
AGARD—Propulsion and Energetics Panel (Member and Former Chairman)
AIDAA, Roma (Member)
Seminario Matematico e Fisico, Milano (Member)
SAE, New York, USA (Member)
Honored Guest, City of Cleveland

Awards

Gold Medal, Benemeriti della Scienza e dell'Arte, Italy, 1969
Gold Medal, Istituto Internazionale delle Comunicazioni, Genova, 1969
Von Karman Medal, AGARD, Bruxelles, 1985

Consultant

U.S. Army, USA; U.S. Air Force, USA; AGARD, Paris; Worthington, USA–Italy; NATO, Bruxelles; CEE, Bruxelles; Ministero dell'Industria, Roma, Italy; Fiat, Italy; Alfa Romeo, Italy; Alfa Avio, Italy; Aeronautica Militare Italiana, Italy; Ansaldo, Italy; Franco Tosi, Italy; Air Force of Turkey; BPD, Italy; OTO Melara, Italy; Montecatini, Italy; Progetto San Marco, Italy; ENI (Ente Nazionale Idrocarburi), Roma, Italy; SNAM-Metanodotti, Milano, Italy.

Publications of Corrado Casci

1. Metodo Teorico per la Ricerca delle Caratteristiche di Progetto dei Giunti Idrodinamici. *Ricerca Scientifica e Ricostruzione*, anno 16, n. 9, Settembre 1946.
2. Sulla Progettazione dei Trasformatori di Coppia Idrodinamici. *Ricerca Scientifica e Ricostruzione*, anno 17, n. 2–3, Marzo 1947.
3. Giunti Idrodinamici *L'Ingegnere*, n. 9, Settembre 1947.
4. Ciclo Combinato a Gas Combusti per Turbine a Combustione Interna. *Atti e rassegna tecnica della Società degli Ingegneri e degli Architetti in Torino*, anni 1, n. 7, Luglio 1947.
5. Sui Cicli Combinati a Gas Combusti per Turbine a Gas. *La Termotecnica*, n. 6, Settembre 1947.
6. Ricerche Teoriche Sulle Tensioni dell'Occhio di Biella. *L'Aerotecnica*, vol. XXVIII, n. 2, 15 Aprile 1948.
7. Su una Proprieta' delle Velocita' Critiche. Rivista A.T.A., n. 6, Settembre 1948.
8. Sui Liquidi di Apporto nei Motori a Carburazione. *La Rivista dei Combustibili*, vol. II, Fasc. 7, Nov.–Dic.
9. Sulla Distribuzione delle Temperature in Regime Permanente di un Anello in Ambienti a Temperatura Diversa. *Rendiconti dell'Accademia dei Lincei, Classe di Scienze Fisiche, Matematiche e Naturali*, serie VIII, vol. VII, fasc. 5.
10. Sulla Distribuzione della Temperatura in un Anello Rotante in Ambienti a Temperatura Diversa. *Rendiconti dell'Accademia Nazionale dei Lincei, Classe di Scienze Fisiche, Matematiche e Naturali*, serie VIII, vol. VII, fasc. 6.
11. L'Impiego dell'Iniezione di Liquidi di Apporto nei Motori in Volo di Crociera. L'Aerotecnica, vol. XXIX, fasc. 6, 1949.
12. Sulla Distribuzione delle Temperature di un Anello in Regime Permanente e Posto in Condotti Percorsi da Correnti Gassose a Temperature Diverse. *Rendiconti dell'Accademia Nazionale dei Lincei, Classe di Scienze Fisiche, Matematiche e Naturali*, serie VIII, vol. IX, fasc. 3–4.
13. Sui Limiti Della Sovralimentazione dei Motori Alternativi a Combustione Interna. *Numero speciale dell'Aerotecnica in onore di Modesto Panetti*, Torino 1950.
14. La Lubrificazione dei Motori a Combustione Interna e le Prove Sperimentali sui Lubrificanti. *Atti e rassegna tecnica della Società degli Ingegneri e degli Architetti in Torino*, luglio 1951.
15. La Sollecitazione a Torsione Nelle Palette di Compressore Assiale. *Ricerche, Rivista A.T.A.*, Gennaio 1951.
16. Un Metodo Teorico-Empirico per la Progettazione delle Turbine con la Teoria

Bidimensionale. *Rendiconti dell'Istituto Lombardo di Scienze e Lettere, Classe di Scienze*, vol. LXXX–IV, 1951.
17. Prove Sperimentali sui Lubrificanti. *La Rivista dei Combustibili*, vol. VI, fasc. 4, aprile 1952.
18. Sull'Iniezione dei Liquidi di Apporto nei Motori a Carburazione. *L'Aerotecnica*, vol. XXXII, fasc. 2, 1952.
19. Sulla Risoluzione delle Equazioni Differenziali del Moto dei Missili. *Studia Ghisleriana, Studi Matematici-Fisici*, serie IV, vol. 1, 1952.
20. Prove sull'Iniezione col Sistema Pilgrim. *Ricerca Scientifica*, anno 22, n. 7, luglio 1952.
21. Essais sur L'emploi des Huiles Lourdes Dans un Moteur Diesel Tosi a' 4 Temps Suralimente' par Turbo-Compresseur. *Atti del Congrès International des moteurs a combustion interne*, IV, 1953.
22. Sul Motore Composito. *L'Aerotecnica*, vol. XXXIII, fasc. 4, 1953.
23. La Turbina a Gas. Stato Attuale, Applicazioni e Possibilita' D'Impiego. *L'Ingegnere*, n. 3–4, anno 1954.
24. Esperienze sull'Impiego della Nafta Pesante in Motori Diesel Tosi Sovralimentati. *L'Ingegnere*, n. 7, anno 1954.
25. Sull'effetto Delle Differenti Sequenze Degli Scarichi nei Motori Diesel a 4 Tempi Sovralimentati con Turbina a Gas di Scarico. *Istituto Lombardo di Scienze e Lettere, Classe di Scienze*, vol. LXXXVII, 1954.
26. L'Evoluzione del Motore a Combustione Interna nel Primo Centenario. Commemorazione Tenuta all'Accademia F. Petrarca di Arezzo. *Centenario dell'Invenzione di Barsanti e Matteucci, Atti e Memorie dell'Accademia Petrarca*, Nuova serie, vol. XXXVI, 1954.
27. Ancora sulla Distribuzione della Temperatura in un Anello Rotante in Ambienti a Temperatura Diversa. *Rendiconti dell'Istituto Lombardo di Scienze e Lettere, Classe di Scienze*, vol. LXXXVIII, 1955.
28. Temperature dello Stantuffo e del Cilindro del Motore Diesel Monocilindro di Media Potenza. *Rendiconti dell'Istituto Lombardo di Scienze e Lettere, Classe di Scienze*, vol. LXXXVIII, 1955.
29. Su un Problema di Misure per la Determinazione delle Caratteristiche dei Turboreattori. *Rendiconti dell'Istituto Lombardo di Scienze e Lettere, Classe di Scienze*, vol. LXXXVIII, 1955.
30. Sulle Prestazioni e Caratteristiche del Missile. *L'Aerotecnica*, vol. XXXV, fasc. 6, 1955.
31. Sul Comportamento dell'Autoreattore al Variare delle Condizioni di Impiego nel Campo Supersonico. *L'Aerotecnica*, vol. XXXV, fasc. 4, 1955.
32. Sulle Possibilita' del Missile in Volo Verticale. Deduzione dei Parametri Fondamentali di Progetto del Missile dalle Caratteristiche del suo Impiego. *Memoria presentata al V Congresso della International Astronautical Federation, Copenaghen, 1955, L'Aerotecnica*, vol. XXXVI, fasc. 1, 1956.
33. Divagazioni Astronautiche. (Conferenza tenuta alla Sezione di Milano dell'AIDA Ass. Italiana d'Aerotecnica). *Rivista d'Ingegneria*, n. 5, 1–15 Maggio 1957.
34. On the Slowing Down of Time. *Jet Propulsion*, June 1957. (in collaborazione con B. Bertotti).
35. Sul Rendimento e sul Lavoro Massico delle Turbine a Gas con Generatori a Gas Combusti a Stantuffi Liberi. *Rendiconti dell'Istituto Lombardo di Scienze e Lettere, Classe di Scienze*, vol. 92, anno 1957.

36. Razzi e Propellenti dalle Origini a Oggi. (Conferenza tenuta su invito dell'Ente Autonomo Fiera di Milano il 21-IV-1958). *Rivista A.T.A., Ricerche* n. 7–8, anno 1958.
37. Nuovi Orientamenti sulla Propulsione a Razzo dei Veicoli Spaziali e dei Satelliti Artificiali. *Rendiconti del Seminario Matematico e Fisico di Milano*, vol. XXIX, anno 1958.
38. Sul Tempo di Vita dei Satelliti Artificiali. *Memoria presentata al IX Congresso International Astronautical Federation, Proceedings IXth International Astronautical Congress, Amsterdam, 1958.* Springer-Verlag, Wien. (in collaborazione con V. Giavotto).
39. Sulla Teoria dei Processi di Combustione e sui loro Confronti con le Trasformazioni delle Correnti Supersoniche. *Studia Ghisleriana, Studi Fisici*, Serie IV, vol. II, 1958.
40. Sui Cicli di Turbine a Gas con Generatore di Gas Combusti. *Rendiconto dell'Istituto Lombardo di Scienze e Lettere*, Classe di Scienze, vol. 92, anno 1958.
41. An Experimental and Indirect Method for Determining High Atmosphere Density. *Il Nuovo Cimento*, vol. XI, n. 2, 16 Gennaio 1959.
42. Sui Sistemi di Propulsione nei Veicoli Spaziali. (Memoria presentata al 70 Convegno Internazionale delle Comunicazioni). *Celebrazione Colombiana*, Genova, 1959.
43. Reattori Nucleari per la Propulsione a Razzo di Veicoli Spaziali. *Memoria presentata al Convegno Internazionale Tecnico-Scientifico dello Spazio, 12–25 Giugno 1961.* Missili, fasc. 6, Dicembre 1962 (in collaborazione con B. Coppi).
44. L'Evoluzione dei Sistemi a Propulsione Chimica. *La Ricerca Scientifica*, anno 33, serie 2, parte I°, Rivista (vol. 3, Gennaio 1963, n. 1). CNPM N.T. n. 1.
45. La Propulsione e i Combustibili. *Relazione generale al Congresso "Combustibili e Propellenti Nuovi" organizzato della F.A.S.T., Milano, 10–14 Giugno 1963.* Atti del Convegno Soc. Pergamon Press—Editrice Politecnica Tamburini, 1963. CNPM N.T. n. 3.
46. Applicazione dei Calcolatori Elettronici allo Studio Degli Effusori Convenzionali e a Spina col Metodo delle Caratteristiche. *Atti del Convegno AIR—IBM, Firenze 1963* (in collaborazione con il prof. Angelino e ing. Janigro). CNPM N.T. n. 7.
47. Turbina a Gas Supersonica quale Mezzo per ridurre la Temperatura Degli Organi Rotanti della Motrice. *Quaderno CCSS* n. 10, 1° Convegno sui Combustibili e Lubrificanti per Uso Navale, Roma, 1964, CNPM N.T. n. 46.
48. Propulsion and Propellants. Ed. Pergamon Press, Tamburini, 1964.
49. Su un Nuovo Metodo per l'Impostazione del Progetto dei Missili Sonda. *XIII Convegno Internazionale delle Comunicazioni*, Genova, 12–16 Ottobre 1965 (in collaborazione con E. Gismodi). CNPM N.T. n. 5.
50. Sui Criteri di Progettazione di un Generatore di Gas a Stantuffi Semiliberi per l'Alimentazione di una Turbina. *La Termotecnica*, n. 1, 1965 (in collaborazione con l'ing. E. Sacchi). CNPM N.T. n. 15.
51. Sui Trasduttori di Pressione per Rilievi Sperimentali di Fenomeni Termodinamici. *La Termotecnica*, Aprile 1965. (in collaborazione con O. Tuzunalp). CNPM N.T. n. 17.
52. Sull'Applicazione di Fluidi non Convenzionali nei Generatori di Potenza. *Atti e rassegna tecnica della Società degli Ingegneri e degli Architetti di Torino*, Nuova Serie, A19, n. 10, Ottobre 1965 (in collaborazione con gli ingg. G. Angelino e A. Ranalletti). CNPM N.T. n. 22.

53. Misurazione della Costante Politropica su Cicli di Macchine Termiche Mediante Calcolatore Analogico. *Atti e rassegna tecnica della Società degli Ingegneri e degli Architetti di Torino*, Nuova Serie, A19, n. 10, Ottobre 1965 (con O. Tuzunalp). CNPM N.T. n. 23.
54. I Vari Aspetti della Combustione nei Motori—Considerazioni Introduttive. *Atti e rassegna tecnica della Società degli Ingegneri e degli Architetti di Torino*, Nuova Serie, A19, n. 10, Ottobre 1965 (in collaborazione con F. Mina). CNPM N.T. n. 24.
55. Sui Fenomeni Transitori degli Scambi di Calore nei Propulsori a Razzo a Propellente Liquido. *Atti e rassegna tecnica della Società degli Ingegneri e degli Architetti di Torino*, Nuova Serie, A19, n. 10, Ottobre 1965 (in collaborazione con gli ingg. F. Chiesi e U. Ghezzi). CNPM N.T. n. 25.
56. Sul Confronto dei Vari Sistemi di Lancio per la Messa in Orbita dei Satelliti. *La Ricerca Scientifica*, anno 1934, vol. 5, serie 2, n. 7–9, 1965 (in collaborazione con L. Corti). CNPM N.T. n. 11.
57. Sul Rendimento di Combustione di uno Statoreattore Supersonico. *L'Aerotecnica*, fasc. 1, Gennaio–Febbraio 1966 (in collaborazione con l'ing. U. Ghezzi). CNPM N.T. n. 28.
58. Una Tecnica di Registrazione dei Dati nell'Analisi Sperimentale dei Processi Termodinamici Periodici. *La Rivista dei Combustibili*, vol. XX, fasc. 4, 1966 (in collaborazione con O. Tuzunalp). CNPM N.T. n. 30.
59. Preliminary Report on the Performance of Plug Nozzles in Solid Propellant Rockets. *VI° E.S.S.T. Brighton*, Maggio 1966 (in collaborazione con gli ingg. E. Gismondi e G. Angelino). CNPM N.T. n. 53.
60. Problems of Auxiliary Propulsion for Satellite Attitude Control. *VI° E.S.S.T. Brighton*, Maggio 1966 (in collaborazione con S. Ricci). CNPM N.T. n. 52.
61. Su una particolare Tecnica Sperimentale per la Determinazione delle Temperature in Organi Mobili di Motori a Combustione Interna. *A.T.A.*, Novembre 1966 (in collaborazione con gli ingg. G. Ferrari e A. Radaelli). CNPM N.T. n. 47.
62. Influence of Some Characteristic Parameters on the Frequency Instability in Bi-Propellant Rocket Engines. *Rivista Meccanica—AIMETA*—n. 2, vol. 11, 1967 CNPM N.T. 57 (in collaborazione con gli ingg. F. Chiesi e U. Ghezzi).
63. Sul Coefficiente di Efflusso dell'Appereto di Alimentazione di un Motore Alternativo a Combustione Interna. *Studia Ghisleriana*, Studi Fisici, Pavia, 1967 (in collaborazione con gli ingg. A. Radaelli e G. Ferrari). CNPM N.T. n. 76.
64. Pressure Dispersion Phenomena in Actual Internal Combustion Engines. *VII° Congresso Mondiale del Petrolio*, Città del Messico, 2–9 Aprile 1967 (in collaborazione con O. Tuzunalp). CNPM N.T. n. 51.
65. Development Report on: An Electronic System for Experimental Investigation of Cycle to Cycle Pressure Dispersion Phenomena in Actual Engine Cycles. *Seventh World Petroleum Congress*, Mexico City, 2nd–8th April 1967 (in collaborazione con O. Tuzunalp).
66. Nuovo Programma di Calcolo delle Prestazioni di un Endoreattore al Variare della Pressione e del Rapporto di Miscela. *XV Convegno Internazionale delle Comunicazioni*, Genova, 12–15 Ottobre 1967 (in collaborazione con l'ing. F. Chiesi). CNPM N.T. n. 60.
67. Applicazione dell'Ugello a Spina a Razzi a Propellente Solido di Media Grandezza. Milano, 1967 (in collaborazione con gli ingg. E. Gismondi, G. Angelino). CNPM N.T. n. 48.
68. Considerazioni sulla Tecnica della Propulsione Navale a Getto Pesante. *Rivista*

A.T.A., Novembre 1967 (in collaborazione con gli ingg. U. Ghezzi e A. Ranalletti). CNPM N.T. n. 58.

69. Impianto di Alimentazione e Regolazione per Prove Endoreattori. *La Rvista dei Combustibili*, vol. XXII, fasc. 6, 1968 (in collaborazione con l'ing. F. Chiesi). CNPM N.T. n. 98.
70. Condizioni Attuali e Prospettive Future della Turbina a Gas nella Trazione Terrestre. *Rivista A.T.A.*, Aprile 1968 (in collaborazione con l'ing. F.M. Montevecchi). CNPM N.T. 77 e 101.
71. Sulla Meccanica di una Serie di Anelli in un Motore Alternativo a Combustione Interna. *Lubrificazione*, n. 12, 1968. CNPM N.T. n. 103 (in collaborazione con gli ingg. A. Radaelli e G. Ferrari).
72. Some Thermodynamic Aspects of Special Fluid Power Plants. *Problems in Fluid Flow Machines*, Warsaw, 1968 (in collaborazione con l'ing. G. Angelino).
73. New Trends for High Speed Helicopter Propulsion. 31st Meeting: Helicopter Propulsion System (AGARD-NATO), 10–14 June 1968, Ottawa, Canada (in collaborazione con E. Bianchi). CNPM N.T. n. 79.
74. The Dependence of Power Cycles Performance on Their Location Relative to the Andrews Curve. *Presented at the ASME Gas Turbine Conf.*, Cleveland, March 1969, ASME Paper 69-GT-65, 1969 (in collaborazione con l'ing. G. Angelino).
75. Organic Fluid and Gas Turbine in Combined Power Cycles. U. Hoepli Ed. Milano, 1969 (in collaborazione con l'ing. G. Angelino).
76. Studio Sperimentale dei Comburenti Solidi—Influenza dei Catalizzatori e della Granulometria sulla Velocita' Lineare di Regressione. *La Rivista dei Combustibili*, vol. XXIII, fasc. 6, 1969 (in collaborazione con l'ing. L. De Luca). CNPM N.T. 106.
77. Studio Sperimentale dei Comburenti Solidi. Influenza di Polvere Metallica sulla Velocita' Lineare di Regressione di NH_4CLO_4. *La Rivista dei Combustibili*, vol. XXIII, fasc. 4, 1969 (in collaborazione con l'ing. L. De Luca). CNPM N.T. 107.
78. La Vitesse Lineaire de Regression du Perchlorate d'Ammonium dans un Ecoulement Gaseux Combustible. *34° Meeting Propulsion and Energetics Panel dell'AGARD*, Dayton, Ohio, 13–17 Ottobre 1969 (in collaborazione con l'ing. L. De Luca).
79. Influenza della Percentuale degli Additivi sulla Velocita' Lineare di Regressione del Perclorato d'Ammonio. *La Rivista dei Combustibili*, vol. XXIV, fasc. 2, 1970 (in collaborazione con gli ingg. L. De Luca e V. Boldrini).
80. Fluid-Dynamic Criteria for Design and Evaluation of Artificial Valves. *AGARD Fluid Dynamics of Blood Circulation and Respiratory Flow*, CP n. 65, Napoli, 1970 (in collaborazione con gli ingg. R. Fumero e F.M. Montevecchi).
81. La Propulsione Spaziale Negli Anni '70, Sviluppo, Indirizzi e Studi. *X Convegno sullo Spazio*, Roma, 9–11 Marzo 1970.
82. Analysis of the Piston Heat Load During Knocking. *The Internal Combustion Engines Conference*, Bucharest, 1970 (in collaborazione con l'ing. G. Ferrari).
83. Analisi Preliminare del Comportamento dei Gas di Scarico di un Motore Volumetrico a Combustione Interna. *Rivista A.T.A.*, Novembre 1970 (in collaborazione con l'ing. U. Ghezzi).
84. Possibilita' d'Impiego dei Molibdati Ridotti nella Decomposizione di Idrazina Monopropellente. *La Rivista dei Combustibili*, vol. XXV, fasc. 9, 1971 (in collaborazione con E. Santacesaria e D. Gelosa).

85. Analisi del Carico Termico dello Stantuffo in Condizioni di Detonazione. *Rivista A.T.A.*, Gennaio 1971 (in collaborazione con l'ing. G. Ferrari).
86. An Experimental Rocket-Engine with Solid Oxidizer in Eutectic of Ammonium Nitrate–Ammonium Perchlorate. *Israel Journal of Technology*, vol. 9, n. 6, 1971 (in collaborazione con gli ingg. F. Chiesi e C. Ortolani). CNPM N.T. n. 161.
87. Proprieta' Termodinamiche dell'Anidride Carbonica fra 0°C e 800°C e fra 1 Bar e 600 Bar: Presentazione del Diagramma di Mollier. *La Termotecnica*, n. 9, 1972 (in collaborazione con gli ingg. E. Macchi e G. Angelino). CNPM N.T. n. 178.
88. Proprieta' Termodinamiche dell'Anidride Carbonica. Ed. Tamburini, 1972 (in collaborazione con gli ingg. E. Macchi e G. Angelino).
89. A Method for Preliminary Analysis of MHD Generator Performance. *39° Meeting AGARD Propulsion and Energetics Panel, Energetics for Aircraft Auxiliary Power Systems*, Colorado Springs, 1972 (in collaborazione con gli ingg. A. Coghe e U. Ghezzi).
90. An Experimental Research on the Behaviour of a Continuous Flow Combustion Chamber. *41° AGARD Meeting on Atmospheric Pollution by Aircraft Engines*, London, 1973 (in collaborazione con gli ingg. A. Coghe, U. Ghezzi e S. Pasini).
91. Gestione Mediante Elaboratore Numerico in Linea dei Rilievi Sperimentali su Motori a Combustione Interna. *La Termotecnica*, n. 7, Luglio 1973 (in collaborazione con gli ingg. B. Abbiati, S. Facchinetti, F.M. Montevecchi e C. Parrella). CNPM N.T. n. 200.
92. Fluidodinamica Simulata dei Processi di Aspirazione e Scarico di un Monocilindro. *Memoria presentata al XXI Convegno Internazionale delle Comunicazioni*, Genova, 8–13 Ottobre 1973 (in collaborazione con l'ing. G. Ferrari). CNPM N.T. n. 202.
93. La Problematica del Consumo di Risorse Energetiche nei Tentativi di Limitazione dell'Inquinamento da Mezzi di Trasporto Terrestri. *Memoria presentata al XXI Convegno Internazionale delle Comunicazioni*, Genova, 8–13 Ottobre 1973 (in collaborazione con l'ing. P. De Marchi). CNPM N.T. 203.
94. Analisi dell'Influenza della Temperatura delle Pareti della Camera di Combustione sui Fenomeni di "Quenching" in Motore a C.I. *La Rivista dei Combustibili*, Marzo 1973 (in collaborazione con l'ing. G. Ferrari). CNPM N.T. n. 212.
95. Energetic Problems in Artificial Hearts: A Survey of the Work Carried on by the Bioengineering Group of the Institute of Machine at the Polytechnic of Milan. *6° Congresso della Società dei Trapianti d'Organo*, Varese, Settembre 1973 (in collaborazione con gli ingg. S. Facchinetti, R. Fumero e F.M. Montevecchi). CNPM N.T. n. 280.
96. Dispersione Ciclica in un Motore Commerciale per Autovettura. *La Rivista dei Combustibili*, 1974, pg. 262 (in collaborazione con gli ingg. B. Abbiati, G. Ferrari e C. Parrella). CNPM N.T. n. 248.
97. Nascita e Sviluppo della Bioingegneria. *Rivista FINA*, n. 51, Marzo 1973.
98. Prosthetic Heart Valves. *Advanced Study Institute "Cardiovascular Flow Dynamics"* NATO, University of Houston, Tex., 6–17 October, 1975 CNPM N.T. n. 288.
99. Behaviour of Nitric Oxide in Continuous Flow Combustion Chambers. *Deuxieme Symposium Europeen sur la Combustion*, Orléans, 1–5 September 1975 (in collaborazione con U. Ghezzi e C. Ortolani).
100. Prospetto e Realizzazione di un Riscaldatore ad Effetto Joule della Potenzialita'

di 250 KW. *XXX Congresso Nazionale ATI*, Cagliari, Settembre 1975 (in collaborazione con gli ingg. E. Bollina e E. Macchi).

101. Centro di Studio per Ricerche sulla Propulsione e Sull'Energetica, Peschiera Borromeo (Milano). *Attività Scientifica svolta nel 1974. La Ricerca Scientifica*, anno 45, n. 6, Nov.–Dic. 1975, pp. 1263–1270. CNPM N.T. n. 281.
102. Experimental Results on High Speed Double Mechanical Seals. *AGARD Conference Proceedings No. 237 on "Seal Technology in Gas Turbine Engines,"* London, April 1989, pp. 11-1–11-10 (in collaborazione con gli ingg. E. Bollina e E. Macchi).
103. Electron Density Measurements in Flames with Microwave Interferometry. *Presented at the Fifth International Colloquium on Gasdynamics of Explosion and Reactive Systems*, Orléans, France, September 1975 (in collaborazione con gli ingg. A. Coghe, U. Ghezzi, N. Gottardi, G. Lisitano).
104. Numerical Analysis and Experimental Data in Continuous Flow Combustion Chamber. *Paper presented at the 54th AGARD Meeting on Combustion Modeling*, Köln, Germany, 3–5 October, 1979 (in collaborazione con gli ingg. F. Gamma, A. Coghe, U. Ghezzi).
105. Control of Alternative Engines by Microcomputer Systems. *54th Specialists Meeting on Advanced Control Systems for Aircraft Powerplants*, Köln, Germany, 1–2 October, 1979.
106. Recent Research on Unsteady Combustion at CNPM. *Paper presented at the VIth International Symposium on Combustion Processes*, Karpacz, Poland, 26–30 August 1979 (in collaborazione con gli ingg. L. De Luca, A. Coghe, G. Ferrari, L. Galfetti, L. Martinelli, C. Zanotti).
107. Heat Transfer and Friction in High Aspect Ratio Rectangular Channels with Repeated-Rib Roughness. *55th (A) Specialists Meeting AGARD on Testing and Measurement Techniques in Heat Transfer and Combustion*, Brussels, Belgium, 5–7 May 1980 (in collaborazione con gli ingg. G. Giglioli e P. Ferrari).
108. Risparmio e Ricupero di Energia nei Moderni Sistemi di Trazione e Propulsione Endotermica. *1° Relazione del Convegno su "Energia e Trasporti," Istituto Internazionale delle Comunicazioni*, Genova, 9–10 Maggio 1980. Atti Istituto Internazionale Communicazioni.
109. Development for New Laboratories for Future Testing. *56th Meeting AGARD on Turbine Engine Testing*, Torino, 29 September—3 October 1980. AGARD-CP 293 pp. 39-1–39-25.
110. Experimental Results and Economics of a Small (40 kW) Organic Rankine Cycle Engine. *15th Intersociety Energy Conversion Engineering Conference*, Seattle, Wash. (U.S.A.), 18–22 August, 1980 (in collaborazione con gli ingg. G. Angelino, P. Ferrari, M. Gaia, G. Giglioli, E. Macchi).
111. Introduzione ai Lavori del Primo Simposio Europeo su "Remote Operation in Fusion Devices," *Proc. 1st European Symposium*, Milano, Italy, 25–26 Maggio 1982. pp. 1–5.
112. La Progettazione Assistita dal Calcolatore (Computer Aided Design) in Relazione alla Applicazione in Macchine e Turbomacchine. *Corso Residenziale "Progettazione di turbine assiali," Giovinazzo* (Ba), 27 Sept. 1983. Atti PFE-2 (in collaborazione con il prof. Alberto Rovetta).
113. Heat Pump Enhanced Gas Turbine Cogeneration. Energy, The International Journal, Pergamon Press, New York, Vol. 9, n. 7, pp. 555–564, July 1984 (in collaborazione con l'ing. Mario Gaia).

114. Introduzione al Volume "*Recent Advances in the Aerospace Sciences*" in Onore del Prof. Luigi Crocco, Plenum Press, New York, 1985, pp. XI–XXIII.
115. Experiments on Solid Propellants Combustion: Experimental Apparatus and First Results. In: "*Recent Advances in the Aerospace Sciences,*" Volume in Onore del Prof. Luigi Crocco, Plenum Press, New York, 1985.
116. La Ricerca sull'Energia e il Futuro dell'Ingegnere, ovvero: la Magnetofluidodinamica, una Scienza Antica per un Insegnamento Nuovo. *Energia Nucleare,* Quadrimestrale Tecnico Scientifico dell'ENEA, anno 2, n. 1, pp. 98–99, Aprile 1985 (in collaborazione con il dr. U. Carretta e il prof. E. Minardi).
117. Introduzione al Convegno "Approccio Multidisciplinare per la Pianificazione e lo Sviluppo del Territorio." *Accademia Petrarca di Lettere Scienze e Arti,* Arezzo, 9–11 Ottobre 1986 (in collaborazione con Prof. M.V. Erba). Atti del Convegno, pp. 39–56.
118. L'Importanza dell'Energia e delle Innovazioni Tecnologiche nello Sviluppo del Territorio. *Convegno 'Approccio Multidisciplinare per la Pianificazione e lo Sviluppo del Territorio,'* Arezzo, 9–11 Ottobre 1986. Atti, pp. 59–104.
119. Osservazioni Conclusive al Convegno "Approccio Multidisciplinare per la Pianificazione e lo Sviluppo del Territorio." *Accademia Petrarca di Lettere Scienze e Arti,* Arezzo, 9–11 Ottobre 1986. Atti del Cocegno, pp. 391–392 (in collaborazione con Prof. M.V. Erba).
120. Pompe e Turbine Reversibili Bistadio Regolanti. *Energia Elettrica,* Giugno 1988, n. 6, pp. 237–238.
121. Sui Problemi Scientifici e Tecnologici del Volo Ipersonico. *Accademia di Lettere e Scienze,* F. Petrarca di Arezzo, 23 Gennaio 1990 (in corso di stampa).

I
Combustion

1
Mechanics of Turbulent Flow in Combustors for Premixed Gases

A.K. OPPENHEIM

ABSTRACT: In order to reveal the mechanism of turbulent flow in a premixed combustor, a numerical technique, using Chorin's random vortex method to solve the Navier-Stokes equations and an interface propagation algorithm to trace the motion of the combustion front, are employed. A successive over-relaxation hybrid method is used as the initial step in the computational scheme to solve the Euler equations for a planar flow field.

Solutions obtained thereby for a backfacing step, the essential element of a planar dump combustor, turn out to be in satisfactory agreement with experimental results especially insofar as the global properties are concerned, such as the average velocity profiles and the reattachment lengths. Velocity fluctuations are found to compare well with experimental data, exhibiting, however, some discrepancies that can be ascribed to the omission of three-dimensional effects and the relatively small size of numerical data sampled for their evaluation.

The combustion field appears to be dominated by the large-scale eddy structure of the turbulent shear layer, whereby the effects of advection overpower those of diffusion—enhancing the entrainment of the fresh mixture into the combustion region. Under such circumstances, the front of the combustion zone acquires the properties of an interface between the unburnt medium and the burnt gases, rather than a flame, while the exothermic regime, being effectively decoupled from it, is confined within the kernel of the large-scale eddy.

Introduction

The next step in the evolution of gas turbine combustors should be associated with the introduction of a premixed working substance to replace direct fuel injection—a system leading, as a rule, to the formation of diffusion-flames with all their well-known deficiencies. Most prominent among them is the generation of pollutants, in particular nitric oxide, that blocked the development of SST, and the detrimental effects of irreversibilities, due to secondary air mixing, upon the thermal efficiency of the system.

Thus, over the last decade combustion of premixed gases in combustors attracted deservedly broad attention, as manifested by the experimental investigations of Ganji and Sawyer (1980), Pitz and Daily (1983), Shepherd et al. (1982), and El-Benhawy et al. (1983), as well as by the numerical studies of Ghoniem et al. (1981, 1982) and Ashurst (1981), all concerned with the flame-holding properties of a wake behind a backfacing step, the typical feature of a dump combustor.

The reason for this lies in a number of advantages offered by this sytem. Firstly, it provides proper means for mass and heat recirculation—a process of particular virtue to lean combustion, enhancing thermal efficiency and reducing pollutant emission. Secondly, it promotes intimate mixing that enhances strong interaction between the turbulent flow field and the combustion process, providing a closed-loop feedback mechanism for its control. Thirdly, it exploits the intrinsic flow instability to stabilize the combustion field, an apparent paradox that in reality is particularly beneficial in spreading out the deposition of exothermic energy throughout the field.

However, by the same token, combustion instabilities are of crucial importance since they are so dependent on various operating conditions, such as equivalence ratio, inlet flow velocity, pressure, and temperature. As the combustion field is stabilized by recirculation, bringing hot products into close contact with the reaction zone and thereby furnishing a continuous supply of ignition sources to the incoming fresh mixture, the turbulent field plays a major role in determining the geometry of the combustion zone. On the other hand, the expansion due to the deposition of the exothermic energy of combustion tends to constrain the turbulent field in that the extent of the recirculation zone (reattachment length) becomes reduced by as much as half of its value in the nonreacting flow. Meanwhile, the free-stream flow and the recirculation vortex interact with the turbulent shear layer—a region dominated by large-scale eddy structure—enhancing the interaction between combustion products and reactants to provide the most conducive conditions for the chemistry to take place. There is a wide range of frequencies detectable in the energy spectrum of the shear layer, corresponding to the size and motion of the eddies forming its elementary components. However, some frequencies can become dominant depending on local properties of the reacting mixture, a feature leading to flow instability that induces oscillations into the flow field (Keller et al. 1982). This process emphasizes the close coupling between fluid mechanics and combustion in the system under study, as well as bringing up the major role of the vortical flow structure that has to be, therefore, properly taken into account by the model.

The combined effects of the fundamental mechanism of a turbulent combustion zone, namely, the combined effects of advection and expansion, are associated with a variety of phenomena governing the motion of its front—an interface popularly referred to as the flame. Most of the computational attempts made so far to reveal the mechanism of these processes have been handicapped by lack of adequate models of turbulent flow, difficulties associated with proper treatment of chemical reactions in a fluctuating field (Mellor and Ferguson 1980), and numerical instabilities that are introduced when dynamic effects of combustion are manifested by expansion across the flame front (Williams 1974). The crux of the problem lies in the effect of the Arrhenius exponential term in the kinetic rate equations that describe the chemical reaction. Thus, attempts to incorporate turbulent fluctuations in the combustion process using statistical decomposition of thermodynamic variables lead to nonconvergent solutions. Moreover, conventional numerical methods used in these studies are influenced by a priori averaging (Chorin 1986), as well as by artificial viscosity that inhibits the amplification of flow instabilities at high Reynolds numbers, and tends to 'laminarize' the flow (McDonald 1979). On the other hand, the use of grids to calculate the flow field imposes a limit on the spatial resolution of the results and may require the added complicity of adaptive modifications around zones of large gradients associated with the concomitant effects of numerical diffusivity, unless an implicit computational scheme is adopted.

Here, conventional artificial modeling of turbulence is avoided by seeking the solution of the basic Navier-Stokes equations without averaging, by the use of a Lagrangian particle technique, the random vortex method of Chorin (1973, 1978). The numerical model of this method is ideally suited to treat the unsteady and highly fluctuating flow field associated with turbulent combustion, as well as the dynamic effects due to the exothermicity of combustion in a flow system (Chorin 1980). It was employed originally to study the evolution of the vorticity field and the development of the flame front behind a step in a relatively short channel (Ghoniem et al. 1981, 1982). The results presented here have been obtained for a long channel incorporating a smooth contraction followed by an abrupt expansion. This relates to essential geometrical features of the combustion tunnel used concomitantly for experimental studies. In order to treat the appreciably large flow field, we used a Cray computer with the numerical procedure properly vectorized for this purpose.

The solutions we thus obtained were averaged to yield mean velocity and turbulence intensity profiles for comparison with the experimentai data of Pitz and Daily (1983). In their preliminary version, some of these results were published in the *Twentieth Symposium on Combustion* (Hsiao et al. 1985).

This paper is based on studies of turbulent combustion fields which the author has been conducting for a number of years (Chen et al. 1983; Dai et al. 1983; Ghoniem and Oppenheim 1984; Ghoniem et al. 1981, 1982, 1986; Hsiao and Oppenheim 1985; Hsiao et al. 1985; Keller et al. 1982; Oppenheim 1982, 1985, 1986, 1987; Oppenheim and Ghoniem 1983; Oppenheim and Rotman 1987; Rotman and Oppenheim 1986; Rotman et al. 1988). In fact, he is taking advantage of this occasion to present a fuller version of the results than published before (Hsiao and Oppenheim 1985), including prominently those obtained in collaboration with Dr. Hsiao, together with updated commentaries and interpretation.

Background

The vorticity field and flame development behind a backfacing step, without taking into account the inlet section and nozzle flow ahead of the step, were studied by Ghoniem et al. (1981, 1982), producing results in essential agreement with schlieren records reported by Ganji and Sawyer (1980).

In order to emphasize the conditions at inlet, the step was moved into the field of view at the test section, while the computational domain was extended to cover the upstream section including the contraction at the step. A cinematographic schlieren record of the combustion field obtained under such circumstances by Vaneveld et al. (1984) is presented in Fig. 1.1, whereas the configuration of the channel adopted for numerical analysis is specified in Fig. 1.2.

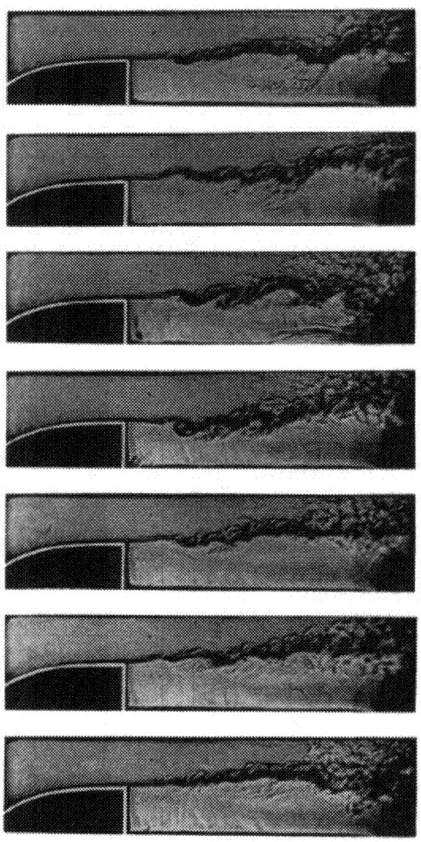

FIGURE 1.1. Extracts of cinematographic schlieren records of combustion field in a back-facing step, dump combustor system. Flowing substance: propane/air mixture at an equivalence ratio of 0.57. Inlet flow velocity: 13.4 m/s (Re = 2.2×10^4). Time interval between frames: 18 milliseconds.

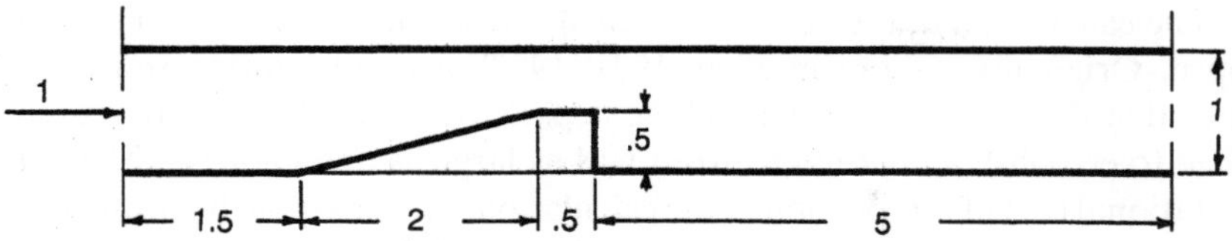

FIGURE 1.2. Combustor configuration used in numerical analysis.

In the computations all the dimensions are normalized with respect to the channel width and the flow velocity at inlet. Since the reattachment length of the shear layer behind a step is about six to eight step heights, the duct length behind the step was taken to be 5, minimizing thereby the influence of the untractable outflow boundary condition upon the recirculation zone. The geometry of the step was construed to match the experimental shape as closely as possible without introducing undue computational complicity. For premixed reactants, a propane-air mixture of equivalence ratio $\phi = 0.57$, the density ratio across the front of the combustion zone ("the flame") is 6, while its own normal speed is assumed to be at a nominal value of 0.01.

The computations were carried out for the Reynolds number $Re = 2.2 \times 10^4$ based on the inlet flow velocity and channel width, while the kinematic viscosity of the reactants, ϒ, matched the experimental conditions of Pitz and Daily (1983). At zero time step, the flow field is initialized by potential flow of uniform velocity at inlet. A vortex sheet layer of thickness $\delta = 3\ \sigma$ is then introduced, where $\sigma = (2\ \delta t/Re)^{1/2}$ is the standard deviation of random displacements used at each time step, δt. For numerical treatment, the layer is discretized into a number of finite-length vortex elements, each $b = 0.2$ in length. A maximum elementary circulation strength $\Gamma_{max} = 0.05$ for each vortex sheet is then used to minimize the error in the diffusion of vorticity. The whole shear layer is, thererore, made out of four sheets whose combined action serves to annihilate the wall velocity, u_w. The motion of vortex sheets, as well as the transformation of sheets to blobs and vice versa, follows then the algorithm of Ghoniem et al. (1982).

The solution or the Euler equations for the specific geometry under study, the first step in the numerical algorithm, was obtained by the use of the successive overrelaxation method (for details see Hsiao and Oppenheim, 1985). The flow domain was discretized for this purpose into square meshes, each $h_f = 0.05$ in size. To reach 10^{-3} accuracy of any function under consideration, the number of iterations for each time step ranged between 20 and 40.

As vortex sheets are continuously created at the walls, the vorticity field of the shear layer behind the step and the boundary layer along the walls grow, while the number of vortex blobs increases at each time step. As vortex blobs leave the computational domain at the exit section of the channel, they are discarded. Eventually then their number reaches a saturation level and the flow field attains thus a stationary state.

The calculation of thc combustion field starts when the flow is fully developed. Originally (Ghoniem et al. 1981, 1982), a point ignition source was introduced for this purpose behind the step. The number of time steps required then to establish a continuous front was as large as one hundred. The computational cost of this portion of the calculation was thus quite high, especially when the number of vortex elements reached several thousand. Hence, instead of expressing the initial condition in terms of a point ignition, the front is introduced as an interface extending along the horizontal centerline throughout the flow field in the combustion section, saving thereby a significant amount of computational time without any detrimental effect upon the eventual stationary state under study.

For the sake of convenience, the mesh size used in the calculation of front kinematics matches that used in nonreacting flow field, i.e., $h_c = h_f = 0.05$. The time step for the reacting flow calculation is then properly reduced to satisfy the Courant stability condition.

Results

Vorticity

To visualize the flow field, all the vortex blobs used in computations are plotted as small circles, while line segments attached to them display vectorial properties of their velocities. The development of vorticity field is presented in Fig. 1.3 up to $T = 10$, exhibiting the onset of the typical large-scale eddy structure of the turbulent shear layer in the course of growth associated with the entrainment of fluid from the surroundings. The coherence is manifested by continuous pairing of eddies having the same sense of rotation, as they

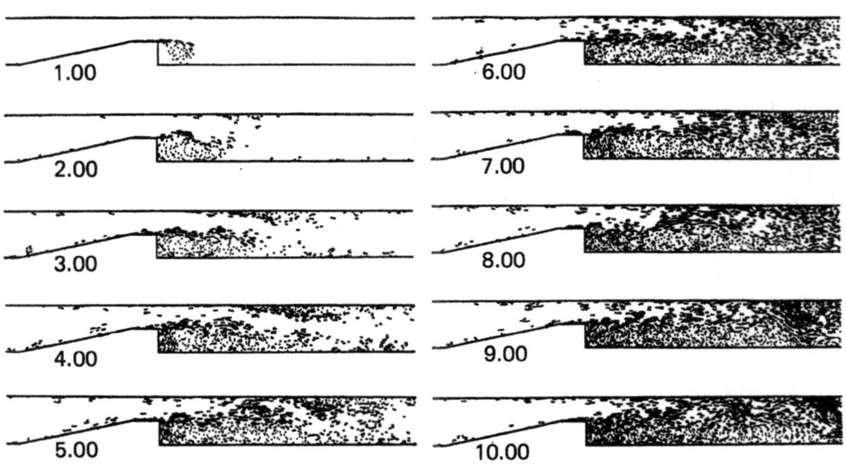

FIGURE 1.3. Development of vorticity field in nonreacting flow.

travel downstream. Thus, mixing takes place between the free stream of the incoming flow and the recirculation zone, associated with an appreciable amount of entrainment. For a step height equal to one-half of the channel width, as is the case here, the shear layer formed behind the apex of the step tends also to interact with the boundary layer developing at the upper wall. This is particularly evident at the fully developed stage of a stationary state betwen $T = 9$ and $T = 10$, portrayed in Fig. 1.4.

Front

Figure 1.5 depicts the front contours and the vorticity field of reacting flow, reached upon the establishment of a stationary state at $T = 10$. The front follows the flow field, bounding the region of concentrated vorticity and manifesting the dominant role of advection.

The reacting portion of the flowing medium is thus essentially confined within the vorticity region where large-scale eddy structures are formed, growing as they move downstream. This indicates that the deposition of the exothermic energy, associated with the expansion of the shear layer, exerts a relatively small effect upon the process of vortex shedding behind the step. At the same time, as a consequence of the action of source blobs, the number of vortex elements forming eddies in the turbulent flow field decreases. This constitutes a mechanism for the "laminarization" observed experimentally in turbulent flames (Takagi et al. 1980) that has been ascribed primarily to the increase in kinematic viscosity of the hot products.

Velocity

A comparison between mean velocity profiles we evaluated and the experimental data of Pitz and Daily (1983) for nonreacting and reacting cases are presented in Figure 1.6. The flow behind the step is dominated by large eddy structure of the turbulent shear layer, a regime well reproduced by our method, whereas the fluctuations are unencumbered by numerical diffusivity. Consequently the agreement between the numerical and experimental profiles is quite satisfactory. The discrepancies of mean velocity profiles near the step are due to the fact that the number of vortex blobs in the mixing layer is too small to simulate the large velocity gradient across it. Thls effect is more pronounced in the case of reacting flow where the number of vortex blobs in the shear layer is reduced by the expansion due to source blobs.

Moreover, as the reacting flow behind the step is dominated by large-scale structures that cause wrinkling and stretching of the interface when the reactants are entrained into the combustion zone, the numerical model also yields quite satisfactory results for mean velocity profiles in the reacting case. Furthermore, the increase of flow velocity in the reaction zone due to thermal expansion is properly modeled by volumetric sources along the flame front.

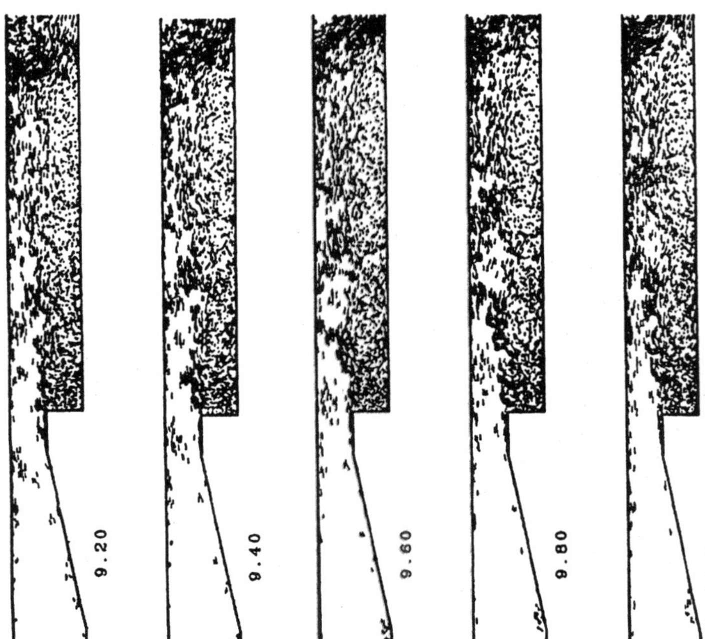

FIGURE 1.4. Vorticity field at stationary state.

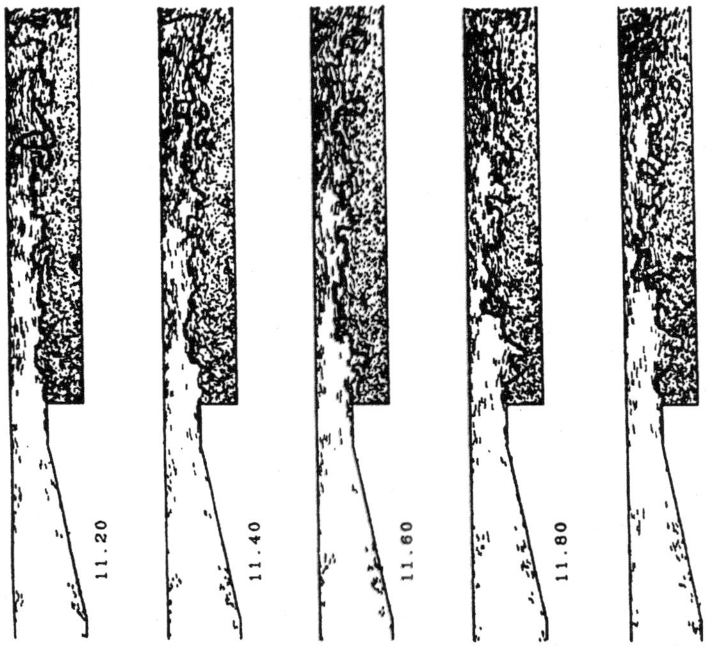

FIGURE 1.5. Vorticity field and combustion front in reacting flow.

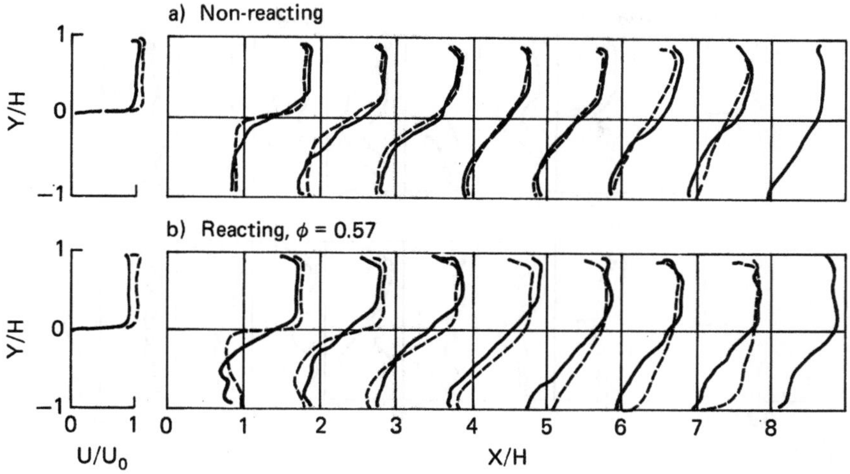

FIGURE 1.6. Average streamwise velocity profiles.

Turbulence

Turbulence intensities are, in general, not easy to evaluate. Even for a rearward-facing step, there is a substantial difference between various experimental measurements of this parameter. The variation is probably caused as much by experimental uncertainty as by real difference in various flow fields. According to Eaton and Johnston (1981), the maximum value of streamwise turbulence intensity $(u'^2)^{1/2}/U_0$ is between 0.16 and 0.21 for most experiments, while the peak value of the shear stress $\overline{u'v'}/U_0^2$ is around 1.25×10^{-3}.

Figure 1.7 depicts the streamwise turbulence intensity $(u'^2)^{1/2}/U_0$, we evaluated for both the nonreacting and reacting cases, in comparison with experimental data denoted by broken lines. Peak values occur, as expected from the velocity field, in three places: at both walls and at the start of the mixing layer. Turbulence due to high shear is continuously generated there, giving rise to maximum levels, followed by a decay in the downstream direction. The maximum value of the streamwise turbulence intensity we obtained in the shear layer region is about 0.16, which is at the lower bound of most experimental data. Large intensities were obtalned by us also, on both walls, in contrast to the decidedly one-sided character of experimental data. Measured intensities appear, moreover, to have smoother profiles within the shear layer. However, the maximum turbulence intensities obtained by Pitz and Dally (1983) of as high values as 0.28, seem to exceed the level one would expect on the basis of other measurements.

While numerical results follow the trend of experimental data, their profiles exhibit more irregular shapes in the region of the shear layer, the discrepancy becoming particularly apparent downstream. The results also show an early decay of the peak value at the distance of about two to three step heights

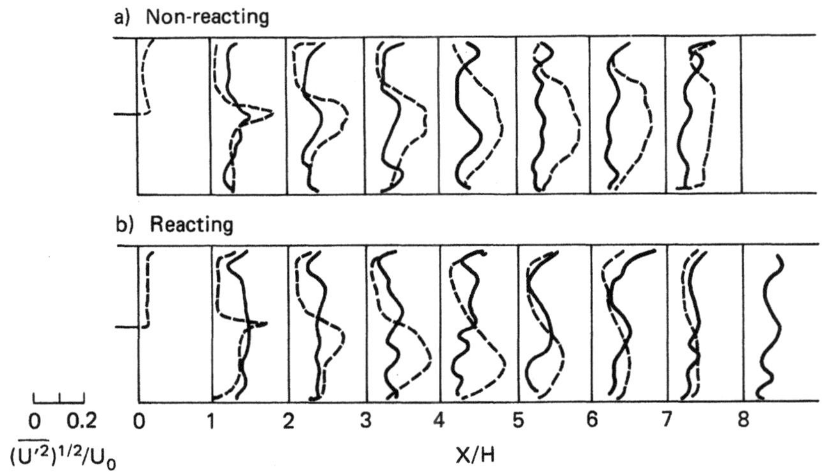

FIGURE 1.7. Streamwise turbulence intensity profiles.

upstream from the reattachment point. These deviations might be due to a number of reasons, namely:

1. Relatively small size of the sample used for the evaluation of numerical averages. This is especially pronounced when higher moments of the solution are evaluated—an operation equivalent to time differentiation emphasizing errors in the deviations from the mean. However, as a consequence of the inadequate size of the sample, there was no point in calculating higher moments, such as skewness and flatness factors.
2. Influences of boundary conditlons at the exit. This is associated with two factors: (a) the omission of vortex blobs at the outflow section made to save computational time and (b) the boundary condition $v = 0$, used to evaluate the potential flow velocity. The fluctuating component of the velocity is thus artificially reduced, a factor producing better agreement between numerical results and experimental data near the step than towards the end of the channel.
3. Less accuracy of the random vortex blob method near the boundaries where a large number of vortex blobs are generated and accumulated. The overlapping of vortex blobs cause large fluctuations of streamwise velocity.
4. Three-dimensional effects, the experimental data indicate, in contrast to numerical results, that turbulence intensity grows around the mixing layer in the vicinity of the separation zone. This phenomenon can be attributed to energy transfer from the main flow into turbulent fluctuations by vortex stretching, an effect beyond the scope of our purely two-dimensional model.

Presented in Fig. 1.8 is a comparison between the transverse turbulence intensities, $(v'^2)^{1/2}/U_0$, we evaluated and the experimental data in both the

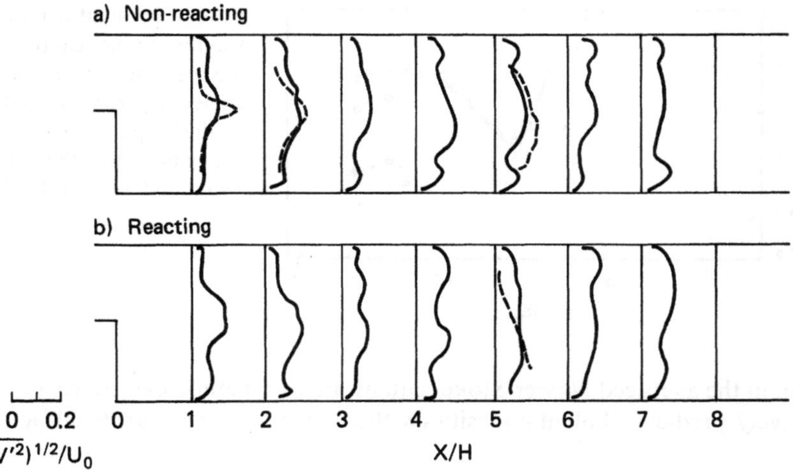

FIGURE 1.8. Transverse turbulence intensity profiles

reacting and nonreacting cases. In this respect our results are evidently in better agreement with the experiment than those of Fig. 1.7. The computed transverse turbulence intensities exhibit similar trends and are of comparable magnitude to their streamwise counterparts, whereas experimental data feature definitely smaller transverse intensities than streamwise (Etheridge and Kemp, 1978).

Since the no-slip boundary condition in our numerical model is satisfied by creating new vortices at every time step, the random displacements of vortex elements and the transformation between blobs and sheets in the region of high vorticity density induce considerable fluctuations in streamwise velocity. The transverse velocity, v, on the other hand, is essentially annihilated in the inviscid flow solution. Therefore one should expect streamwise turbulence intensities, obtained by methods of vortex dynamics, to have larger values near the walls (Ashurst et al. 1980; Dai et al. 1983) and the transverse turbulence intensities to be smaller, decaying faster to zero.

Similar discrepancy between experimentally observed turbulence intensities and the corresponding numerical results was also reported by Briggs et al. (1976), who used a Reynolds stress closure model and a law-of-the-wall boundary conditions to compute a similar flow field. The streamwise component of turbulence intensity they computed depended drastically on the rate of velocity generation at the step. On the other hand, Walterick, et al. (1984) obtained a very good agreement between the results of numerical calculations and the experimentally determined turbulence intensities for flow behind a back-facing step. They used the $k - \varepsilon$ model with special modeling technique for pressure-velocity correlation to match the experimental data. While most of the existing turbulence model requires artificial modeling of some correla-

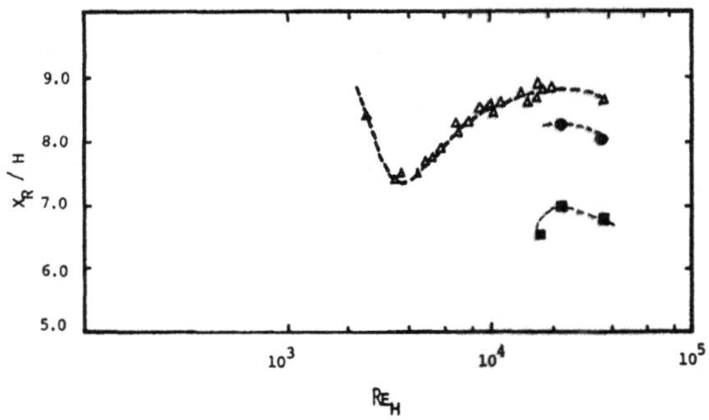

FIGURE 1.9. Reattachment lengths. Numerical results are presented by dark circles, experimental data by dark squares, and triangles denote the data of Durst and Tropea (1981).

tion in the averaged Navier-Stokes equations, our method can, at least qualitatively, predict turbulent intensities without resorting to any artificial factors.

Reattachment

The reattachment length is the distance from the point the stream separates from the boundary to where the dividing streamline hits the wall. Its value is usually expressed in terms of a number of step heights and varies as a function of the geometrical expansion ratio of the channel and the Reynolds number, as well as being influenced by the structure of the boundary layer at the step (Eaton and Johnston, 1981).

As shown in Fig. 1.9, the reattachment lengths deduced from the mean velocity profiles for the nonreacting case, are around eight step heights for both Reynolds numbers, slightly larger than those measured by Pitz and Daily (1983). However, when compared with other experimental data, such as those of Durst and Tropea (1981), our results are even more satisfying, especially with respect to the dependence on the Reynolds number and expansion ratio.

For reacting flow, as a consequence of the expansion due to combustion, the length is decreased to between five and six step heights. The recirculation is concomitantly reduced, while the maximum reverse velocity and recirculation rates are increased.

Conclusions

This paper presented a numerical solution of Navier-Stokes equations that is in remarkably good agreement with experimental results. This is so in spite of the fact that the model is two-dimensional (planar) and consequently devoid of the essential three-dimensional properties of turbulence. Evidently then, the particular results under consideration are relatively little affected by these properties. The solution furnishes, therefore, quite a good insight into the mechanism of a turbulent combustion field, complementing our previous study (Ghoniem et al. 1981, 1982), to provide a rational background for most of the concepts pointed out in concomitant publications (Oppenheim 1982, 1985, 1986, 1987).

The major conclusion they emphasize is that in a turbulent fleld, such as that attainable by flow in a channel at an inlet Reynolds number in excess of 10^4, the advective effects are predominant to such an extent that the influence of diffusion is reduced to practically negligible proportions. Under such circumstances, the interface formed when the unburnt medium gets in contact with burnt gases—a front associated with a particularly significant temperature and concentration gradient—acquires the role of a contact discontinuity rather than that of a flame front—an interface coupled with the evolution and deposition of exothermic energy referred to popularly as *heat release*. The particularly conspicuous influence of the now familiar large-scale structure of turbulent shear layer (vid., e.g., Peters and Williams 1974) is in causing the exothermic region to be decoupled from the front of the combustion zone.

All the properties of the so-called highly turbulent flames are due to this effect. Thus the turbulent combustion zone is capable of consuming the reactants at a relatively high rate (i.e., at a significantly large normal burning speed) primarily as a consequence of entrainment produced by eddies—the elementary components of the large-scale structure of turbulent shear layers—a mechanism that today should be well understood (vid., e.g., Papilou and Lykoudis, 1974). Concomitantly, the region of exothermic reaction is distributed over the kernels of these eddies, so that, on the average, the exothermic power density pulse becomes spread out and thus the residence time of the reacting species in the zone of essential chemical reaction (the zone of appreciable concentration of active radicals serving as chain carriers) is prolonged. The benefits of this effect are self-evident. It is then up to the designers and manufacturers of combustion systems to maximize its influence in order to improve their performance.

Acknowledgment. Work on this paper was supported by the director, Office of Energy Research, Office of Basic Energy Science, Engineering and Geosciences Division, and the Office of Energy Conversion and Utilization Technologies of the U.S. Department of Energy, under Contract No. DE-AC03-76SF0098, and by the U.S. Army Research Office under Contract No. DAAL03-87-K-0123.

References

Ashurst, W.T., 1981, "Vortex Simulation of a Model Turbulent Combustor," *Prog. Astronaut. Aeronaut.*, *76*, 259–273.

Ashurst, W.T., Durst, F., and Tropea, C., 1980, "Two-Dimensional Separated Flow Experiment and Discrete Vortex Dynamics Simulation," *AGARD Report*, *24*, 1–13.

Briggs, M., Mellor, G., and Yamada, Y., 1976, "A Second-Moment Turbulence Model

Applied to Fully Separated Flows," *Turbulence in Internal Flows* (S.N.B. Murthy, ed.), Hemisphere Publlshing Co.. Washington.

Chen, D.-Y., Ghoniem, A.F., and Oppenheim, A.K., 1983, "Experimental and Theoretical Study of Combustion Jet Ignition," *NASA Report CR-168139*, 130 pp.

Chorin, A.J., 1973, "Numerical Studies of Slightly Viscous Flow," *J. Fluid Mech.*, *57*, 785–796.

Chorin, A.J., 1978, "Vortex Sheet Approximation of Boundary Layers," *J. Comput. Phys.*, *27*, 428–442.

Chorin, A.J., 1980, "Flame Advection and Propagation Algorithms," *J. Comput. Phys.*, *35*, 1–11.

Chorin, A.J., 1986, "Vortex Methods for the Study of Turbulent Combustion at a Low Mach Number," *Progress in Astronautics and Aeronautics*, *105*, 14–21.

Dai, Y.W., Ghoniem, A.F., Sherman, F.S., and Oppenheim, A.K., 1983, "Numerical Modeling of Turbulent Flow in a Channel," *NASA Report CR-168278*, 63 pp.

Durst, F., and Tropea, C., 1981, 'Turbulent, Backward-Facing Step Flows in Two-Dimensional Ducts and Channels," *Proceedings of the Third International Symposium of Turbulent Shear Flows*, Davis, Calif., 18.1–18.6.

Eaton, J.K., and Johnston, J.P., 1981, "A Review of Research on Subsonic Turbulent Flow Reattachment," *AIAA J.*, *19*, 9, 1093–1100.

El-Benhawy, Y., Sivasegram, S., and Whitelaw, J.H., 1983, "Premixed, Turbulent Combustion of a Sudden-Expansion Flow," *Combustion and Flame*, *50*, 153–165.

Etheridge, D.W., and Kemp, P.H., 1978, "Measurement of Turbulent Flow Downstream of a Rearward-Facing Step," *J. Fluid Mech.*, *86*, 3, 545–566.

Ganji, A.R., and Sawyer, R.F., 1980, "An Experimental Study of the Flow Field of a Two-Dimensional Premixed Turbulent Flame," *AIAA J.*, *18*, 817–824.

Ghoniem, A.F., and Oppenheim, A.K., 1984, "Numerical Solution for the Problem of Flame Propagation by the Random Element Method," *AIAA J.*, *22*, 10, 1429–1435.

Ghoniem, A.F., Chen, D.Y., and Oppenheim, A.K., 1986, "Formation and Inflammation of a Turbulent Jet," *AIAA J.*, *24*, 2, 224–229.

Ghoniem, A.F., Chorin, A.J., and Oppenheim, A.K., 1981, "Numerical Modeling of Turbulent Combuation in Premixed Gases," *Eighteenth Symposium (International) on Combustion*, The Combustion Institute, Pittsburgh, Pa., 1375–1383.

Ghoniem, A.F., Chorin, A.J., and Oppenheim, A.K., 1982, "Numerical Modelling of Turbulent Flow in a Combustion Tunnel," *Phil. Trans. R. Soc. Lond.*, *A 304*, 303–325.

Hsiao, C.C., and Oppenheim, A.K., 1985, "Aerodynamic Properties of Turbulent Combustion Fields," *NASA Contractor Report 175005*, 126 pp.

Hsiao, C.C., Ghoniem, A.F., Chorin, A.J., and Oppenheim, A.K., 1985, "Numerical Simulation of a Turbulent Flame Stabilized Behind a Rearward-Facing Step," *Twentieth Symposium (International) on Combustion*, The Combustion Institute, Pittsburgh, Pa., 495–504.

Keller, J.O., Vaneveld, L., Krochlet, D., Hubbard, G.L., Ghoniem, A.F., Daily, J.W., and Oppenheim, A.K., 1982, 'Mechanisms of Instabilities Leading to Flashback,' *AIAA J.*, *20*, 2, 254–262.

McDonald, H., 1979, "Combustion Modeling in Two and Three Dimensions—Some Numerical Considerations," *Prog. Energy Combustion Science*, 5, 97–122.

Mellor, A.M., and Ferguson, C.R., 1980, "Practical Problems in Turbulent Reacting Flow," *Turbulent Reacting Flows* (P. Libby and F. Williams, eds.), Springer-Verlag, 45-64.

Oppenheim, A.K., 1982, "Dynamic Effects of Combustion," *Proceedings of the Ninth U.S. National Congress of Applied Mechanics*, ASME, New York, 29–40.

Oppenheim, A.K., 1985, "Dynamic Features of Combustion." *Phil. Trans. R. Soc. Lond., A 315*, 471–508.

Oppenheim, A.K., 1986, "The Beauty of Combustion Fields and Their Aerothermodynamic Significance." *Dynamics of Reactive Systems Part 1: Flames and Configurations. Progress in Astronautics and Aeronautics*, AIAA, New York, 105, 3–13.

Oppenheim, A.K., 1987, "Fluid Mechanical Control of Combustion," Plenary Lecture, *11th Canadian Congress of Applied Mechanics*, Edmonton, Canada, 31—44.

Oppenheim, A.K., and Ghoniem, A.F. 1983, "Aerodynamic Features of Turbulent Flames,"*AIAA Paper 83-0470*, 10 pp.

Oppenheim, A.K., and Rotman, D.A., 1987, 'Fundamental Features of Ignition and Flame Propagation in Engines," *ASME Paper 87-ICE-21*, 8 pp.

Papilou, P.S., and Lykoudis, P.S., 1974, "Turbulent Vortex Streets and the Entrainment Mechanism of the Turbulent Wake," *J. Fluid Mech., 62*, 11—31.

Peters, N., and Williams, F.A., 1980, "Coherent Structures in Turbulent Combustion," *The Role of Coherent Structures in Modeling Turbulence and Mixing* (J. Jimenez, ed.), Lecture Notes in Physics, 136, Springer-Verlag.

Pitz, R.W., and Daily, J.W., 1983, "Experimental Study of Combustion in a Turbulent Free Shear Layer Formed at a Rearward-Facing Step," *AIAA J., 21*, 11, 1565—1570.

Rotman, D.A., and Oppenheim, A.K., 1986, "Aerothermodynamic Properties of Stretched Flames in Enclosures," *Twenty-first Symposium (International) on Combustion*, The Combustion Institute, Pittsburgh, Pa., 1303—1312.

Rotman, D.A., Pindera, M.Z., and Oppenheim, A.K., 1988, "Fluid Mechanical Properties of Flames in Enclosures," *XIth International Colloquium on Dynamics of Explosions and Reactive Systems*, Warsaw, Poland, 15 pp.

Shepherd, I.G., Moss, J.B., and Bray, K.N.C., 1982, "Turbulent Transport in a Confined Premixed Flame," *Nineteenth Symposium (International) on Combustion*, The Combustion Institute, Pittsburgh, Pa., 423—431.

Takagi, T., Shin, H., and Ishio, A., 1980, "Local Laminarization in Turbulent Flames," *Combustion and Flame, 37*, 163—170.

Vaneveld, L., Hom, K., and Oppenheim, A.K., 1984, "Secondary Effects in Combustion Instabilities Leading to Flashback," *AIAA J., 22*, 1, 81—82.

Walterick, R.E., Jagoda, J.E., Richardson, C.R.J., de Groot, W.A., Strahle, W.C., and Hubbartt, J.E., 1984, "Experimental and Computation on Two-Dimensional Turbulent Flow Over a Backward Facing Step," *AIAA-84-0013*, AIAA 22d Aerospace Sciences Meeting, Reno, Nev.

Williams, F.A., 1974, "A Review of Some Theoretical Consideration of Turbulent Flame Structure," AGARD REP 43d Meetings, Lieye, Belgium, II, 1—125.

2
A Pore-Structure-Independent Combustion Model for Porous Media with Application to Graphite Oxidation

M.B. RICHARDS AND S.S. PENNER

ABSTRACT: This paper deals with combustion in porous media under conditions when reactions may be described by volume-averaged equations that are independent of pore-structure details. The generally applicable governing equations are developed. Available kinetic data are then used for gas-phase and solid-phase conversions in graphite oxidation in order to define the temperature regime ($T \lesssim 1250°C$) for which the model may be applied with acceptably small errors. An extension of the model to higher temperatures is discussed.

Our pore-structure-independent model may be used provided there are three greatly different and well-defined characteristic lengths, as is known to be the case for nuclear graphite before extensive burn-up.

Introduction

Combustion in porous media is of considerable practical importance but usually not properly allowed for in such diverse applications as coal oxidation and conversion, in situ recovery of shale oil, and oxidation of graphite in nuclear reactors. Quantitative analyses have been severely impeded by the general lack of necessary information concerning pore structure and reactivities in chemical conversions that occur within pore structures. Even if all of the physicochemical details and the morphology of the porous medium were accurately defined, it is not clear that the solution of the associated combustion problem could be obtained with the use of modern-day supercomputers. For this reason, it is important to search for useful simplified models that apply under well-defined conditions.

In this paper, we first derive a set of generally valid governing equations for the special conditions when pore-structure details do not significantly affect observable overall reaction rates. It is intuitively obvious that a reaction regime of this type will exist provided heat-absorbing and heat-releasing reactions occur relatively slowly compared with intra- and interparticle diffusion. Locally volume-averaged parameters should then be usable for the chemical species and the temperature because the ratios of deviations from the means to the means for these quantities are relatively small (i.e., $\lesssim 0.1$).

An approach of this type has been used in recent years for drying by Whitaker (1981), catalytic conversions by Ryan et al (1980), and basic studies on heat-and mass transfer by Carbonell and Whitaker (1983), Zanotti and Carbonell (1984), and Levec and Carbonell (1985). We make extensive use of the methodology developed for these applications in this first analysis dealing with the high local heat release that may be encountered during combustion in porous media. Following development of the general governing equations, we estimate for graphite oxidation the temperature limit below which the simplified model involving volume-averaged quantities may be applied with acceptably small errors. Graphite oxidation under the specified conditions is of considerable importance in nuclear reactor safety and appears to have been involved, in particular, in the Chernobyl accident (Wilson 1987).

In the course of the derivation, quantitative relations are developed between void fraction, tortuosity, diffusion coefficients, thermal conductivity, and specific heat of the porous medium. The needed approximations and assumptions in the conservation equations to assure pore-structure independence are defined and the conditions for pore-structure independence are then estimated for graphite oxidation. Model extension to large heat-release rates is indicated briefly. Numerical exploitation of our results and comparisons with experimental measurements are described in a sequel.

Governing Equations

The generalized conservation equations for reacting, multicomponent gas mixtures are given in standard references (Chapman and Cowling 1953; Hirschfelder et al. 1954; Penner 1957; Williams 1985). We make the following assumptions to generate simplified forms of the conservation equations: (1) Fick's law may be used to estimate a single diffusion velocity for all of the species, i.e., the binary diffusion coefficients for all pairs of species are taken to be equal, transport caused by thermal and pressure diffusion is assumed to be negligibly small, and the body force is taken to be the same for all species. (2) The mass-average velocity and mixture viscosity are so small that the contribution of viscous dissipation to internal energy may be neglected. (3) The momentum and internal energy associated with compressibility effects are negligibly small. (4) The net transfer of internal energy with species diffusion is negligibly small. (5) The gas mixture behaves as a Newtonian fluid with negligibly small components of the stress tensor associated with species diffusion. Using the specified assumptions, the conservation equations may be written in vector form as follows.

Mass conservation:

$$\partial \rho / \partial t + \underline{\nabla} \cdot (\rho \underline{v}) = 0; \tag{1}$$

species conservation:

$$\rho \partial Y_K/\partial t + \rho \underline{v} \cdot \underline{\nabla} Y_K = \underline{\nabla} \cdot (\rho D \underline{\nabla} Y_K) + \omega_K \, (K = 1, 2, \ldots, N-1); \tag{2}$$

momentum conservation:

$$\rho \partial \underline{v}/\partial t + \rho \underline{v} \cdot \underline{\nabla} \underline{v} = -\underline{\nabla} p + \underline{\nabla} \cdot \mu \underline{\nabla} \underline{v} + \rho \underline{g}; \tag{3}$$

energy conservation:

$$\rho c_p \partial T/\partial t + \rho c_p \underline{v} \cdot \underline{\nabla} T = \underline{\nabla} \cdot \lambda \underline{\nabla} T - \underline{\nabla} \cdot \underline{q}_R - \sum_{K=1}^{N} h_K \omega_K. \tag{4}$$

The symbols appearing in Eqs. (1) through (4) have the following meanings: ρ = mass-average density, $\underline{v}$ = mass-average velocity, Y_K = mass fraction of species K, D = an assumed representative diffusion coefficient, ω_K = mass-production rate of species K per unit volume associated with homogeneous chemical reactions, N = total number of chemical species, p = pressure, μ = viscosity coefficient, $\underline{g}$ = gravitational acceleration, c_p = mass-average specific heat at constant pressure, T = temperature, λ = thermal conductivity, $\underline{q}_R$ = radiant heat-flux vector, and h_K = specific enthalpy for species K. By definition,

$$\sum_{K=1}^{N} Y_K = 1, \tag{5}$$

which completes the required number of governing equations. The ideal-gas equation of state is

$$\rho = p\overline{W}/RT, \tag{6}$$

where $\overline{W}$ = mixture molecular weight and R = universal gas constant. The mixture molecular weight is obtained from

$$\overline{W} = \left(\sum_{K=1}^{N} Y_K/W_K \right)^{-1}, \tag{7}$$

where W_K is the molecular weight of species K. We represent the specific enthalpy of species K as the sum of the enthalpy of formation h_K^0 at the reference temperature T^0 and the change $\Delta h_K(T)$ in h_K between T^0 and T, viz.

$$h_K(T) = h_K^0 + \Delta h_K(T). \tag{8}$$

Volume-Averaged Equations for a Porous Medium

The porous medium has two phases: an impermeable solid phase and a gas-filled pore phase. We assume that the mass-average velocity components are negligibly small compared with the diffusion-velocity components and that the pressure may be treated as constant within the porous medium. The constant-pressure approximation represents an integral of the overall momentum-conservation equation. It has been shown empirically that these assumptions are reasonable for oxidation in air of a typical nuclear-grade graphite with low permeability (Hewitt 1965).

If the pore structure of the two-phase medium were accurately known, we could, in principle, solve the governing equations within each phase, subject to appropriate interphase boundary conditions. However, for most porous media, including graphite, the pore geometry is not known. The pore geometry will also change with time since there are heterogeneous chemical reactions occurring within the porous medium. For this reason, we use a volume-averaging technique for an averaging volume that is small compared with the total volume of the two-phase medium (see Fig. 2.1). In using volume-averaging procedures, it will be convenient to express a priori unknown point values of the dependent variables in terms of the sum of a phase average and phase-average deviation. The precise procedures for evaluating integrals over V_p and A_p (see Fig. 2.1) will be discussed in the following analysis for unknown pore structure.

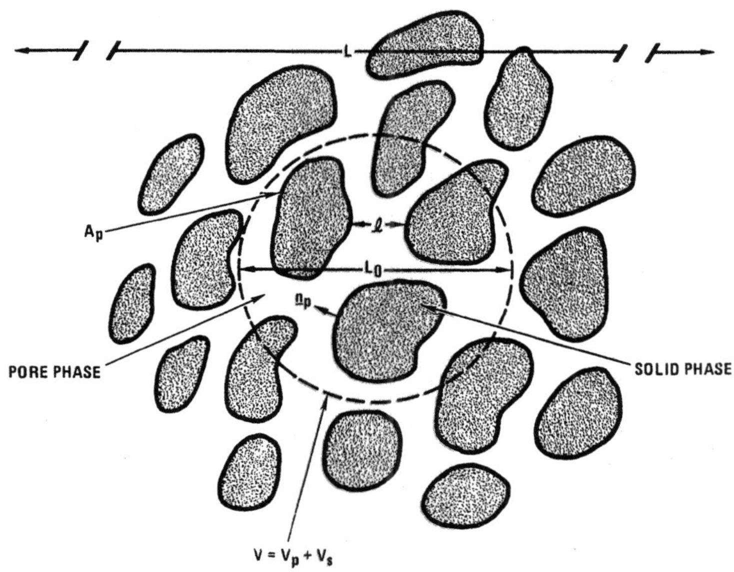

FIGURE 2.1. Schematic of a two-phase porous medium. The dotted enclosure represents a planar projection of the sum V of the pore-phase (V_p) and solid-phase (V_s) volumes over which we average; A_p = total surface over V_s within V; l = effective mean pore diameter; L_0 = diameter of a sphere with volume equivalent to V; $\underline{n}_p$ = unit outward normal from the solid phase to the gaseous phase within the pores; $L \gg L_0 \gg l$, where L represents the characteristic dimensions of the total two-phase medium. Representative values are $L \approx 10 - 10^3$ cm, $L_0 \approx 10^{-2} - 1$ cm, $l \approx 10^{-5} - 10^{-3}$ cm.

Species-Conservation Equations

With the assumption of negligibly small mass-average velocity components, the species-conservation equations for gases in the pores may be written as

$$\rho \partial Y_K/\partial t = \underline{\nabla}\cdot(\rho D \underline{\nabla} Y_K) + \omega_K. \tag{9}$$

We next determine volume averages for each term in Eq. (9) over the averaging volume V defined in Fig. 2.1, viz.

$$(1/V)\int_V \rho(\partial Y_K/\partial t)\,dv = (1/V)\int_V \underline{\nabla}\cdot(\rho D\underline{\nabla} Y_K)\,dv + (1/V)\int_V \omega_K\,dv. \tag{10}$$

The volume integrals in Eq. (10) may be replaced by integrals over the pore volume V_p since the species mass fractions are zero within the impermeable solid phase, viz.

$$(1/V)\int_{V_p} \rho(\partial Y_K/\partial t)\,dv = (1/V)\int_{V_p} \underline{\nabla}\cdot(\rho D\underline{\nabla} Y_K)\,dv + (1/V)\int_{V_p} \omega_K\,dv. \tag{11}$$

The subsequent derivation is based on procedures developed by Carbonell and Whitaker (1983) and Zanotti and Carbonell (1984). The following notation will be used for the phase and intrinsic phase averages of ψ, respectively:

$$\langle\psi\rangle = (1/V)\int_{V_p} \psi\,dv = \text{phase average of } \psi, \tag{12}$$

$$\langle\psi\rangle_p = (1/V_p)\int_{V_p} \psi\,dv = \langle\psi\rangle/\varepsilon = \text{intrinsic phase average of } \psi, \tag{13}$$

where ψ may represent an arbitrary scalar, vector (with single underline), or tensor (with double underline) quantity and $\varepsilon = V_p/V$ is the volume fraction occupied by the pore phase. When averaging over the solid volume V_s, the subscript p is replaced by s in Eqs. 12 and 13 and ε by $1 - \varepsilon$ in Eq. 13, where $1 - \varepsilon = V_s/V$.

The minimum acceptable value for L_0 (see Fig. 2.1) is such that $\langle\psi\rangle$ and $\langle\psi\rangle_p$ remain nearly independent of position over distances of the order of $L_0(\langle\langle\psi\rangle\rangle = \langle\psi\rangle, \langle\langle\psi\rangle_p\rangle_p = \langle\psi\rangle_p)$. Whitaker (1969) has shown that a sufficient condition for averaging is $L \gg L_0 \gg l$, which is intuitively consistent with the diagram shown in Fig. 2.1. This last condition constitutes an acceptable approximation for nuclear-grade graphite (Baker 1970).

Using the notation of Eq. 12, we may rewrite Eq. 11 as

$$\rho\langle\partial Y_K/\partial t\rangle = \langle\underline{\nabla}\cdot(\rho D\underline{\nabla} Y_K)\rangle + \langle\omega_K\rangle, \tag{14}$$

where it has been assumed in the first term that ρ is constant within V_p. Next, we apply a Leibnitz theorem to the transient term in Eq. 14, with the result

$$\rho\partial\langle Y_K\rangle/\partial t = \langle\underline{\nabla}\cdot\rho D\underline{\nabla} Y_K\rangle + \langle\omega_K\rangle + (\rho/V)\int_{A_p} \underline{n}_p\cdot\underline{v}_p Y_K\,dA, \tag{15}$$

where $\underline{v}_p$ is the rate of regression of A_p caused by heterogeneous chemical reactions. For the graphite-oxidation problem, the last term in Eq. 15 may be shown to be negligibly small. Equation 15 then becomes

$$\rho\partial\langle Y_K\rangle/\partial t = \langle\underline{\nabla}\cdot\rho D\underline{\nabla}Y_K\rangle + \langle\omega_K\rangle. \tag{16}$$

Using next the spatial averaging theorem (Slattery 1981) for a two-phase system, we find

$$\langle\underline{\nabla}\cdot\underline{\psi}\rangle = \underline{\nabla}\cdot\langle\underline{\psi}\rangle - (1/V)\int_{A_p}\underline{n}_p\cdot\underline{\psi}\,dA \tag{17}$$

and

$$\langle\underline{\nabla}\psi\rangle = \underline{\nabla}\langle\psi\rangle - (1/V)\int_{A_p}\underline{n}_p\psi\,dA. \tag{18}$$

Applying Eqs. 17 and 18 to Eq. 16 and assuming that the product ρD is constant within V_p yields

$$\rho\partial\langle Y_K\rangle/\partial t = \underline{\nabla}\cdot(\rho D\underline{\nabla}\langle Y_K\rangle) - \underline{\nabla}\cdot\left[(\rho D/V)\int_{A_p}\underline{n}_p Y_K\,dA\right] - (1/V)\int_{A_p}\underline{n}_p\cdot(\rho D\underline{\nabla}Y_K)\,dA + \langle\omega_K\rangle. \tag{19}$$

The third term on the right-hand side of Eq. 19 may be expressed in terms of the boundary condition at the pore surface, which is given by

$$-\underline{n}_p\cdot\rho D\underline{\nabla}Y_K = \Gamma_K \text{ on } A_p, \tag{20}$$

where Γ_K is the mass-production (Γ_K positive) or mass-consumption (Γ_K negative) rate of species K per unit area of A_p caused by heterogeneous chemical reactions or by adsorption or desorption. Substituting Eq. 20 into Eq. 19 and using the definition of ε yields

$$\rho\partial(\varepsilon\langle Y_K\rangle_p)/\partial t = \underline{\nabla}\cdot[\rho D\underline{\nabla}(\varepsilon\langle Y_K\rangle_p)] - \underline{\nabla}\cdot\left[(\rho D/V)\int_{A_p}\underline{n}_p Y_K\,dA\right] + (1/V)\int_{A_p}\Gamma_K\,dA + \langle\omega_K\rangle. \tag{21}$$

In the subsequent analysis, it will be convenient to express the point value of the mass fraction Y_K as the sum of the intrinsic phase average $\langle Y_K\rangle_p$ and a deviation from the average $\bar{Y}_K$, where

$$\bar{Y}_K = Y_K - \langle Y_K\rangle_p. \tag{22}$$

By definition, $\langle\bar{Y}_K\rangle_p = 0$. Next we note that the area integral in the second term on the right-hand side of Eq. 21 becomes, with the use of Eq. 22,

$$\int_{A_p}\underline{n}_p(\bar{Y}_K + \langle Y_K\rangle_p)\,dA = \int_{A_p}\underline{n}_p\bar{Y}_K\,dA + \langle Y_K\rangle_p\int_{A_p}\underline{n}_p\,dA. \tag{23}$$

According to Gray's theorem (1975), which relates area integrals of the unit normal to gradients in volume fraction,

$$\underline{\nabla}\varepsilon = (1/V)\int_{A_p} \underline{\mathrm{n}}_p\, dA. \tag{24}$$

We will approximate ε as a constant over the entire porous medium (and hence delete the second term on the right-hand side of Eq. 23), i.e., we assume that small variations in V_p/V occur over the medium compared with variations in $\langle Y_K\rangle_p$. For low levels of burn-up, this approximation is reasonable for the kinetically and boundary-layer-controlled regimes, and may also be acceptable for the diffusion-controlled regime. Modifications will be introduced in a subsequent paper that lead to a relation that is analogous to Eq. 25 when gradients in ε are not neglected. For constant ε, Eq. 21 becomes

$$\rho\partial(\varepsilon\langle Y_K\rangle_p)/\partial t = \underline{\nabla}\cdot(\varepsilon\rho D\underline{\nabla}\langle Y_K\rangle_p) - \underline{\nabla}\cdot\left[(\rho D/V)\int_{A_p} \underline{\mathrm{n}}_p \bar{Y}_K\, dA\right]$$
$$+ (1/V)\int_{A_p} \Gamma_K\, dA + \langle\omega_K\rangle, \tag{25}$$

or, after dividing each term by ε and noting that $\varepsilon V = V_p$ and $\langle\omega_K\rangle/\varepsilon = \langle\omega_K\rangle_p$,

$$\rho\partial\langle Y_K\rangle_p/\partial t = \underline{\nabla}\cdot(\rho D\underline{\nabla}\langle Y_K\rangle_p) - \underline{\nabla}\cdot\left[(\rho D/V_p)\int_{A_p} \underline{\mathrm{n}}_p \bar{Y}_K\, dA\right]$$
$$+ (1/V_p)\int_{A_p} \Gamma_K\, dA + \langle\omega_K\rangle_p. \tag{26}$$

The corresponding equation for $\bar{Y}_K$ is obtained by subtracting Eq. 26 from Eq. 9 and using the definition of $\bar{Y}_K$ given by Eq. 22. The result is

$$\rho\partial\bar{Y}_K/\partial t = \underline{\nabla}\cdot(\rho D\underline{\nabla}\bar{Y}_K) + \underline{\nabla}\cdot\left[(\rho D/V_p)\int_{A_p} \underline{\mathrm{n}}_p \bar{Y}_K\, dA\right] - (1/V_p)\int_{A_p} \Gamma_K\, dA$$
$$+ \omega_K - \langle\omega_K\rangle_p. \tag{27}$$

Equations 26 and 27 are two coupled equations for the unknowns $\langle Y_K\rangle_p$ and $\bar{Y}_K$. If the source terms were not present, the equations would be uncoupled and we could solve Eq. 27 for $\bar{Y}_K$ and substitute this solution into Eq. 26 to obtain a single equation for $\langle Y_K\rangle_p$. However, this approach requires that A_p is a known function of position within the porous medium. Ryan et al. (1980) and Eidsath et al. (1983) have successfully performed calculations of this type for the transport of a single phase in a spatially periodic porous medium. Zanotti and Carbonell (1984) have extended this development and have considered the transport of two phases in one-dimensional coaxial flow.

The second term on the right-hand side of Eq. 26 accounts for the tortuosity of the porous medium. The conventional approach (Hewitt 1965) used for graphite and other materials with unknown pore structure is to account for tortuosity by multiplying the diffusion coefficient by an empirical tortuosity coefficient. In Eq. 26, the dimensionless tortuosity coefficient ϕ is defined according to

$$\underline{\nabla}\cdot(\phi\rho D\underline{\nabla}\langle Y_K\rangle_p) = \underline{\nabla}\cdot(\rho D\underline{\nabla}\langle Y_K\rangle_p) - \underline{\nabla}\cdot\left[(\rho D/V_p)\int_{A_p}\underline{n}_p\bar{Y}_K\,dA\right]. \quad (28)$$

For graphite, ϕ is typically measured (Hewitt 1965) by flowing gas streams of different composition past opposite faces of a graphite specimen and measuring the concentration changes in one or both gas streams. The transport equations are then solved by varying ϕ until good agreement is obtained with the experimental data. Typical measured values for ϕ are about 0.01 for nuclear-grade graphite (Hawtin et al. 1967).

We assume that ϕ is small, as is done customarily and known to be the case from experiments. Using this empirical information, an order-of-magnitude estimate may be obtained for the ratio $|\bar{Y}|/\langle Y_K\rangle_p$, as will now be shown. We use Eq. 28 and note that averaged quantities vary over distances of order L_0, whereas the ratio A_p/V_p is of order l^{-1}. We then find that

$$\phi = O[1 - (L_0/l)|\bar{Y}_K|/\langle Y_K\rangle_p]. \quad (29)$$

From Eq. 29, we see that ϕ is a small number and agrees with measured values if the ratio $|\bar{Y}_K|/\langle Y_K\rangle_p$ is of order l/L_0.† Subsequently, we will apply this type of methodology also to the temperature. In view of Eq. 29, we may evaluate properties and source or sink terms using averaged quantities without introducing significant errors and thus justify our previous assumptions that ρ and ρD are constants within V_p. Using also Eq. 28, we may write Eq. 26 as

$$\rho\partial\langle Y_K\rangle_p/\partial t = \underline{\nabla}\cdot(\phi\rho D\underline{\nabla}\langle Y_K\rangle_p) + \Gamma_K a_p + \omega_K, \quad (30)$$

where $a_p = A_p/V_p$ is the internal surface area per unit volume and ρ, D, Γ_K, and ω_K are evaluated by using averaged quantities. The measured values for ϕ are usually obtained under non-reactive or low-temperature reactive conditions. Later, we will discuss limitations on using Eq. 30 for high-temperature graphite oxidation.

Energy-Conservation Equation

For negligibly small mass-average velocity components, the energy-conservation equation (Eq. 4) reduces to

$$\rho c_p\partial T/\partial t = \underline{\nabla}\cdot\lambda\underline{\nabla}T - \underline{\nabla}\cdot q_R + q, \quad (31)$$

where $q = -\sum_{K=1}^{N} h_K\omega_K$ is the net volumetric heat-release rate from homogeneous chemical reactions. For the solid and pore phases, Eq. 31 yields,

† The tortuosity coefficient ϕ in graphite increases with burn-up and our empirical adjustment $|\bar{Y}_K|/\langle Y_K\rangle_p$ will then ultimately fail.

respectively,

$$\rho_s c_{p,s} \partial T_s/\partial t = \underline{\nabla} \cdot \lambda_s \underline{\nabla} T_s, \tag{32}$$

$$\rho_p c_{p,p} \partial T_p/\partial t = \underline{\nabla} \cdot \lambda_p \underline{\nabla} T_p + q_p, \tag{33}$$

where the subscript s has been introduced to denote the solid phase. Heat release or heat absorption at the pore surface will be allowed for by introducing appropriate boundary conditions for Eqs. 32 and 33. In Eq. 33, we have assumed that transport due to thermal radiation is either negligibly small or else may be accounted for by including a Rosseland thermal conductivity in the total thermal conductivity λ_p.

We now proceed in the same manner as in the derivation of Eq. 30 for species conservation. Again referring to Fig. 2.1, Eqs. 32 and 33 are integrated over the solid and pore phases, respectively, with the results

$$\rho_s c_{p,s} \langle \partial T_s/\partial t \rangle = \langle \underline{\nabla} \cdot \lambda_s \underline{\nabla} T_s \rangle, \tag{34}$$

$$\rho_p c_{p,p} \langle \partial T_p/\partial t \rangle = \langle \underline{\nabla} \cdot \lambda_p \underline{\nabla} T_p \rangle + \langle q_p \rangle, \tag{35}$$

where we have assumed that $\rho_s c_{p,s}$ and $\rho_p c_{p,p}$ are constants within their respective averaging volumes. Application of a Leibnitz theorem to the transient terms transfers Eqs. 34 and 35, respectively, to

$$\rho_s c_{p,s} \partial \langle T_s \rangle/\partial t = \langle \underline{\nabla} \cdot \lambda_s \underline{\nabla} T_s \rangle, \tag{36}$$

$$\rho_p c_{p,p} \partial \langle T_p \rangle/\partial t = \langle \underline{\nabla} \cdot \lambda_p \underline{\nabla} T_p \rangle + \langle q_p \rangle, \tag{37}$$

where we have assumed that the term involving motion of the surface is negligibly small, as we did previously in the species-conservation equations. Application of the spatial averaging theorem (see Eqs. 17 and 18), with the assumptions that λ_s and λ_p are constant within their averaging volumes and that variations in ε are also negligibly small, yields

$$(1-\varepsilon)\rho_s c_{p,s} \partial \langle T_s \rangle_s/\partial t = \underline{\nabla} \cdot [(1-\varepsilon)\lambda_s \underline{\nabla} \langle T_s \rangle_s] + \underline{\nabla} \cdot \left[(\lambda_s/V) \int_{A_p} \underline{\underline{n}}_p T_s \, dA \right] + (1/V) \int_{A_p} \underline{\underline{n}}_p \cdot \lambda_s \underline{\nabla} T_s \, dA \tag{38}$$

and

$$\varepsilon \rho_p c_{p,p} \partial \langle T_p \rangle_p/\partial t = \underline{\nabla} \cdot (\varepsilon \lambda_p \underline{\nabla} \langle T_p \rangle_p) - \underline{\nabla} \cdot \left[(\lambda_p/V) \int_{A_p} \underline{\underline{n}}_p T_p \, dA \right] - (1/V) \int_{A_p} \underline{\underline{n}}_p \cdot \lambda_p \underline{\nabla} T_p \, dA + \langle q_p \rangle. \tag{39}$$

We now express the point values of temperature in each phase as the sum of the intrinsic phase average and a spatial deviation as follows:

$$T_s = \langle T_s \rangle_s + \bar{T}_s \tag{40}$$

$$T_p = \langle T_p \rangle_p + \bar{T}_p. \tag{41}$$

Using Eq. 40 in Eq. 38 and Eq. 41 in Eq. 39 and applying Gray's theorem (see Eq. 24) yields

$$(1-\varepsilon)\rho_s c_{p,s}\partial\langle T_s\rangle_s/\partial t = \underline{\nabla}\cdot[(1-\varepsilon)\lambda_s\underline{\nabla}\langle T_s\rangle_s] + \underline{\nabla}\cdot\left[(\lambda_s/V)\int_{A_p}\underline{n}_p\bar{T}_s\,dA\right] + (1/V)\int_{A_p}\underline{n}_p\cdot\lambda_s\underline{\nabla}T_s\,dA \tag{42}$$

and

$$\varepsilon\rho_p c_{p,p}\partial\langle T_p\rangle_p/\partial t = \underline{\nabla}\cdot(\varepsilon\lambda_p\underline{\nabla}\langle T_p\rangle_p) - \underline{\nabla}\cdot\left[(\lambda_p/V)\int_{A_p}\underline{n}_p\bar{T}_p\,dA\right] - (1/V)\int_{A_p}\underline{n}_p\cdot\lambda_p\underline{\nabla}T_p\,dA + \langle q_p\rangle. \tag{43}$$

We next introduce the assumption of local thermal equilibrium (LtE) between the solid and pore phases through the relations

$$\langle T_s\rangle_s = \langle T_p\rangle_p = \langle T\rangle. \tag{44}$$

Whitaker (1981) has shown that this assumption is valid provided that

$$(1-\varepsilon)(l/L_0)^2[1+(\lambda_s/\lambda_p)] \ll 1. \tag{45}$$

Since the ratio l/L_0 is of the order of 10^{-3} (see Fig. 2.1) and the ratio λ_s/λ_p is of the order of 10^3 for oxidation of nuclear-grade graphite in air, the assumption of LtE will be justified for the present application. Using this assumption and adding Eqs. 42 and 43, we find

$$(\rho c_p)_e\partial\langle T\rangle/\partial t = \underline{\nabla}\cdot\{[(1-\varepsilon)\lambda_s+\varepsilon\lambda_p]\underline{\nabla}\langle T\rangle\} + \underline{\nabla}\cdot\left[(\lambda_s/V)\int_{A_p}\underline{n}_p\bar{T}_s\,dA - (\lambda_p/V)\int_{A_p}\underline{n}_p\bar{T}_p\,dA\right] + (1/V)\int_{A_p}(\underline{n}_p\cdot\lambda_s\underline{\nabla}T_s - \underline{n}_p\cdot\lambda_p\underline{\nabla}T_p)\,dA + \langle q_p\rangle, \tag{46}$$

where $(\rho c_p)_e$ is the effective volumetric heat capacity of the porous medium and is defined according to the relation

$$(\rho c_p)_e \equiv (1-\varepsilon)\rho_s c_{p,s} + \varepsilon\rho_p c_{p,p}. \tag{47}$$

The boundary conditions at the pore-solid interface are

$$T_s = T_p \quad\text{and}\quad \bar{T}_s = \bar{T}_p \text{ on } A_p \tag{48}$$

and

$$\underline{n}_p\cdot\lambda_s\underline{\nabla}T_s = \underline{n}_p\cdot\lambda_p\underline{\nabla}T_p - \sum_{K=1}^{N}\Gamma_K h_K \text{ on } A_p, \tag{49}$$

where the second equality in Eq. 48 expresses the assumption of LtE. Using Eqs. 48 and 49 in Eq. 46 and the fact that $\lambda_s \gg \lambda_p$ leads to

$$(\rho c_p)_e \partial \langle T \rangle / \partial t = \underline{\nabla} \cdot [(1 - \varepsilon)\lambda_s \underline{\nabla} \langle T \rangle] + \underline{\nabla} \cdot \left[(\lambda_s / V) \int_{A_p} \underline{n}_p \bar{T}_s \, dA \right] + (1/V) \int_{A_p} Q \, dA + \langle q_p \rangle, \tag{50}$$

where $Q = -\sum_{K=1}^{N} \Gamma_K h_K$ is the net heat-release rate per unit area from heterogeneous chemical reactions on the pore surface. The second term on the right-hand side of Eq. 50 is referred to as a tortuosity term. It is convenient to define an effective thermal conductivity λ_e for the porous medium through the relation

$$\underline{\nabla} \cdot \lambda_e \underline{\nabla} \langle T \rangle = \underline{\nabla} \cdot [(1 - \varepsilon)\lambda_s \underline{\nabla} \langle T \rangle] + \underline{\nabla} \cdot \left[(\lambda_s / V) \int_{A_p} \underline{n}_p \bar{T}_s \, dA \right]. \tag{51}$$

We make an order-of-magnitude estimate for λ_e by following the procedure which we used previously to estimate ϕ (see Eq. 29). The result is

$$\lambda_e = O[(1 - \varepsilon)\lambda_s + \varepsilon \lambda_s (L_0 / l) | \bar{T}_s | / \langle T \rangle]. \tag{52}$$

We observe that since $\lambda_s \gg \lambda_p$, the maximum value for λ_e is λ_s, i.e., the presence of the gas-filled pore structure reduces λ_e to values less than λ_s. For this result to hold, we see from Eq. 52 that the ratio $|\bar{T}_s|/\langle T \rangle$ must be of order l/L_0. The ratio $|\bar{T}_p|/\langle T \rangle$ must also be of order l/L_0 since the assumption of LtE requires $\bar{T}_p$ to be of the same order as $\bar{T}_s$. We conclude that the absolute values of spatial deviations in temperature are much smaller than the corresponding intrinsic phase-averages of the temperatures. Since we previously showed that a similar result was also valid for the species mass fractions, we may evaluate properties and source or sink terms by using averaged quantities without introducing significant errors (see Eq. 30 and the previous discussion). Using this result, Eq. 51 and the definition of ε, we may write Eq. 50 as

$$(\rho c_p)_e \partial \langle T \rangle / \partial t = \underline{\nabla} \cdot \lambda_e \underline{\nabla} \langle T \rangle + \varepsilon Q a_p + \varepsilon q_p, \tag{53}$$

where $(\rho c_p)_e$, λ_e, Q, and q_p are evaluated by using averaged quantities. In the next section, we will discuss limitations on using Eq. 53 for high-temperature graphite oxidation.

Model Limitations

The previously-derived governing equations for the phase averages of species mass fractions (Eq. 30) and temperature (Eq. 53) may be solved without requiring knowledge of either the pore structure or spatial deviations in mass fractions and temperature. However, to arrive at Eqs. 30 and 53, the assumptions $|\bar{Y}_K|/\langle Y_K \rangle \ll 1$ and $|\bar{T}_s|/\langle T \rangle \ll 1$ are required. The conditions under which these assumptions are valid will now be discussed for oxidation of nuclear-grade graphite in air.

An order-of-magnitude estimate for the ratio $|\bar{Y}_K|/\langle Y_K \rangle_p$ may be obtained from Eq. 26. For sufficiently long times (of the order of milliseconds for the present application), the contribution to $|\bar{Y}_K|/\langle Y_K\rangle_p$ from the transient term is negligibly small and this ratio is given by

$$|\bar{Y}_K|/\langle Y_K\rangle \simeq L_0|\Gamma_K|/\rho D\langle Y_K\rangle_p + lL_0|\omega_K|/\rho D\langle Y_K\rangle_p + l/L_0. \tag{54}$$

The first two terms on the right-hand side of Eq. 54 are the contributions to $|\bar{Y}_K|/\langle Y_K\rangle_p$ from heterogeneous and homogeneous reactions, respectively, and the term l/L_0 is the contribution from diffusion. A similar estimate for temperature using Eq . 50 yields

$$|\bar{T}_s|/\langle T\rangle \simeq L_0|Q|/\lambda_s\langle T\rangle + lL_0|q_p|/\lambda_s\langle T\rangle + l/L_0. \tag{55}$$

The terms on the right-hand side of Eq. 55 are analogous to the terms in Eq. 54. Next we apply Eqs. 54 and 55 to the graphite-oxidation problem and estimate the limits on temperature for using Eqs. 30 and 53.

For oxidation of nuclear-grade graphite in air, we consider the heterogeneous reaction

$$C + [1 - (f_{CO}/2)]O_2 \rightarrow f_{CO}CO + (1 - f_{CO})CO_2 \tag{56}$$

and the homogeneous reaction

$$CO + (1/2)O_2 \xrightarrow{H_2O} CO_2. \tag{57}$$

We may neglect the Boudouard reaction,

$$C + CO_2 \rightarrow 2\,CO, \tag{58}$$

since it is about five orders of magnitude slower (Walker et al. 1959) than graphite gasification by oxygen at the temperatures of interest. We note that the rate for the reaction given by Eq. 57 is negligibly small in the absence of moisture (Dixon-Lewis and Williams 1977). In the reaction given by Eq. 56, f_{CO} is the number of moles of CO formed per mole of C reacting. The heat release for this reaction is an appropriateiy weighted sum of the heat released for two separate heterogeneous reactions, one producing only CO and the other producing only CO_2.

The graphite reaction rate is usually determined experimentally in terms of a reaction frequency k defined by the rate of mass loss as follows:

$$dM_C/dt = -kM_{C,o}, \tag{59}$$

where M_C and $M_{C,o}$ are the mass and original mass of graphite, respectively. Since the mass does not change significantly during the observation time, we may approximate the mass-loss equation by the relation

$$d\rho_C/dt = -k\rho_C, \tag{60}$$

where ρ_C is the observed mass density of the two-phase porous graphite. It is apparent that the quantity $\Gamma_C a_p$ is dimensionally and physically equivalent to $-k\rho_C/\varepsilon$, i.e., it has the units g of C/cm^3 of pore volume-time. In a similar manner, we may write Γ_K for O_2, CO, and CO_2, respectively, as

$$\Gamma_{O_2} = -[1 - (f_{CO}/2)](k\rho_C/\varepsilon a_p)(W_{O_2}/W_C), \tag{61}$$

$$\Gamma_{CO} = (f_{CO}k\rho_C/\varepsilon a_p)(W_{CO}/W_C), \tag{62}$$

$$\Gamma_{CO_2} = (1 - f_{CO})(k\rho_C/\varepsilon a_p)(W_{CO_2}/W_C). \tag{63}$$

The quantity a_p is approximately $1.5 \times 10^4\,cm^{-1}$ for graphite with representative measured nitrogen BET surface areas of about $2200\,cm^2/g$ (Baker 1970). This value for a_p corresponds to an equivalent cylindrical pore diameter of about 2.7×10^{-4} cm that is in good agreement with measured pore-size distributions (Baker 1970). Typical values for ρ_C and ε are $1.7\,g/cm^3$ and 0.2, respectively (Baker 1970). A correlation for the graphite-oxygen reaction frequency is (Jensen et al. 1973)

$$k = 7.78 \times 10^4 \exp(-20,129/T)p(\overline{W}/W_{O_2})Y_{O_2}, \tag{64}$$

where k is in sec^{-1}, T in K, and p in atm. To estimate f_{CO}, we use the correlation developed by Arthur (1951), viz.

$$X_{CO}/X_{CO_2} = f_{CO}/(1 - f_{CO}) = 2512\exp(-6240/T), \tag{65}$$

where X_K is the mole fraction of K. Arthur performed his experiments over the temperature range 480–1000°C under conditions in which gas-phase reactions were blocked by injecting $POCl_3$, even in the presence of moisture.

In the presence of moisture, an empirically determined reaction frequency in sec^{-1} for the reaction given by Eq. 57 is (Howard et al. 1973)

$$k' = 1.3 \times 10^{14}\rho(Y_{O_2}/W_{O_2})^{0.5}(Y_{H_2O}/W_{H_2O})^{0.5}\exp(-15,097/T), \tag{66}$$

where ρ is in g/cm^3, We may then evaluate ω_K for O_2, CO, and CO_2 as follows.

$$\omega_{O_2} = (-1/2)\rho Y_{CO}k'(W_{O_2}/W_{CO}), \tag{67}$$

$$\omega_{CO} = -\rho Y_{CO}k', \tag{68}$$

and

$$\omega_{CO_2} = \rho Y_{CO}k'(W_{CO_2}/W_{CO}). \tag{69}$$

To estimate D, we use the following correlation for binary diffusion of O_2 in N_2 (Edwards et al. 1979):

$$D = (0.325/p)(T/400)^{1.673} \text{ in } cm^2/sec. \tag{70}$$

We make the following additional assumptions to evaluate Eq. 54 for $|\overline{Y}_K|/\langle Y_K\rangle_p$ and Eq. 55 for $|\overline{T}_s|/\langle T\rangle$: (1) $l = 2.7 \times 10^{-4}$ cm; (2) $L_0 = 0.27$ cm; (3) $Y_{O_2} = 0.233$; (4) $Y_{H_2O} = 0.01$ (48 percent relative humidity); (5) ratios of different species mass fractions, e.g., Y_{O_2}/Y_{CO}, are taken to be unity, (6) $\overline{W} =$ 28.85 g/mole; (7) $p = 1$ atm; (8) $\lambda_s = 0.2$ W/cm-°C; (9) the heat release for graphite gasification to CO = 110,529 J/mole of CO; (10) the heat release for graphite gasification to CO_2 = 393,522 J/mole of CO_2; and (11) the heat release for CO combustion = 282,993 J/mole of CO. Assumptions (3) through (6) yield estimates of $|\overline{Y}_K|/\langle Y_K\rangle_p$ and $|\overline{T}_s|/\langle T\rangle$ without actually solving the

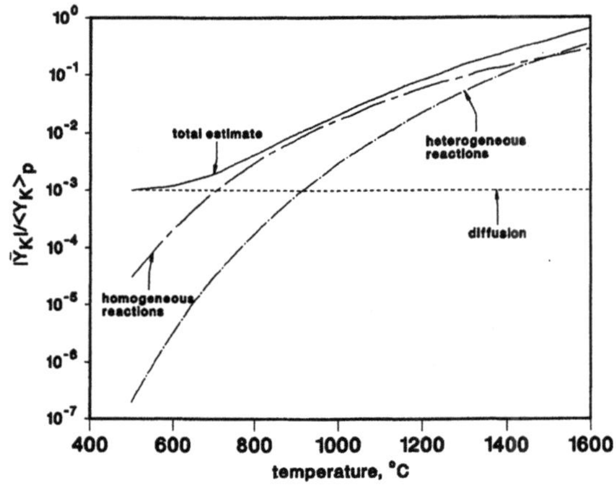

FIGURE 2.2. Estimates for species-averaged values of $|\bar{Y}_K|/\langle Y_K\rangle_p$ for graphite oxidation in air using Eq. 54. The four curves represent averages for $K = O_2$, CO, and CO_2 and refer to the individual contributions made by diffusion, heterogeneous reactions, homogeneous reactions and the sum of these.

conservation equations. It is not certain that these assumptions will lead to upper-bound estimates over the entire temperature range considered in the present analysis. In a sequel, the present model is compared with experimental data and more rigorous estimates are made for $|\bar{Y}_K|/\langle Y_K\rangle_p$ and $|\bar{T}_s|/\langle T\rangle$ in order to obtain good fits to the observed results.

Figure 2.2 shows species-averaged estimates for $|\bar{Y}_K|/\langle Y_K\rangle_p$ derived from Eq. 54 for the specified numerical values. Differences among the calculated ratios for $K = O_2$, CO, or CO_2 were within a factor of 3 over the temperature range (500°–1600°C) considered. Variations within a factor of 3 fall within the ranges of uncertainties associated with the empirical correlations and assumptions employed. Also shown in Fig. 2.2 are species-averaged values for the individual contributions made by diffusion, heterogeneous reactions, and homogeneous reactions. Reference to Fig. 2.2 shows that the contribution to $|\bar{Y}_K|/\langle Y_K\rangle_p$ made by chemical reactions is negligible for $T \lesssim 600°C$. For $600 \lesssim T$, $°C \lesssim 800$, both diffusion and homogeneous reactions make significant contributions. For $800 \lesssim T$, $°C \lesssim 1100$, homogeneous reactions are dominant. Above about 1100°C, both homogeneous and heterogeneous reactions are important, with the fractional changes in mass fractions caused by heterogeneous reactions exceeding those made by the homogeneous reactions at about 1500°C. The ratio $|\bar{Y}_K|/\langle Y_K\rangle_p$ is less than 0.01 for $T \lesssim 900°C$ and is less than 0.1 for $T \lesssim 1250°C$.

Estimates for $|\bar{T}_s|/\langle T\rangle$ using Eq. 55 are shown in Fig. 2.3 for the same numerical values. The contributions to $|\bar{T}_s|/\langle T\rangle$ made by chemical reactions are negligible for $T \lesssim 850°C$. The estimated value of $|\bar{T}_s|/\langle T\rangle$ is less than 0.01 at 1600°C and the criterion $|\bar{T}_s|/\langle T\rangle \ll 1$ is seen to be satisfied at higher temperatures than the corresponding criterion for the species-averaged mass fractions. Therefore, our pore-structure-independent combustion model may be used for graphite oxidation provided $T \lesssim 1250°C$. Extensions of the model to higher temperatures are discussed in the next section.

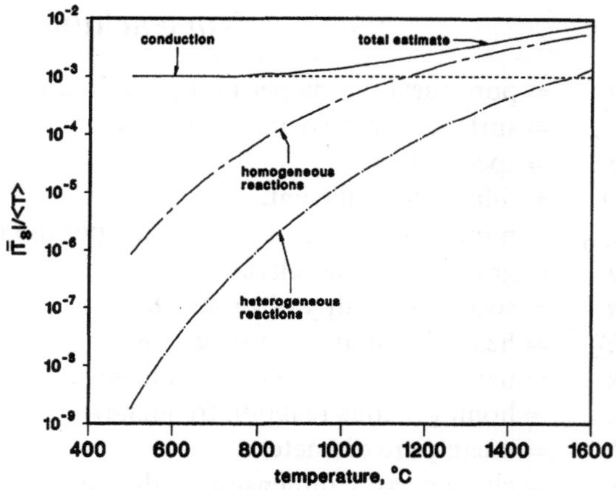

FIGURE 2.3. Estimates of $|\bar{T}_s|/\langle T\rangle$ for graphite oxidation in air using Eq. 55. The four curves represent, respectively, the total estimate for $|\bar{T}_s|/\langle T\rangle$ and the individual contributions made by conduction, homogeneous reactions, and heterogeneous reactions.

Model Assessment and Extension

We have shown that order-of-magnitude estimates of diffusion and reaction rates in porous graphite allow the construction of a pore-structure-independent model for graphite oxidation at temperatures up to about 1250°C, for which reaction rates and in-pore diffusion jointly control combustion. At appreciably higher temperatures, the constraints imposed by the magnitudes of the source or sink terms invalidate the assumptions required for pore-structure-independent combustion models of the type we have discussed in this paper. However, in these high-temperature regimes, a different type of simplification will become useful. At high temperatures, the maximum diffusion length over which the oxidizer persists is confined to a thin zone near the surface. Since this zone is difficult to define, measured surface-reaction rates usually refer to the external surface area. Therefore, at high temperatures, effective chemical conversions may be modeled at the geometric boundary of the porous graphite. In this limit, the oxidation rate becomes controlled by boundary-layer transport. The source and sink terms may then be described in terms of boundary conditions external to the entire porous medium, i.e., in the mathematical formulation, they no longer occur in the conservation equations for the porous medium.

For graphite oxidation and most combustion problems, chemical reaction rates increase so rapidly with temperature that the transition between the two specified oxidation regimes occurs over a narrow temperature range. For this reason, we expect to be able to use the pore-structure-independent model for $T \lesssim T_1$ whereas the boundary-layer-controlled model applies for $T \gtrsim T_2$. Furthermore, $T_2 - T_1$ covers a relatively small temperature range for which we obtain solutions by extrapolating the high- or low-temperature results. In a sequel (Richards and Penner 1990), a model of this type is applied to the graphite-oxidation problem.

Nomenclature

a_p = pore surface area per unit pore volume.
A_p = surface area associated with the pore structure.
c_p = specific heat.
D = diffusion coefficient.
f_{CO} = number of moles of CO formed per mole of C reacting.
$\underline{g}$ = gravitational acceleration.
h_K = specific enthalpy of species K.
h_K^0 = heat of formation of species K.
k = heterogeneous reaction frequency.
k' = homogeneous reaction frequency.
l = mean pore diameter.
L = characteristic dimension of the porous medium.
L_0 = characteristic dimension of the averaging volume.
M_C = mass of carbon.
$M_{C,0}$ = initial mass of carbon.
N = number of chemical species.
$\underline{n}_p$ = unit normal pointing from the solid phase to the pore phase.
p = pressure.
q = net volumetric heat-release rate from homogeneous chemical reactions.
Q = net heat-release rate per unit area from heterogeneous chemical reactions on the pore surface.
$\underline{q}_R$ = radiant heat-flux vector.
R = universal gas constant.
T = temperature.
T^0 = reference temperature.
$\tilde{T}$ = spatial deviation of temperature.
$\underline{v}$ = mass-average velocity.
$\underline{v}_p$ = velocity of pore surface.
V = averaging volume.
W_K = molecular weight of species K.
$\bar{W}$ = molecular weight of gas mixture.
X_K = mole fraction of species K.
Y_K = mass fraction of species K.
$\tilde{Y}_K$ = spatial deviation of mass fraction of species K.
Δh_K = change in enthalpy of species K from the reference value.
ε = void fraction of the porous medium.
Γ_K = mass-generation rate of species K per unit area of pore surface.
λ = thermal conductivity.
μ = coefficient of viscosity.
ω_K = volumetric mass-generation rate of species K.
ϕ = tortuosity coefficient.
ρ = density.

ρ_C = density of porous graphite.
ψ = arbitrary scalar.
$\underline{\psi}$ = arbitrary vector.
$\langle\psi\rangle$ = phase average of ψ.
$\langle\psi\rangle_p$ = intrinsic phase average of ψ.

Subscripts

e = effective value for the porous medium.
p = pore phases.
s = solid phase.

References

Arthur, J.R., 1951, "Reactions Between Carbon and Oxygen," *Trans. Faraday Soc.*, *47*, 164–178.

Baker, D.E., 1970, "Graphite as a Neutron Moderator and Reflector Material," *Nuc. Engrg. Des.*, *14*, 413–444.

Carbonell, R.G., and Whitaker, S., 1983, "Dispersion in Pulsed Systems—II. Theoretical Developments for Passive Dispersion in Porous Media," *Chem. Engrg. Sci.*, *38*, 1795–1802.

Chapman. S., and Cowling, T.G., 1953, *The Mathematical Theory of Non-Uniform Gases*, Cambridge University Press, Cambridge.

Dixon-Lewis, G., and Williams, D.J., 1977, "The Oxidation of Hydrogen and Carbon Monoxide," *Comprehensive Chemical Kinetics*, C.H. Bamford and C.F.H. Tipper, eds., Elsevier Scientific Publ. Co., Amsterdam, The Netherlands, Vol. 17, 1–248.

Edwards, D.K., Denny, J.E., and Mills, A.F., 1979, *Transport Processes, An Introduction to Diffusion, Convection, and Radiation*, 2d ed., Hemisphere Publ. Corp., New York.

Eidsath, A., Carbonell, R.G., Whitaker, S., and Herrmann, L.R., 1983, "Dispersion in Pulsed Systems—III. Comparison Between Theory and Experiments for Packed Beds," *Chem. Engrg. Sci.*, *38*, 1803–1816.

Gray, W.G., 1975, "A Derivation of the Equations for Multiphase Transport," *Chem. Engrg. Sci.*, *30*, 229–233.

Hawtin, P., Hewitt, G.F., and Roberts, J., 1967, "Flow of Gases in Porous Solids," *Nature*, *215*, 1415–1416.

Hewitt, G.F., 1965, "Gaseous Mass Transport Within Graphite," *Chemistry and Physics of Carbon*, P.L. Walker, Jr., ed., Marcel Dekker, Inc., New York, 1, 73–120.

Hirschfelder, J.O., Curtiss, C.F., and Bird, R.B., 1954, *Molecular Theory of Gases and Liquids*, Wiley, New York.

Howard, J.B., Williams, G.C., and Fine, O.M., 1973, "Kinetics of Carbon Monoxide Oxidation in Postflame Gases," *14th Symposium (International) on Combustion*, The Combustion Institute, Pittsburgh, Pa., 975–986.

Jensen, D., Tagami, M., and Velasquez, C., 1973, "Air/H-327 Graphite Reaction Rate as a Function of Temperature and Irradiation," *Report Gulf-GA-A12647*, General Atomics, San Diego, Calif.

Levec, J., and Carbonell, R.G., 1985, "Longitudinal and Lateral Thermal Dispersion in Packed Beds. Part I: Theory," *AIChE J.*, *31*, 581–590.

Penner, S.S., 1957, *Chemistry Problems in Jet Propulsion*, Pergamon, New York.

Richards, M.B., and Penner, S.S., 1990, "Oxidation of a Porous Graphite Cylinder with Airflow through a Coaxial Hole," *12th Intl. Colloqium on Dynamics of Explosions and Reactive Systems*, Progress in Astronautics and Aeronautics, AIAA, Washington, D.C. (in press).

Ryan, D., Carbonell, R.G., and Whitaker, S., 1980, "Effective Diffusivities for Catalyst Pellets Under Reactive Conditions," *Chem. Engrg. Sci.*, *35*, 10–16.

Slattery, J.C., 1981, *Momentum, Energy, and Mass Transfer in Continua*, 2d ed., Robert E. Krieger Publ. Co., Huntington, N.Y.

Walker, P.L., Jr., Rusinko, F., Jr., and Austin, L.G., 1959, "Gas Reactions of Carbon," *Adv. Catal.*, *11*, 133–221.

Whitaker, S., 1969, "Advances in Theory of Fluid Motion in Porous Media," *Ind. Engrg. Chem.*, *61*, 12, 14–28.

Whitaker, S., 1981, "Heat and Mass Transfer in Granular Porous Media," *Advances in Drying* (A.S. Mujumdar, ed.), Hemisphere Publishing Corp., New York, 23–61.

Williams, F.A., 1985, *Combustion Theory*, 2d ed., Benjamin/Cummings Publishing Co., Menlo Park, Calif.

Wilson, R., 1987, "A Visit to Chernobyl," *Science*, *236*, 1636–1640.

Zanotti, F., and Carbonell, R.G., 1984, "Development of Transport Equations for Multiphase Systems—I. General development for Two Phase Systems," *Chem. Engrg. Sci.*, *39*, 263–278.

3
Stabilization of Hydrogen-Air Flames in Supersonic Flow

G. WINTERFELD

ABSTRACT: The operating range of scramjet engines for supersonic flight includes combustor inlet conditions, particularly in the lower flight Mach number range, which does not allow thermal self-ignition of hydrogen-air flames. Therefore, other methods have to be applied in order to insure safe and reliable ignition of supersonic combustion. Besides other methods, a very efficient tool for supersonic flame stabilization at low temperatures and pressures is the use of recirculation zones. These zones can also be generated in various ways in supersonic flows, for example, by bluff-body-type flame holders, by suitable shaping of fuel injectors or can be observed in connection with fuel jets injected from the sidewalls.

The Damköhler number for flame stabilization by recirculation zones, introducing the fuel characteristics via the laminar burning velocity, is derived and was checked experimentally for hydrocarbon fuels and hydrogen. It has been applied to hydrogen-air diffusion flames stabilized by cylindrical flame holders in supersonic flows up to Mach numbers of 2.1. Characteristic fluid mechanic times for hydrogen-air flames have been measured for a broad range of equivalance ratios. These results can be used for the sizing of recirculation zones or flames holders for supersonic and subsonic flame stabilization of hydrogen-air flames.

Introduction

In recent years interest in air-breathing propulsion for hypersonic flight speeds has been stimulated again by proposals for second-generation space transportation systems. Projects like the United States' National Aerospace Plane, the German "Sänger II," or the British "Hotol" will be equipped with combined propulsion systems, which are composed of turbine engines for the lower flight speed range and ramjet engines for the higher flight Mach number range. In the latter, subsonic and/or supersonic combustion will be employed according to the flight Mach number at which the rocket-propelled flight into orbit is to be started.

As is well known, hydrogen is the most suitable fuel for such propulsion systems, the efficiency of which is—to a large extent—dependent on the combustion characteristics of hydrogen, particularly at high flow velocities in the combustors. Therefore, numerous research activities on hydrogen combustion in supersonic flow were pursued in the 1960s and 70s, part of which extended until the 80s. The German aerospace research establishment, DFVLR (formerly DVL) has participated in these research activities, mainly concentrating on problems of thermal self-ignition in supersonic flow (Suttrop 1972, 1973), reaction kinetics of hydrogen-air ignition under conditions relevant for supersonic combustion (Schmalz 1971), and flame stabilization of hydrogen-air mixtures in supersonic flow by means of recirculation zones (Winterfeld 1968, 1976). Particularly, the results of the latter work are of interest also for hydrogen combustion in subsonic flow, as it prevails in turbine engines and conventional ramjets for Sänger-type vehicles. Therefore, it seems to be appropriate to review this work, which was described in an earlier German publication (Winterfeld 1976). In the literature, a number of similar studies can be found; Wilhelmi (1972) and Baev et al. (1971) may be mentioned as examples.

Basic Consideration for the Stabilization of Flames by Means of Recirculation Zones

It is well known that supersonic combustion in scramjets at high flight Mach numbers can be achieved by thermal self-ignition of the combustible mixture. This requires that the mixture temperature in the ignition region be above a certain level, the actual value of which depends on the aerothermodynamic state of the fuel and the air at the combustor inlet as well as on the permitted induction length of the flame (Suttrop 1973). The induction length is the distance between the fuel injector and the flame, that the combustible gases travel in the downstream direction during the ignition delay or induction time of the chemical reaction. The pertaining minimum temperature level fixes a flight Mach number below which thermal self-ignition cannot be reliably used.

If, however, supersonic combustion is to be applied below this threshold value, for example, in order to simplify the propulsion system by elimination of the conventional ramjet, ignition and flame stabilization must be achieved by other means (Suttrop 1972). One possibility is the dual-mode combustion using strong shock waves; another possibility is the use of recirculation zones, which will be discussed later. Recirculation zones will be present or can be easily generated in supersonic flow, for example, by fuel injection from side-walls, by rearward-facing steps in the combustor wall, or simply by a suitable combination of fuel injectors and flame holders.

The principles of flame stabilization by recirculation zones are well known from earlier work in subsonic flow, e.g. by Zukoski and Marble (1956). They will be discussed by means of Fig. 3.1, which shows a test setup used for

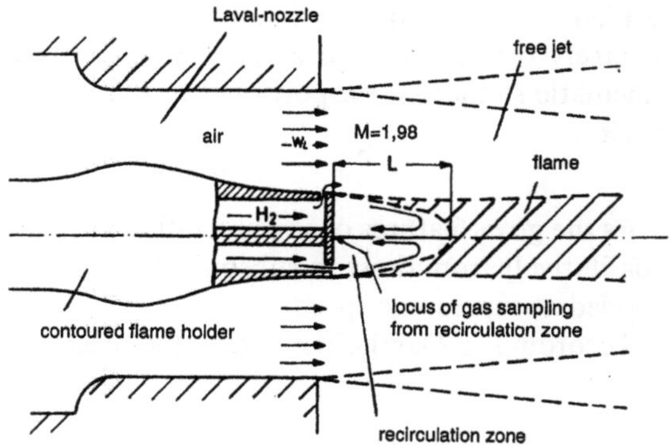

FIGURE 3.1. Schematic illustration of the test setup for flame stabilization experiments in supersonic flow, using a contoured flame holder in a casing with cylindrical cross section. Two different fuel-injection angles shown.

supersonic flame stabilization experiments described below. A contoured flame holder with cylindrical cross section is situated within a cylindrical casing; they form a Laval nozzle. Air leaves the nozzle as a supersonic free jet, generating a recirculation zone downstream of the flame holder's base after it has separated from the flame holder. Close to the separation point gaseous hydrogen fuel is injected into the adjacent airflow under different directions; hence a combustible mixture is formed in the wake and in the adjacent supersonic stream tubes. In this region a diffusion flame can be ignited. The stabilization process depends on the recirculating flow, which transports heat and radicals upstream toward the separation region where it mixes with the unburnt combustibles. The energy per unit mass available for continuous ignition depends on the enthalpy of the burnt gases, which must be as high as possible if reliable ignition is to be achieved. This, in turn, demands complete combustion in the critical zone. The whole process is therefore governed by two characteristic times, namely, a reaction time, t_R, which is needed for complete combustion, and a residence time, t_A, which must be provided by the flow process downstream of the flame holder. Reliable flame stabilization therefore requires that, throughout the whole burning range of the flame, the residence time t_A must always be larger than—or at the burning limit just equal to—the needed reaction time t_R:

$$t_A \geq t_R \tag{1}$$

The ratio t_A/t_R of these two characteristic times is known as Damköhler's number; it represents the similarity parameter for flow processes with chemical reaction. Therefore, it can be used to derive a stabilization criterion that enables the selection of flame holder dimensions if a certain burning range for different fuels is required. The actual values of these times can only be determined by experiments, as will be described later. However, the influence of important parameters on flame stabilization can be investigated if the characteristic times are expressed by the governing physical quantities.

Zukoski and Marble (1956) found that for subsonic flame stabilization the residence time t_A is proportional to the length L of the recirculation zone and inversely proportional to a velocity w_1 at its outer edge, or in the case of kinematic similarity, proportional to w_0, the velocity upstream of the flame holder:

$$t_A \sim L/w_1 \sim L/w_0 \tag{2}$$

As the gross features of a recirculation zone in supersonic flow do not differ qualitatively from those in subsonic flow, this proportionality (2) can also be applied at supersonic speeds.

According to Damköhler (1936), the reaction time t_R can be expressed by

$$t_R \sim c_B/\bar{r} \tag{3}$$

where c_B is, for example, the fuel concentration of the fresh mixture, and $\bar{r}$ a global reaction rate of the fuel-air mixture employed. Eq. 3 can be reshaped using the simplified theory of laminar flame propagation. From the latter one can deduce the relationship (Bartlmä 1975):

$$S_L \sim \left(\frac{\bar{k}/\rho_0\bar{c}_p}{t_R}\right)^{1/2} \tag{4}$$

where S_L is the laminar burning velocity and $\bar{k}/\rho_0\bar{c}_p$ is an average thermal diffusivity in the preheating zone of a laminar flame ($\bar{k}$, $\bar{c}_p$ average thermal conductivity and specific heat, ρ_0 density of the unburnt mixture far upstream).

Eqs. 2, 3, and 4 can be combined to yield a stabilization criterion based on Damköhler's number:

$$\mathrm{Dam} \sim \frac{L \cdot S_L^2}{w_0 \cdot \bar{k}/\rho_0\bar{c}_p}. \tag{5}$$

Eq. 5 is a dimensionless parameter, and it contains other non-dimensionless stabilization criteria, as they have been derived in early work on flame stabilization. From Eq. 5 the influence of the fuel properties on the burning range of a flame can be deduced once the geometry of the recirculation zone is fixed (Löblich 1962). If Eq. 5 is evaluated with respect to the maximum flow speed at the burning limits of stoichiometric propane-air and hydrogen-air flames, one has to consider the laminar flame speeds S_L of the mixtures, which is approximately 40 cm/s for propane-air and 200–250 cm/s for hydrogen-air. A reasonable value for the thermal diffusivity of a hydrogen-air mixture would be twice that of propane-air. Therefore, the minimum residence time at the burning limit of a hydrogen-air flame should be 1/15 to 1/20 the value for a propane-air flame. In other words, with hydrogen as a fuel the maximum flow speed at which flame stabilization by a given flame holder is possible should be much higher than with propane (as a typical hydrocarbon fuel). Therefore, it was concluded that recirculation zones can be very well used to stabilize hydrogen flames at supersonic speeds. Townend's (1963) experiments on base-burning in supersonic flow were taken as a first confirmation.

Experimental Results

Experiments have been carried out at DFVLR in order to quantitatively evaluate Eq. 5 and to determine numerical values of reaction times and residence times. First tests have been carried out with premixed gases, using propane, ethylene, and hydrogen, where, due to the limitations of hydrogen supply, a rather small test setup with a cylindrical flame holder of 6-mm diameter had to be used. These experiments have been described elsewhere (Winterfeld 1968). Here, only Fig. 3.2 is reproduced in order to show, by means of the lean burning limits, the smooth transition from subsonic to supersonic outer flow that was observed in these tests. With the limitations of the hydrogen supply, the maximum flow speed at the lean burning limit observed was 530 m/s, corresponding to a flow Mach number of 1.6.

Further experiments have been carried out in order to determine the burning limits of hydrogen diffusion flames as well as numerical values of t_A and t_R, respectively, using the condition $t_A = t_R$ at the burning limit. This was done by measuring the length L of the recirculation zone as well as by determining the flow velocity at the separation point at the flame holder base, w_1. Different geometries of axisymmetric flame holders within supersonic flows have been employed: a contoured body within a straight-walled outer casing as illustrated in Fig. 3.1, a parallel-walled body within a conical Laval nozzle, and a short cone-cylinder flame holder in a parallel supersonic flow (Winterfeld 1976). Flame holder diameters of 10, 20, and 30 mm have been used. The gaseous hydrogen fuel was injected from the base of the flame holder normal or parallel to the surrounding supersonic flow. Thus, hydrogen-air diffusion flames were stabilized by the flame holders. Since the fuel supply from a battery of conventional pressure bottles was rather limited, the resulting flames were confined to a very narrow region downstream of the flame holder, i.e., to the subsonic wake and the immediately neighboring supersonic stream tubes. This, however, corresponds to a real case, because it covers the regions where

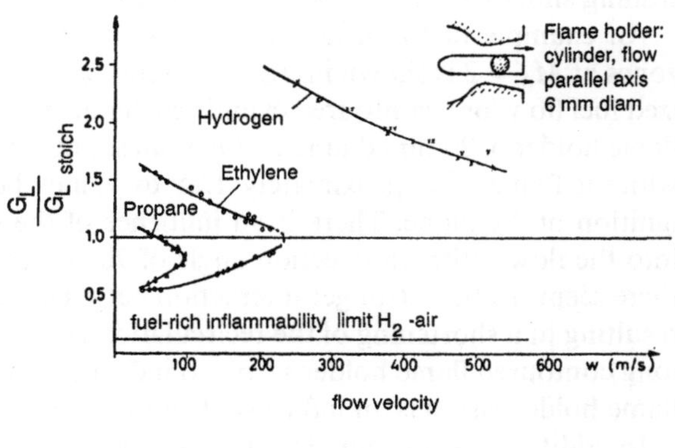

FIGURE 3.2. Burning limits of premixed flames stabilized by a 6-mm diameter cylindrical flame holder, for different fuel-air mixtures. (G_L/G_L stoich = air equivalence ratio λ).

the important stabilization processes occur. Thus, the flame is burning in a mixed subsonic-supersonic flow region.

The tests have been conducted in the free jet downstream of the nozzle at a pressure of 1 bar at the nozzle exit, at Mach numbers between 1.5 and 2.1 in the plane of the flame holder base, and at a total temperature of 375 K. The static temperatures of the outer airflow were in the range between 260 and 210 K; the recovery temperatures in the boundary layer of the flame holder amounted to 357 to 365 K.

The length of the recirculation zone L with flame was determined by coloring the flame with sodium particles from the tip of a probe that was introduced into the wake from downstream. Due to the highly unsteady flow conditions downstream of the flame holder the point of flow reversal fluctuates over some distance and the center of that interval was taken as the average end of the recirculation zone. The resulting measurement error inherent in this method was estimated to about 5 to 8 percent. For more details see Winterfeld (1976).

The flow field structure immediately downstream from the base of the flame holder is completely changed by the presence of the flame. In a flow field without flame the direction of the surrounding supersonic flow is changed in an expansion fan originating from the edge of the base. The amount of flow deviation depends on the pressure within the recirculation zone, i.e., the base pressure, which in turn is determined by the characteristics of the turbulent shear layer between the wake and the surrounding flow. The latter is turned back to the direction parallel to the axis by means of the trailing shock waves. Thus, the length of the recirculation zone as well as the base pressure are functions of the flow Mach number. The presence of the flame produces a large change of the density in the wake, and the base pressure increases considerably, approximately to values very close to the static pressure in the surrounding flow. This has already been reported by Townend (1963). As a consequence the expansion fan at the flame holder base is greatly reduced and only a weak flow turning can be observed. Consequently, the length of the recirculation zone increases compared to the case without flame. Also, the trailing shocks are considerably weakened or disappear totally.

An example of the measured results for the length of the recirculation zones at $M_0 = 2$ is shown in Fig. 3.3. Here, L/d is plotted against the normalized fuel flow for a contoured cylindrical flame holder and for a cone-cylinder flame holder of 20-mm diameter. One can see that L/d increases from its value without flame of approximately 1.25 to values between 2 and 3 after the ignition of the flame. There is an influence of the direction of fuel injection into the flow. With an injection angle of 90 degrees to the surrounding flow there seems to be a stronger interaction between the outer flow and the wake resulting in a shortening of the recirculation zone. The lower diagram for the long contoured flame holder seems to indicate that the boundary layer of the flame holder also has an influence. For more details see Winterfeld (1976).

In addition, the burning limits of the hydrogen diffusion flames have been

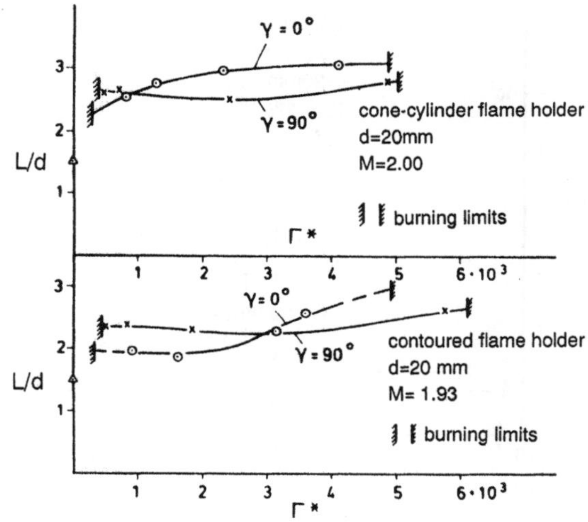

FIGURE 3.3. Relative lengths of recirculation zones behind contoured flame holders and short cone-cylinder flame holders, for different relative fuel mass flows and different fuel injection angles γ.

determined for different Mach numbers of the airflow. The fuel flow has been either increased or decreased until the flame was blown off. Some of the measured results are shown in Fig. 3.4, where the normalized fuel flow (normalized with the airflow through the flame holder cross section) at the burning limits is plotted against the flow Mach number, for a cylindrical flame holder of 20-mm diameter in a conical Laval nozzle (Winterfeld 1968). Whereas at the lean burning limit almost no influence of the flow velocity could be found, the fuel flow at the rich burning limit decreased with increasing velocity. Major differences in the normalized fuel flow at the rich burning limit are caused by the angle of fuel injection. Theoretical estimations based on mass exchange rates measured in subsonic flow have shown that with a fuel injection angle of 90 degrees to the surrounding flow, 30 to 25 percent of the total fuel flow is blown into the external supersonic flow, whereas with injection at 0 degrees, i.e., parallel to the outer flow, this share is rather small, about 5 percent. This explains why the fuel flow at the rich burning limit of these diffusion flames depends sensibly on the fuel injection angle.

Additional measurements of the gas composition within the recirculation zone by means of gas sampling from the flame holder base were used to determine the air-equivalence ratio λ at or close to the blow-off limits. Based on these results the characteristic times L/W_1 at the burning limits have been evaluated for flow Mach numbers between 1.4 and 2.1. The flow velocity and Mach number have been determined by the usual relationships for flow expanslon in a Laval nozzle. The results are shown in Fig. 3.5, where L/W_1 is plotted against the air equivalence ratio. Values for L/W_1 have been found between $0.6 \cdot 10^{-4}$ and $1.4 \cdot 10^{-4}$ sec, where there is still a considerable range of mixture ratios between the burning limits. For a rough comparison, the characteristic times L/W_1 at the lean burning limits of the premixed hydrogen-

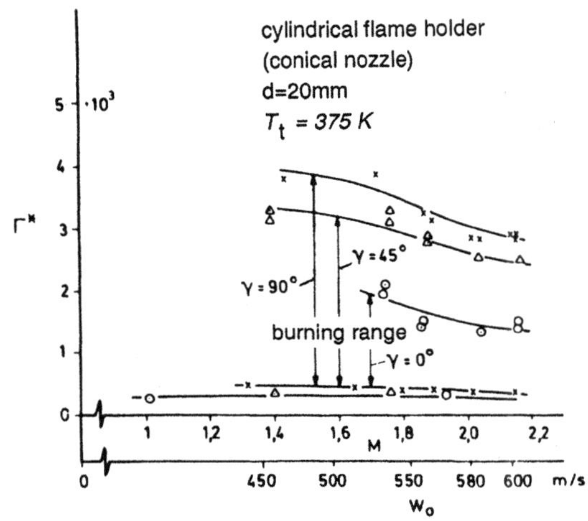

FIGURE 3.4. Burning range of hydrogen-air diffusion flames behind cylindrical flame holders for different airflow Mach numbers and different fuel injection angles γ.

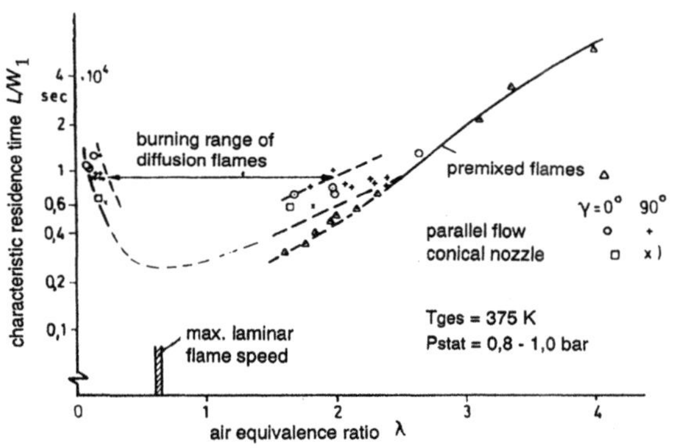

FIGURE 3.5. Characteristic time L/d, measured at the burning limits of supersonic hydrogen-air flames, as a function of the air-equivalence ratio λ within the recirculation zone, for cylindrical flame holders.

air flames stabilized by the 6-mm flame holder have been also entered, although in this case the determination of the length L by the coloring method is subject to larger scatter. The corresponding times are of the same order of magnitude, although at somewhat higher air-equivalence ratios than with the diffusion flames. This seems to indicate that in the premixed case the average air-equivalence ratio in the recirculation zone at the lean limit is higher than that value which has been measured for the diffusion flames at the flame holder base.

Although there are basic difficulties in the measurement method, i.e., determination of the length L and of the fuel-air ratio at blow-off, the measured values for the characteristic time could be used for the design of the supersonic flame holders. According to the definition of the Damköhler number the measured residence times correspond to characteristic reaction times, from which the dimensions of a flame holder can be determined provided the length

L of the recirculation zone is known. The values L/d given earlier correspond to bodies with circular cross section. In the case of two-dimensional or annular wall steps the lengths L should be larger, as can be expected from experiences in subsonic flow. Here, further measurements with flames are necessary for confirmation.

For the measured values of L/W_1 of $0.6 \cdot 10^{-4}$ to $1.4 \cdot 10^{-4}$ sec there is a rather large range of fuel-air ratios in which the fuel flow can change without causing blow-off. Expressed in terms of the fuel flow injected from the flame holder it corresponds to a ratio of approximately 1:15. This is of course much smaller than required for safe operation for turbine engine combustors; however, it should be possible to operate the flame holders in scramjets as pilot flame holders with a fixed fuel flow at all operating conditions of the engine. According to Zukoski and Marble (1956), the characteristic residence time for a hydrocarbon-air flame having the same range of mixture ratios would be on the order of 2 milliseconds or more. Thus, the preceding rough estimation of residence times for hydrogen and propane fuel by means of Eq. 4 is confirmed, at least on an order-of-magnitude basis.

Combustion efficiency was only measured within the recirculation zone, where the extraction of gas samples did not disturb the flow field. For cone-cylinder flame holders actual values found for combustion efficiency ranged between 90 and 98%, for dimensionless fuel flows Γ^* between 1 and $5 \cdot 10^{-3}$, 141.

Concluding Remarks

The experiments described herein have shown that flame stabilization in supersonic combustors using hydrogen as a fuel is possible if the flame holder dimensions are chosen appropriately. Thus operation of scramjets can be extended to lower hypersonic flight speeds, where thermal self-ignition cannot be safely applied. However, the pressure and temperature dependence of the reaction time, and consequently of the residence time, have yet to be determined. A first estimation should be possible by using a global rate equation for hydrogen-air combustion, as it is, for example, given in Lezberg and Lancashire (1961), or as it may be eventually deduced from reaction kinetic calculations. Experiments to verify such estimations are necessary.

However, as is mentioned in Winterfeld (1976), interactions between shock waves in the supersonic flow and the subsonic flame holder wake can become important once a flame is burning in the wake. They can, on one hand, be used to improve flame stability; on the other hand, permanent blockage effects can be generated that are detrimental for the supersonic flow processes in the combustor. Thus, careful flow examinations within the combustor are necessary if flame stabilization by recirculation zones is applied in scramjet combustors.

Nomenclature

C_B = fuel concentration, kg/m^3.
$\bar{C}_p$ = average specific heat of combustible mixture in the preheating zone of a laminar flame, kJ/kg K.
d = flame holder diameter (at base), m.
Dam = Damköhler's number.
$\bar{k}$ = average thermal conductivity of combustible mixture in the preheating zone of a laminar flame, kJ/m·s·K.
L = length of recirculation zone, m.
$\dot{m}_B$ = fuel mass flow, kg/s.
$\dot{m}_L$ = air mass flow, which would pass through the flame holder cross section with velocity w_1, kg/s.
M = flow Mach number.
P_{stat} = static pressure, N/m^2.
$\bar{r}$ = global reaction rate, Mol/m^3 s.
S_L = laminar flame speed, m/s.
T_{ges} = total temperature, k.
t_A = residence time, sec.
t_R = reaction time, sec.
w_0, w_1 = flow velocity upstream of flame holder along the contours of the recirculation zone, m/s.
Γ^* = $\dot{m}_B/\dot{m}_L$.
γ = fuel-injection angle, measured against the axis of flame holder test setup, degree.
ρ_0 = density of unburnt mixture, kg/m^3.
λ = air-equivalence ratio.

References

Baev, V.K., Triet'yakov, P.K., and Konstantinovsky, V.A., 1981, "The Study of Hydrogen Combustion in a Ducted Supersonic Flow with a Sudden Expansion," *Archivum Combustionis, 1*, 251–259,

Bartlmä, F., 1975, *Gasdynamik der Verbrennung*, Springer, Wien, New York.

Damköhler, G., 1936, "Einflüsse der Strömung, Diffusion und des Wärmeübergangs auf die Leistung von Reaktionsöfen," *Z. Elektrochemie, 42*, 846.

Lezberg, E.A., and Lancashire, R.B., 1961, "Expansion of Hydrogen-Air Combustion Products Through a Supersonic Exhaust Nozzle; Measurements of Static Pressure and Temperature Profiles," *Fourth AGARD Colloquium on Combustion and Propulsion*, Pergamon Press, 286 pp.

Löblich, K.R., 1962, "Zur Kinetik der Stabilisierung Turbulenter Flammen durch Flammenhalter," *Diss. Th.* Hannover.

Schmalz, F., 1971, "Messung und Theoretische Berechnung von Zündverzugszeiten in Wasserstoff-Luft-Gemischen bei Temperaturen um 1.000 K und Drücken unter 1 at," *Diss. TH.*, Aachen; DLR-FB71-08.

Suttrop, F., 1972, "Ignition of Gaseous Hydrocarbon Fuels in Hypersonic Ramjets," paper presented at the 1st ISABE, Marseille.

Suttrop, F., 1973, "Experimentelle Untersuchungen über Thermische Selbstzündung Turbulenter Diffusionsflammen bei Hoher Geschwindigkeit," *Diss. TH.*, Aachen (D82), DLR-FB74-10.

Townend, L.H., 1963, "Some Effects of Stable Combustion in Wakes Formed in a Supersonic Stream," *RAE. Tech. Note Aero 2872.*

Wilhelmi, H., 1972, Fourteenth Symposium, International, on Combustion, 585–593.

Winterfeld, G., 1968, "On the Burning Limits of Flame Holder Stabilized Flames in Supersonic Flow," *AGARD-CP 34*, paper 28.

Winterfeld, G., 1976, "Untersuchungen über die Stabilisierung von Wasserstoff-Diffusionsflammen durch Flammenhalter in Überschallströmungen," *DLR-FB76-35.*

Zukoski, E.E., and Marble, F.E., 1956, "Experiments Concerning the Mechanism of Flame Blow-off from Bluff Bodies," *Proc. Gas Dyn. Symp.*, Northwestern Univ., Evaston, I11., 205 pp.

4

Thermodynamics of Refractory Material Formation by Combustion Techniques

I. GLASSMAN, K. BREZINSKY, AND K.A. DAVIS

ABSTRACT: Equilibrium thermodynamic calculations of adiabatic flame temperatures and species concentration for the reactions of Ti, Si, and Al with nitrogen at a variety of pressures and stoichiometries have revealed interesting characteristics similar to metal-oxygen reactions. The results suggest the feasibility of continuous formation of the refractory nitrides in a self-sustaining rocket process approaching stirred reactor conditions.

Introduction

Spurred primarily by the use of high-temperature combustion techniques to form pure solar-grade silicon (Dickson et al. 1981) and silica optical fibers, interest has developed in the possibility of using various combustion techniques to create high-temperature materials. Ulrich (1971, 1984) has reviewed most of the work related to the gas phase combustion reactions in which optical-grade silica and metal oxides can be formed. Merzhanov and Borovinskaya (1975) and Holt (1983, 1986) and Holt and Munir (1986) have written about the formation of refractory materials by reactions of solids similar to the classical "thermite" reactions used to create heat for welding. Thermite-type reactions were also used (Glassman et al. 1987) to disperse barium vapor into the atmosphere for geophysical observations. Both the Russian investigators and Holt also explored solid-liquid (nitrogen) reactions as a process for the synthesis of titanium nitride. They have termed the process self-propagating high-temperature synthesis (SHS). In chemical process terms, these are batch techniques and are not continuous, such as those typically used for the formation of silicon compounds. This paper offers a new approach for the continuous process formation of refractory materials.

The ultimate objective is to evaluate the feasibility of creating refractories such as metallic nitrides by the self-sustained reaction of the metal and gaseous nitrogen in a rocket motor designed to approach stirred reactor conditions. The rocket motor technique should be more economical than those proposed by other researchers and may permit, through regulation of the cooling rate, some control over refractory particle size and structure. Although little is currently known about the high-temperature reaction kinetics of titanium and gaseous nitrogen, calculations show the rocket stirred reactor concept to be thermodynamically feasible (Glassman et al. 1987). The results reveal that the reaction between nitrogen and titanium is quite exothermic at 298 K, about −265.8 kJ/mole for the formation of liquid TiN. As important, however, is the thermodynamic result that the calculated adiabatic flame temperature for the Ti-N_2 reaction is limited by the unique dissociation characteristics of the TiN over a range of pressures. This result illustrates a concept similar to one discovered a number of years ago (Glassman 1959, 1987); the oxidation reaction temperatures of metals are limited (equivalent) to the saturation temperature of the metal oxides formed. It appears that the ramifications and importance for combustion processes of the saturation temperature limitation have rarely been explicitly stressed in the literature. Stimulated by the interesting results for the Ti-N_2 system, the authors have been exploring many other metal nitride systems using the same type of thermochemical approach that has been very successful in evaluating which metals have the greatest performance potential for solid, liquid, and hybrid rockets (Glassman 1965; Glassman and Sawyer 1970). This preliminary exploration has led to other process concepts for formation of many types of refractories and suggests that the thermochemical approach should be explored in detail. In order to elucidate the thermodynamic and combustion concepts alluded to, the next section is a review of the investigators' earlier and current thinking about metal combustion processes and how this thinking reflects on the analytical and experimental objectives discussed in this paper.

Metal Combustion Processes

Many metals and higher periodic table atomic number species in their natural state exist with a protective oxide coat. Ignition of these metallic species requires deterioration of this coat. The protective value of the oxide depends to a great degree on the Pilling-to-Bedworth ratio, the ratio of the volume of oxide formed to the volume of metal consumed. When this ratio is greater than one, the coating scales are not very protective and reaction and ignition can proceed most readily. When the ratio is much less than one, some metal surface remains exposed. However, when the ratio is near one, the coating is very protective and generally is not destroyed by cracking, melting, or evaporation until a certain temperature is reached. This transition temperature and its relation to ignition have been discussed at length (Glassman et al., 1970; Mellor and Glassman 1965). Fortunately, because of the high temperature and other experimental conditions to be discussed later, ignition, though of general importance in many metal combustion processes, is of no particular relevance to the generation of refractories by continuous combustion methods.

For evaluating metal combustion processes, the adiabatic flame temperature of the process has been found to be an important controlling parameter. The calculation of the adiabatic (flame) temperature for Reaction 1,

$$xM + (y/2)O_2 \rightarrow M_xO_y, \tag{1}$$

the formation of a metallic oxide from its elemental constituents, at some reference state, can be considered to proceed by the transformation of the reactants to products at the reference state temperature and then the subsequent absorption by the products of all the heat released. So much heat can be absorbed that there can be phase changes and dissociation of the products themselves. The final temperature reached is defined as the adiabatic reaction temperature (T_f). Essentially the statement of reaction at the reference state, say 298 K, can be written as

$$-\Delta H_R = Q_p = -\Delta H^0_{f,M_xO_y}\,(298\text{ K}) \tag{2}$$

where ΔH_R is the enthalpy change of the reaction, Q_p is the heat of reaction and $\Delta H^0_{f,M_xO_y}$ is the standard state heat of formation of the metal oxide at 298 K. As depicted in Fig. 4.1, equilibrium thermodynamics permits the system to proceed by another path, i.e., the reactants are heated to T_f and then reacted at this value to form the products. It is important to note for the concepts to be discussed later that if there is no dissociation of products at T_f then Eq. 2 will also hold if the various thermodynamic properties ΔH_R and ΔH_f are evaluated at T_f.

For systems represented by Reaction 1, there is never sufficient heat released and subsequently absorbed by the oxide product to completely vaporize it (Glassman 1959, 1987). There is, however, always enough heat released to raise the oxide to its saturation temperature (boiling point); i.e.,

$$|\Delta H^0_{f,298M_xO_y}| > (H^0_{T_{sat}} - H^0_{298})_{M_xO_y} \tag{3}$$

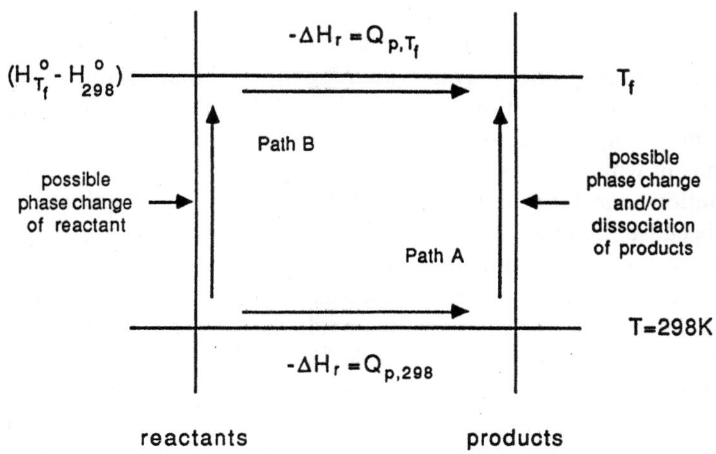

FIGURE 4.1. Thermodynamic Route to Flame Temperature Calculation.

but

$$(H^0_{T_{sat}} - H^0_{298})_{M_xO_y} + \Delta H^0_{vap, M_xO_y} > |\Delta H^0_{f,298,M_xO_y}| \tag{4}$$

where $H^0_{T_{sat}}$ is only the standard state enthalpy at the saturation temperature and does not include any heat of vaporization or dissociation. The inequalities represented by Eqs. 3 and 4 specify that the flame temperature of a metal oxidation system must be equal to the saturation temperature of the metal oxide:

$$T_f = T_{sat, M_xO_y}. \tag{5}$$

This result has important ramifications for the combustion of any condensed phase metal particle. Droplet burning rate theory reveals that during steady-state combustion the droplet approaches a temperature near its saturation value (Glassman 1959, 1987). For metal particles, it would be the metal saturation temperature. For low latent heat of vaporization condensed phases, such as liquid hydrocarbons, the droplet temperature is very close to its saturation temperature. For large latent heat of vaporization substances such as metals, the droplet temperature can be 100–200 K lower than its saturation point (Brzustowski and Glassman 1964). With this theoretical result and Eq. 5, the criterion (Glassman 1959, 1987) for a metal particle, liquid or solid, to burn in the vapor phase as liquid hydrocarbons do, becomes

$$T_{sat, M_xO_y} > T_{sat, M}; \tag{6}$$

the saturation temperature or boiling point of the metal oxide must be greater than that of the metal. If the opposite condition holds, i.e.,

$$T_{sat, M_xO_y} < T_{sat, M} \tag{7}$$

the metal burns (oxidizes) by a surface (heterogeneous) process, and a molten surface oxide generally inhibits the particle burning rate (Glassman 1959, 1987; Glassman et al. 1984). If the oxide is heated by a source other than the heat of Reaction 1, then the constraint of Eq. 7 can be relieved. Similar considerations come into play in evaluating the possibility of the Ti-N_2 rocket combustion system mentioned. This point will be discussed subsequently.

The preceding discussion of metal combustion processes has dealt with reactions in which at least one metallic component is gaseous. For example, the reaction zone, for the condition of Eq. 6, is a completely gaseous diffusion flame because the metal vaporizes. When the constraint of Eq. 7 is relieved because of external heating, the heterogeneous solid-gas oxidation can continue depending on the amount of external heat but the product will now be a gas. In either case, the solid product oxide that would eventually form, in a propulsion system, for example, would be the result of a physical, gaseous nucleation step occurring away from the primary reaction zone.

There is another class of metal combustion reactions quite different from the gaseous processes described earlier. These are exchange reactions between one metal and another metal oxide. Such systems have been called thermite reactions and are represented by the iron oxide–aluminum reaction frequently used as a heat source for welding. There are insufficient data to indicate the type of physical state in which such reactions actually take place. Liquid-solid, liquid-liquid, or even gas-solid reactions are possible since the reaction temperatures achieved can be above one or both constituent's fusion and sometimes vaporization temperatures. One's intuition would allow one to conclude that the conversion efficiency of condensed phase thermite-type systems would not be as large as for the gaseous systems because intimate mixing and contact between reactant particles can be very difficult to obtain.

Nevertheless, various investigators (Holt 1983, 1986; Holt and Munir 1986; Merzanov and Borovinskaya 1975) have made substantial progress in evaluating these thermite-type (SHS) systems, not as convenient solid heat sources but, as possible economical means for fabricating high-temperature refractory materials such as nitrides, borides and carbides. This progress has resulted from the use of metal-liquid nitrogen (Holf 1983; Merzanov and Borovenskaya 1975) or metal-sodium azide systems (Holf 1983) to provide solid-gaseous reactants. The azide provides a higher mass density than liquid nitrogen. These processes, despite approaching very attractive efficiencies, are batch processes unlike the continuous gas phase reaction systems to be discussed in the sections to follow.

Although pure thermite reactions are not the primary focus of this paper, an understanding of the thermodynamics of this unusual class of reactions provides the foundation for understanding their potential applications. The "classical" thermite reaction is represented by Reaction 8:

$$Fe_2O_3 + 2Al \rightarrow Al_2O_3 + 2Fe \qquad Q_p = 850\ \text{kJ/mole.} \tag{8}$$

The large heat release of this and other thermite reactions leads to high temperatures, which can cause melting of one component. Melting can prevent the propagation through the compressed granular rod of reactants of the hot vapors necessary to sustain the propagation of the reaction front. To alleviate this inhibiting condition, the initial reactants can be mixed with a sufficient amount of the product oxide to lower the temperature below the controlling melting point. For the system represented by Reaction 8, the reaction then becomes

$$Fe_2O_3 + 2Al + xAl_2O_3 \rightarrow (x + 1)\ Al_2O_3 + 2Fe \qquad Q_p = 850\ \text{kJ/mole.} \tag{9}$$

Although Reaction 9 has the same heat release per mole of Fe_2O_3 as Reaction 8, its adiabatic flame temperature must be lower due to the larger molar content of the Al_2O_3.

A similar system was used during the International Geophysical Year and in a variety of space programs to create metallic ions (Foppi et al. 1967) for the evaluation of stratospheric and ionospheric phenomena. The ion-producing reaction is

$$Ba + CuO \rightarrow BaO + Cu \qquad Q_p = 392\ \text{kJ,} \tag{10}$$

which is quite exothermic. Although the heat release is not sufficient to ionize any of the products of Reaction 10, it is sufficient to ionize barium which has a low ionization potential. Therefore, more than a stoichiometric amount of barium is added to the reactant mixture so that some unreacted barium will be available for vaporization and ionization.

Surprisingly, a review of the literature revealed no paper emphasizing that the adiabatic stoichiometric reaction (flame) temperature of Reactions 8 and 10 and similar thermite reactions is, in many instances, limited (equal) to the saturation temperature of the metal product. The limitation occurs because the saturation temperature of the metal product will always be less than that of the product oxide. Furthermore, the heat of vaporization of all the higher atomic number metal products is generally so large that the heat of reaction is absorbed entirely by the boiling metal. One could even conceive of situations in which the reaction temperature is limited to the melting point of the metallic product.

Venturini (1953) has compiled a list of other metal–metal-oxide systems that should have highly exothermic, and possibly self-sustaining reactions:

	Q_p (kJ/mole Al_2O_3)
$(3/5)\,V_2O_5 + 2\,Al \rightarrow Al_2O_3 + (6/5)\,V$	782
$WO_3 + 2\,Al \rightarrow Al_2O_3 + W$	887
$3\,CuO + 2\,Al \rightarrow Al_2O_3 + 3\,Cu$	1208
$Cr_2O_3 + 2\,Al \rightarrow Al_2O_3 + 2\,Cr$	541

The preceding systems are all exothermic; however some metal–metal oxide reactions can be endothermic:

	Q_p (kJ)
$3\,MgO + 2\,Al \rightarrow Al_2O_3 + 3\,Mg$	−127
$3\,CaO + W \rightarrow WO_3 + 3\,Ca$	−1116
$3\,CaO + 2\,Al \rightarrow Al_2O_3 + 3\,Ca$	−229
$2\,BaO + Si \rightarrow SiO_2 + 2\,Ba$	−193

A good screening measure for selecting which among the numerous metal–metal-oxide systems will be exothermic is the heat of formation of the oxide in kJ per atom of oxygen (or as will be discussed later, per atom of the oxidizing element) (Venturini 1953). Using the heat of formation per atom accounts for the varying stoichiometries of the different metal-oxide reactions. An update of this type of data is shown in Table 4.1. To obtain an exothermic system, one chooses a metal whose oxide heat of formation per atom of oxygen is high on the list to react with an oxide that is lower on the list. Since there are small entropy changes for most of the reactions discussed earlier, the free energy change is very close to the enthalpy change and the reaction thus will proceed exothermically.

TABLE 4.1. Heats of formation of certain oxides.

Oxide	Heat, f, @298 (kJ/mol)	Per O atom
CaO	−635	−635
ThO_2	−1222	−611
BeO	−608	−608
MgO	−601	−601
Li_2O	−599	−599
SrO	−592	−592
Al_2O_3	−1676	−559
ZrO_2	−1097	−549
BaO	−548	−548
UO_2	−1084	−542
CeO_2	−1084	−542
TiO_2	−945	−473
SiO_2	−903	−452
B_2O_3	−1272	−424
Cr_2O_3	−1135	−378
V_2O_5	−1490	−298
Fe_2O_3	−826	−275
WO_3	−789	−263
CuO	−156	−156

Data from 3d edition of the JANAF tables.

TABLE 4.2. Heats of formation of certain chlorides and nitrides.

Chloride	Heat, f, @298 (kJ/)	Per Cl atom
$CsCl$	−443	−443
KCl	−437	−437
$BaCl_2$	−859	−430
$RbCl$	−430	−430
$SrCl_2$	−829	−415
$NaCl$	−411	−411
$LiCl$	−408	−408
$CaCl_2$	−796	−398
$CeCl_3$	−1088	−362
$AlCl_3$	−706	−235
$TiCl_4$	−815	−204
$SiCl_4$	−663	−166

Nitride	Heat, f, @298 (kJ)	per N atom
HfN	−369	−369
ZrN	−365	−365
TiN	−338	−338
AlN	−318	−318
BN	−251	−251
Mg_3N_2	−461	−231
Si_3N_4	−745	−186
Li_3N	−165	−165
NaN_3	22	7

Similar lists to that in Table 4.1 can be prepared for the various metallic halides. Table 4.2 lists the data collected for the chlorides and the nitride refractories of major concern here.

Postulated Metal Refractory Combustion Technique

The previous section provides the background to understand the reasoning that has led to the suggested combustion system for the economic synthesis of titanium, aluminum, and silicon nitride. As stated in the Introduction, the reaction between Ti and N_2 is quite exothermic. The heat release of this reaction should be sufficient to sustain the formation of TiN. As mentioned previously, investigators (Holt 1983; Merzanov and Borovinskaya 1975) have reported some success by reacting Ti with liquid nitrogen and with sodium azide. The processes generally involve solid columns ignited at the circular face of one end by a heated tungsten wire. The Ti/NaN_3 system appears to give the best yields of TiN (Holt 1983) through a readily attained self-propagating reaction. This self- sustaining reaction has been reproduced in this laboratory. There are two major improvements, however, that can be made to the process: making the process continuous and eliminating the expensive azide.

Examination of the thermodynamics of the Ti/N_2 reaction suggests a means to achieve these two improvements. For the stoichiometric ratio of 0.5 moles N_2/mole Ti and without considering dissociation of the TiN liquid product, the reaction has sufficient energy release to raise the Ti to its saturation temperature, but not sufficient energy to vaporize all the Ti on a mole-per-mole of TiN basis, i.e.,

$$|\Delta H_R|_{\mathrm{Ti+N_2},T_f} > (H^0_{T_f} - H^0_{298})_{\mathrm{Ti,N_2}} \tag{11}$$

but

$$|\Delta H_R|_{\mathrm{Ti}+N_2,T_f} < \Delta H_{\mathrm{vap,Ti},T_f} \tag{12}$$

Essentially this observation, which is similar to that expressed in Equations 3 and 4, signifies that the adiabatic combustion temperature for this stoichiometric ratio can be considered to be limited to (or equal to) the saturation temperature of the Ti. Recall, due to dissociation there will be a partial pressure effect exerted by the nitrogen. It would follow then that a range of off-mixture ratio conditions would give the same temperature as the various partial pressures and mole fractions reequilibrate, and that all these temperatures limited by the saturation condition would be a function of the total pressure. The conceptual ideas are very similar to the concepts that led to Glassman's criterion for vapor phase combustion of metals (1959, Glassman et al. 1987), as stated by Eq. 6.

These considerations provided the motivation for performing detailed calculations using the Gordon-McBride (Gordon and McBride 1976) program, which would account for the effects of the vapor phase dissociation of the

TABLE 4.3. Stoichiometric Ti-N_2 equilibrium mole fractions of combustion products and temperature as a function of total pressure.

P_t(atm)	T_f(K)	N_2	Ti(v)	Ti(l)	TiN(l)	TiN(s)	p[Ti(v)] (atm)
0.2	3155	0.1091	0.1838	0.0343	0.0000	0.6728	0.126
1.0	3451	0.0819	0.0775	0.0864	0.7541	0.0000	0.486
2.0	3596	0.0796	0.0609	0.0983	0.7612	0.0000	0.867
5.0	3804	0.0728	0.0421	0.1035	0.7816	0.0000	1.831
10.0	3975	0.0645	0.0301	0.0990	0.8064	0.0000	3.183

TABLE 4.4. Temperature and mole fraction product distribution of Ti-N_2 reaction at various pressures.

Mole N_2 / Mole T_i	T_f(K)	Ti(v)	Ti(l)	TiN(l)	TiN(s)	N_2	p[Ti(v)] atm
			$P_T = 0.2$ atm				
0.50	3155	0.1838	0.0343	0.0000	0.6728	0.1090	0.1255
0.52	3155	0.1951	0.0010	0.0000	0.6880	0.1158	0.1255
0.54	3154	0.1903	0.0000	0.0000	0.6797	0.1300	0.1188
0.56	3152	0.1852	0.0000	0.0000	0.6708	0.1440	0.1125
0.58	3150	0.1803	0.0000	0.0000	0.6622	0.1575	0.1067
0.60	3147	0.1755	0.0000	0.0000	0.6538	0.1707	0.1014
0.70	3133	0.1540	0.0000	0.0000	0.6151	0.2308	0.0800
0.80	3117	0.1356	0.0000	0.0000	0.5815	0.2829	0.0648
0.90	3101	0.1197	0.0000	0.0000	0.5519	0.3284	0.0534
1.00	3085	0.1057	0.0000	0.0000	0.5258	0.3685	0.0446
			$P_T = 1$ atm				
0.50	3451	0.0775	0.0864	0.7541	0.0000	0.0819	0.4860
0.52	3451	0.0850	0.0584	0.7667	0.0000	0.0899	0.4860
0.54	3451	0.0923	0.0308	0.7791	0.0000	0.0977	0.4860
0.56	3451	0.0996	0.0038	0.7913	0.0000	0.1053	0.4860
0.58	3445	0.0973	0.0000	0.7836	0.0000	0.1191	0.4496
0.60	3437	0.0771	0.0000	0.7729	0.0000	0.1334	0.4123
0.70	3397	0.0633	0.0000	0.7240	0.0000	0.1988	0.2796
0.80	3358	0.0514	0.0000	0.6816	0.0000	0.2551	0.1988
0.90	3320	0.0514	0.0000	0.6445	0.0000	0.3041	0.1446
1.00	3282	0.0412	0.0000	0.6117	0.0000	0.3471	0.1063
			$P_T = 5$ atm				
0.50	3804	0.0421	0.1035	0.7816	0.0000	0.0728	1.831
0.52	3804	0.0475	0.0802	0.7902	0.0000	0.0822	1.831
0.54	3804	0.0528	0.0573	0.7985	0.0000	0.0914	1.831
0.56	3804	0.0580	0.0348	0.8068	0.0000	0.1004	1.831
0.58	3804	0.0631	0.0128	0.8148	0.0000	0.1092	1.831
0.60	3798	0.0648	0.0000	0.8148	0.0000	0.1204	1.751
0.70	3730	0.0510	0.0000	0.7610	0.0000	0.1879	1.068
0.80	3666	0.0394	0.0000	0.7146	0.0000	0.2459	0.6909
0.90	3604	0.0297	0.0000	0.6740	0.0000	0.2963	0.4556
1.00	3542	0.0217	0.0000	0.6378	0.0000	0.3400	0.2990

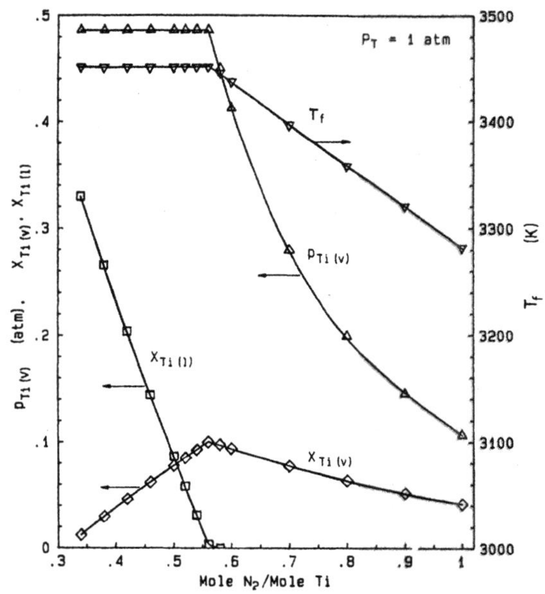

FIGURE 4.2. Variation of reaction temperature (T_f), titanium partial pressure (p[Ti(v)]), vapor mol fraction (X[Ti(v)]), and liquid mol fraction (X[Ti(l)]) as a function of N_2/Ti molar mixture ratio.

liquid TiN formed. Not only did the results confirm the concepts postulated, but they also revealed some trends that one would not at first intuitively expect. Table 4.3 lists the calculated temperature and composition results for the stoichiometric condition at various total pressures. Table 4.4 lists similar results for various mixture ratios at total pressures of 0.2, 1.0, and 5.0 atms. The combustion temperature, mole fraction of liquid and vapor Ti, and the vapor pressure of Ti are plotted in Fig. 4.2 for the 1 atm condition as a function of mixture ratio. The trends shown in this figure are the same for all the total pressure conditions considered in Table 4.4.

Table 4.4 and Fig. 4.2 reveal, as postulated, that over a range of mixture ratios the adiabatic combustion temperature remains constant, as does the Ti vapor pressure. If the vapor pressure in this mixture ratio range is plotted as a function of temperature in a Clausius-Clapeyron form, as shown in Fig. 4.3, then one obtains a latent heat of vaporization of 411 kJ/mole from the slope, corresponding to the values of the most recent JANAF Tables (1986) and used in the Gordon-McBride thermochemical data bank (Gordon and McBride 1976). Obviously this agreement must be found. Note also from Fig. 4.2 that as one moves to the left from a Ti lean mixture condition all the way on the right the mole fraction of Ti(v) increases and at the first value where the maximum temperature is reached the maximum mole fraction is obtained. Thereafter the mole fraction of Ti(v) decreases and Ti(l) begins to form and increase in mole fraction . The trend appears almost counterintuitive. In the maximum temperature range as one leans out the mixture, moving from the far left of the figure to the right, the mole fraction of Ti(v) increases and that

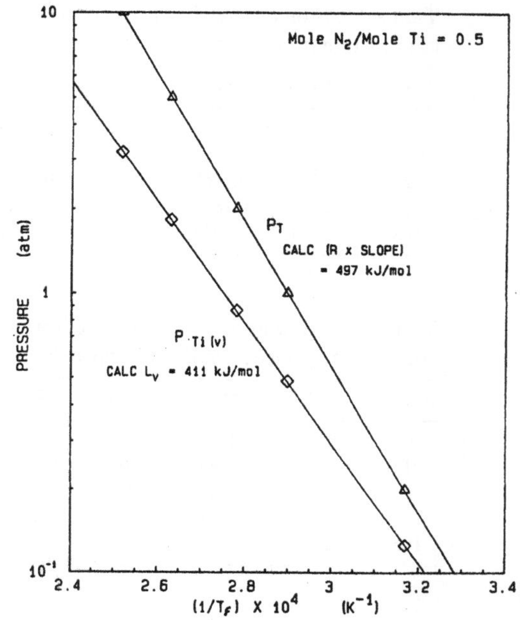

FIGURE 4.3. Combustion calculation of titanium vapor pressure as function of reaction temperature and calculated reaction temperature as function of total pressure for a stoichiometric mixture of titanium and nitrogen.

of Ti(l) decreases! This trend arises because of the equilibrium condition [Ti(l) → Ti(v)] + N_2 → TiN(l). Adding N_2 shifts the equilibrium to form more TiN(l) and the Ti(v) required is supplied by the equilibrium with Ti(l). Note as well in Fig. 4.2 that on the lean side no Ti(l) forms until the Ti(v) reaches its maximum mole fraction or saturation pressure. Obviously there can be no Ti(l) at lean conditions where Ti(v) is below its saturation value.

Plotted on Fig. 4.3 are also the maximum combustion temperatures obtained as a function of the total pressure in the form $\log P_T$ versus $(1/T_f)$. All values fall on a straight line! This trend is certainly not fortuitous, but as yet no acceptable physical explanation has been formulated. The Ti saturation temperatures shown in Fig. 4.3 impose a difficulty in freezing the TiN during an expansion. If one expands from a high pressure to 1 atm, the final temperature at 1 atm will always be that value reported in Table 4.4, which for most conditions is above the 3220 K fusion temperature of TiN. During the expansion the temperatures are again controlled by the saturation condition of Ti(v) and one does not obtain a major drop in temperature as one does when expanding a noncondensible gas mixture. Thus to freeze the TiN(l) one would have to expand to conditions of about 0.2 atm or cool the mixture by further nitrogen addition after reaction had been completed.

The overall production of TiN indicated by the thermodynamic results of Tables 4.3 and 4.4 is encouraging. The optimum conditions show TiN(l) mole fractions of about 0.80, which corresponds to a mass fraction of about 0.83. An 83 percent conversion would appear attractive. Tables 4.3 and 4.4 also

TABLE 4.5. Stoichiometric Al-N_2 equilibrium mole fractions of combustion products and temperature as a function of total pressure.

P_t(atm)	T_f(K)	N_2	Al(v)	Al(l)	AlN(s)	p[Al(v)] atm
0.2	2491	0.1456	0.2882	0.0	0.5647	0.1324
1.0	2707	0.1401	0.2729	0.0	0.5834	0.6550
2.0	2807	0.1662	0.1839	0.1426	0.5044	1.042
5.0	2934	0.1873	0.1061	0.2642	0.4404	1.795
10.0	3028	0.1955	0.0696	0.3181	0.4152	2.610

TABLE 4.6. Stoichiometric Si-N_2 equilibrium mole fractions of combustion products and temperature as a function of total pressure.

P_t(atm)	T_f(K)	N_2	Si(v)	Si(l)	Si_3N_4(s)	p[Si(v)] atm
0.2	2017	0.3422	8.77E-5	0.5131	0.1446	0.51E-4
1.0	2151	0.3368	7.36E-5	0.5051	0.1580	2.18E-5
2.0	2214	0.3341	6.83E-5	0.5011	0.1647	4.08E-4
5.0	2304	0.3301	6.18E-5	0.4951	0.1747	9.35E-4
10.0	2378	0.3276	5.73E-5	0.4899	0.1833	17.5E-4

demonstrate that an increase in total pressure provides higher yields of TiN(l). Similar results are found for Al/N_2 and Si/N_2 as shown in Tables 4.5 and 4.6, except that for Al/N_2 a maximum yield of AlN is obtained at one atmosphere pressure.

Because the flame temperature of the Ti/N_2 system is not significantly higher and, in fact is even lower than the atmospheric pressure saturation temperature of the Ti, 3631 K, pure vapor phase burning of a titanium particle in nitrogen cannot be attained. To circumvent this constraint, which would lead to poor conversion efficiencies, a stirred reactor procedure is proposed. Under stirred conditions there should be sufficient heat feedback and mixing through recirculation to create the necessary metal vapors for good product yields. The exit orifices in a normal stirred reactor, however, must necessarily be kept small. Such orifices would easily clog if a condensed phase is formed. Since rockets, particularly near the injector face plate stabilization zone, behave very much like stirred reactors, it appears feasible to use them to form refractory materials. The suggested approach is to build a rocket tripropellant injector system with ports for gaseous hydrogen, gaseous oxygen, and titanium-laden gaseous (and possibly liquid) nitrogen. The rocket would be ignited with H_2-O_2 fuel. After steady-state H_2-O_2 operation, the Ti/N_2 mixture would be injected. The temperature of the system would then rise due to the high heat release from the formation of TiO_2. When this tripropellant system reaches its steady state, the hydrogen flow is gradually terminated. The reaction should sustain itself and form the thermodynamically favored major product, TiO_2. Again after steady state, the oxygen flow rate would gradually be reduced and the product mixture would consist of TiO_2, TiN, and excess

N_2. The oxygen flow is then terminated leaving sufficient continued release of heat to sustain the Ti-N_2 reaction. If this approach should prove unsuccessful, a modification modeled on one of the processes of Holt (1983) may be feasible. Sodium azide would be dispersed in the Ti particle–laden nitrogen streams of the rocket. Although no thermodynamics calculations of the Ti/NaN_3/N_2 combination have been performed, the endothermic decomposition of NaN_3 could lower the temperature to the point where the nitride product immediately freezes. Such a system would have the advantage of less agglomeration, but the disadvantages of the higher cost of the azide and the inconvenience of sodium vapor as a product.

Acknowledgments. The authors wish to gratefully acknowledge the support of the National Science Foundation under award No. CBT-8813890.

References

Brzustowski, T.A., and Glassman, I., 1964, "Vapor-Phase Diffusion Flames in the Combustion of Magnesium and Aluminum," In *Heterogeneous Combustion*, H.G. Wolfhard, I. Glassman, and H.L. Green, Jr., Eds., Academic Press, New York, p. 117.

Dickson, C.R., Dickson, C.R., Felder, W., Gould, R.K., 1981, "Development of Processes for Production of Solar Grade Silicon from Halides and Alkali Metals," *AeroChem TP-410*.

Foppl, H., et al., 1967, *Planet. Space Science, 15*, 357.

Glassman, I., 1959, "Metal Combustion Processes," *American Rocket Society Preprint 938–59*.

Glassman, I., 1965, "The Chemistry of Propellants and the Space Age," *American Scientist, 53*, 508.

Glassman, I., 1987, *Combustion*, 2d ed., Academic Press, Orlando.

Glassman, I., Ronney, P.D., Takahashi, F., and Brezinsky, K., 1987, "The Thermodynamics of TiN Formation by Combustion Techniques," *Paper No. 39*, Eastern States Section/The Combustion Institute, Nov.

Glassman, I., and Sawyer, R.F., 1970, "The Performance of Chemical Propellants," *AGARDograph 129*, Technivision, Slough, England.

Glassman, I., Williams, F.A., and Antaki, P.A., 1984, "A Physical and Chemical Interpretation of Boron Particle Combustion," *22d Symp. (Int'l.) on Combustion*, The Combustion Institute, 2057.

Glassman, I., et al., 1970, "A Review of Metal Ignition and Flame Models," *AGARD Conf. Proc. No. 52*, AGARD/NATO, Neuilly, France.

Gordon, S., and McBride, B.J., 1976, *NASA SP-273*, Interim Rev., NASACET86.

Holt, J.B., 1983, "Exothermic Process Yields Refractory Nitride Materials," *Industrial Research and Development*, April.

Holt, J.B., 1986, "Combustion Synthesis," *Report LLL-TB-84*, Lawrence Livermore National Lab., May.

Holt, J.B., and Munir, Z.A., 1986, "Combustion Synthesis of Titanium Carbide: Theory and Experiment," *J. Material Sci., 21*, 251.

JANAF Thermochemical Tables, 1985, 3d ed., Amer. Chem. Soc., Washington.
Mellor, A.M., and Glassman, I., 1965, "A Physical Criterion for Metal Ignition," Pyrodynamics, *3*, 43.
Merzanov, A.G., and Borovinskaya, I.P., 1975, "A New Class of Combustion Processes," *Comb. Sci. and Tech.*, *10*, 195.
Ulrich, G.D., 1971, "Theory of Particle Formation and Growth in Oxide Synthesis Flames," *Comb. Sci. and Tech.*, *4*, 47.
Ulrich, G.D., 1984, "Flame Synthesis of Fine Particles," *C&E News*, Aug. 6.
Venturini, J., 1953, *Metaux and Corrosion*, *28*, 293.

5
Catalytic Combustion Processes

A.P. GLASKOVA

ABSTRACT: This work presents experimental data on the effect of catalytic additives on the combustion characteristics of ammonium nitrate and perchlorate and the explosives of different classes. Burning rates are determined photographically in a constant pressure bomb within a broad range of pressure: from the lower limit up to 1000 atmospheres.

The ratio, K, of the catalyzed, R_m, to the uncatalyzed, R_0, mass burning rate under the same conditions is used as a criterion of the efficiency of the additive.

The upper and lower limits of manifestation of the catalytic effect of inorganic and metallorganic additives are determined and the corresponding diagrams, $K(R_0)$, are drawn. An equation of catalysis in the form of $K = AR_0^n$ is used to present the data.

The general aspects and principles of positive and negative catalysis during the combustion of condensed and gaseous systems and the arbitrary nature of the "combustion catalyst" concept are considered.

Introduction

The problem of chemically regularing the combustion rate within broad limits with the aid of catalysts and inhibitors is particularly interesting. This interest is due to the fact that compositions whose rate of combustion can be altered from zero or a few millimeters to hundreds of centimeters per second are needed for different technical applications. Thus, an ideal safety consideration for explosives employed under mining conditions (which are hazardous with respect to gas and dust) would be their incapacity for combustion, which would make it possible to solve the extremely pressing problem of the burning out of such explosives if they fail to detonate.

Catalysts for the acceleration or retardation of the combustion process are also needed in other technical fields, e.g., for developing fuels with required combustion parameters, for reducing the content of poisonous gases in combustion and explosion products such as the exhaust of automobiles, and for extinguishing fires.

Experimental data on the effect of catalysts on the combustion characteristics of ammonium perchlorate and nitrate and explosives of different clases within a wide range of pressures (from the lower limit up to 1000 atmospheres) will be reviewed herein.

The following questions will be examined:

1. the arbitrary nature of the concept of the "combustion catalyst;"
2. the upper and lower limits of the catalytic effect; and
3. certain general aspects and principles of positive and negative catalysis.

Experimental Results and Discussion

Burning rates were determined photographically in a constant pressure bomb (Glaskova and Tereshkin 1961) The additives used in the work had a particle size of less than 100 μ. The ratio, K, of the catalyzed (R_m) to the uncatalyzed (R_0) mass burning rates under the same conditions was used as a criterion of the efficiency of the additive. When $K > 1$, the additive acts as catalyst and where $K < 1$, it acts as an inhibitor.

The dependence of the burning rate on pressure (p) expressed by the equation $R_m = Bp^{\nu}$.

The Arbitrary Nature of the Combustion Catalyst Concept

Effects of Additives on the Combustion of Ammonium Perchlorate

It has been shown (Glaskova 1952) that during the deflagration of mixtures based on ammonium nitrate the optimal content of the catalyst is 5 to 7 percent by weight.

Let us consider at first the dependence of $R_m(p)$ for pure ammonium perchlorate (Friedman et al. 1957; Glaskova 1963) The $R_m(p)$ curve for ammonium perchlorate can be divided into three characteristic segments: 50–150 atm, the region of stable combustion with a linear relationship $R_m(p)$; 160–500 atm, a transitional region characterized by unsteady, pulsating combustion; and finally the 500–1000 atm region, where exponent ν becomes significantly greater than 1. In this case the reactions determining the rate of combustion in the first region occur in the condensed phase, and at pressures above 500 atm in the gas phase (Glaskova 1970). Unstable combustion in the transition region is due to the autoinhibiting process of combustion by water (Glaskova 1968b) and by interaction of the different combustion stages.

Inasmuch as the experiments dealing with the combustion of catalyzed ammonium perchlorate were conducted in Plexiglas tubes, and, as shown in Glaskova (1963), the spread of the flame over the surface of the Plexiglas–ammonium perchlorate boundary is the leading one during the combustion of pure perchlorate at pressures above 500 atm; it is very important which values of the rates of combustion of perchlorate are taken for comparison: those obtained during the combustion of specimens in Plexiglas tubes or those obtained without a container.

The latter was taken as the true rate of combustion of pure perchlorate (Glaskova 1963).

Experiments were conducted to determine the effect of the container material on the rate of combustion of ammonium perchlorate with the catalyst (5 percent $K_2Cr_2O_7$). The results of these experiments are given in Fig. 5.1, which shows that Plexiglas and perchlorvinyl-coated Plexiglas tubes have no effect on the burning rate during the combustion of catalyzed ammonium perchlorate.

Therefore K values for combustion of pure ammonium perchlorate were calculated from experimental data obtained from the combustion of specimens without a container, while values for ammonium perchlorate with a catalyst were taken from rates of combustion in Plexiglas tubes.

It was established in a study (Boggs et al. 1972) that adding the $K^{=+}$ ion to ammonium perchlorate in the pressure range of 40 to 140 atm in the amount of 0.05 to 0.3 percent increased the rate of combustion of the perchlorate, while adding it in the amount of 0.5 to 0.8 percent, on the contrary, reduced it. The authors explained this interesting phenomenon (Boggs et al. 1971, 1972). We studied this phenomenon in greater detail and within a broad range of pressure.

A study was made on the effect of the content of potassium dichromate (0.05–10 percent) on the relationship $R_m(p)$ for rectangular specimens of ammonium perchlorate (7 × 8 × 24 mm) without a container. Figure 5.2 and Table 5.1 give the obtained results, from which it follows that when 0.05 and 0.1 percent potassium dichromate are added, the upper limit of combustion relative to pressure shifts from 150 atm to the region of higher pressures

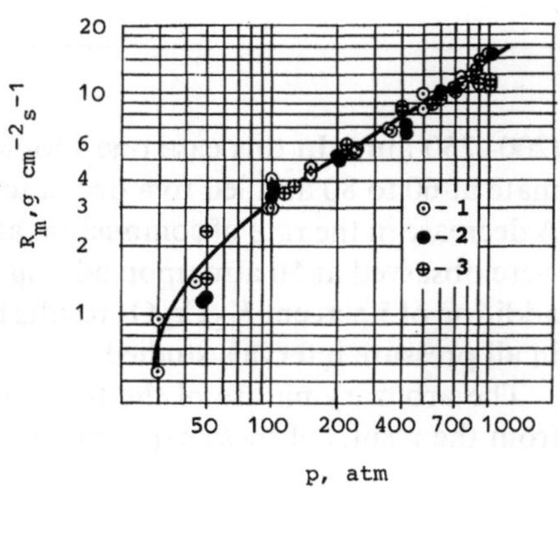

FIGURE 5.1. Absence of an effect of the container on the rate of combustion of ammonium perchlorate with 5 percent potassium dichromate: (1) specimens in Plexiglas tubes; (2) rectungular pellets (7 × 8 × 24 mm) without containers; and (3) specimens in glass tubes 7 mm in diameter, $\delta = 1.3 - 1.4$ g/cm^3.

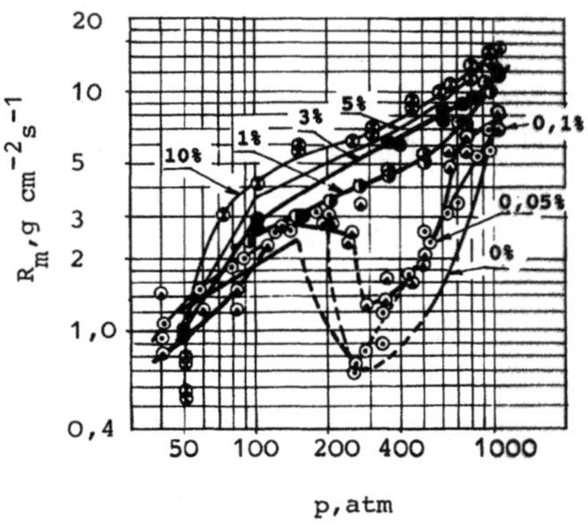

FIGURE 5.2. Effect of the content of potassium dichromate on the relationship of the rate of combustion of ammonium perchlorate with pressure. The numbers next to the curves represent the content of potassium dichromate (percentage by weight).

TABLE 5.1. Effect of the content of potassium bichromate on the combustion characteristics of ammonium perchlorate.

$K_2Cr_2O_7$ (%)	p_{lim} (atm)	K (atm) 50	150	300	600	1000	B	ν	p (atm)
0	50	—	—	—	—	—	0.22	0.49	50– 150
							∞	−3.84	150– 200
							$5.2 \cdot 10^{-7}$	2.30	500–1000
0.05	40	1.0	1.2	1.1	2.4	0.9	0.068	0.75	40– 200
							∞	−5.9	200– 250
							$6.4 \cdot 10^{-5}$	1.68	250–1000
0.1	40	1.0	1.1	3.4	2.4	1.0	0.0134	1.08	40– 125
							1.07	0.18	125– 275
							∞	−4.16	300– 375
							$4.8 \cdot 10^{-5}$	1.734	400–1000
1.0	75	—	1.3	6.0	4.5	1.6	0.26	0.49	100– 500
							0.008	1.058	500–1000
3.0	50	0.9	1.5	7.9	6.1	1.6	0.003	1.50	50– 100
							0.15	0.63	100–1000
5.0	20	1.3	1.9	9.0	7.7	2.0	0.011	1.25	20– 100
							0.142	0.67	100–1000
10.0	50	0.6	2.2	10.9	7.9	1.8	$3 \cdot 10^{-5}$	2.55	50– 100
							0.484	0.485	100–1000

(200–250 atm). In this case one should note that adding 0.1 percent dichromate at 60 to 80 atm led to a decrease in the rate of perchlorate combustion. A decrease in the rate of combustion and a pulsating character of combustion were observed at 50 atm upon adding 10 percent potassium dichromate. The addition of 5 percent $K_2Cr_2O_7$ results in the increase of the rate of combustion in all pressure intervals studied.

The arbitrary nature of the term "combustion catalyst" is visibly evident from the results of these experiments when precisely the same additive acce-

FIGURE 5.3. The effect of the content of sodium salicylate (figures next to curves) on the relationship of the rate of combustion of ammonium perchlorate with pressure for rectangular specimens without armor.

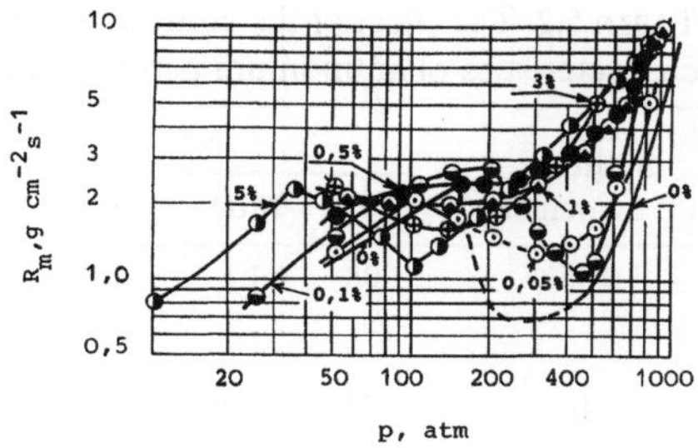

lerated the combustion process in small amounts and inhibited it in large amounts.

The effect of the content of sodium salicylate on the rate of combustion of ammonium perchlorate, shown in Fig. 5.3, serves as still another convincing example of the arbitrary nature of concepts of "catalyst" and combustion "inhibitor." It is evident from the figure that adding 1 to 5 percent sodium salicylate to ammonium perchlorate led to an increase in the rate of combustion at pressures up to 70 to 100 atm and above 160 atm, and reduced it in the region of pressures of 100 to 150 atm. It is interesting to note that not only the rate of combustion, but also the character of the $R_m(p)$ curves depends upon the amount of additive: the curves for perchlorate with 0.05, 0.1, and 5 percent sodium salicylate are reminiscent of the curve for pure ammonium perchlorate, but the maximum of the rate of combustion is not reached at 150 atm—it is displaced to 100, 200, and 50 atm, respectively. The curve for ammonium perchlorate with 0.5 percent sodium salicylate has a plateau at pressures of 100 to 200 atm, but the curves with 1 and 3 percent have a minimum, which is particularly sharply pronounced for perchlorate with 3 percent sodium salicylate.

The values of B and ν in the combustion equation of ammonium perchlorate with sodium salicylate are shown in Table 5.2. Although sodium salicylate can participate in combustion as an organic additive as both a catalyst and a fuel, it is evident from Table 5.2 that the maximum catalytic effect is reached upon adding 3 percent, not 5 percent, for example, at 50 atm, which indicates the predominant role of the catalytic properties of the additive in combustion.

The specificity of the effect of combustion catalysts appears in that precisely the same substance can catalyze combustion in one range of pressure but not influence it on another and can inhibit it in a third, as happened when silicon dioxide was added to ammonium perchlorate. Ammonium perchlorate with 5 percent SiO_2 burns in the pressure range studied according to the law: $R_m = 0.025\,p^{0.878}$. It was shown in Glaskova (1968) that adding silicon dioxide does not influence the burning rate at pressures up to 150 atm; in the higher pressure range it increases it, but at still higher pressures it decreases burning rate. This is associated with the transfer of the leading role of chemical reactions which determine the rate of combustion from one reaction to an-

TABLE 5.2. The effect of the content of sodium salicylate on the combustion characteristics of ammonium perchlorate.

Additive	K (atm)						
(% by weight)	50	100	300	600	B	ν	p (atm)
0.05	1.08	1.1	1.86	1.85	0.07	0.742	50– 100
					14	−0.425	100– 300
					$4.2 \cdot 10^{-5}$	1.77	300– 800
0.1	1.25	1.26	2.28	2.08	0.12	0.622	25– 200
					4600	−1.42	200– 400
					$16 \cdot 10^{-11}$	4.04	500– 800
0.5	1.50	1.21	4.0	3.61	0.70	0.242	50– 400
					$6.4 \cdot 10^{-4}$	1.41	400– 800
1.0	1.75	1.05	3.43	3.46	Plateau		50– 200
					0.0015	1.29	300– 800
3.0	1.92	0.84	3.29	—	20	−0.555	50– 100
					0.01	0.978	200– 400
5.0*	1.67	0.63	4.40	5.00	0.26	0.606	1– 40
					40	−0.785	40– 100
					0.0123	1.0	100–1000

* The experiments were performed in Plexiglas tubes.

other with the change in pressure. The fact that the effectiveness of the catalytic action of different catalysts can change with the increase in pressure is also associated with the same circumstance; it can increase, pass through a maximum, or decrease.

The increase in rate of combustion of ammonium perchlorate upon adding silicon dioxide (in the pressure range 150 to 550 atm) is due to the fact that the latter binds the water (Glaskova 1968b).

The Upper and Lower Limits of Manifestation of the Catalytic Effect

Within what limits can one increase or decrease the rate of combustion of explosives and on what factors does the effectiveness of manifestation of the catalytic effect depend?

We shall examine the relationship of catalytic effectiveness with the initial rate of combustion with the example of the explosives studied in general detail: ammonium perchlorate and ammonium nitrate and nitroguanidine.

Table 5.3 gives the values of K, B, and ν in the combustion equation and shows the range of pressure in which it is applicable for the ammonium perchlorate with inorganic catalysts. Figure 5.4 shows the relationship of the catalysis coefficient with the initial rate of combustion R_0 in the form of a diagram. In this diagram the vertical lines pertain to the equal rates of

TABLE 5.3. The effect of inorganic catalysts (5 percent by weight) on ammonium perchlorate combustion characteristics.

Catalyst	K (atm)						B	ν	p (atm)
	50	150	300	600	800	1000			
Copper	0.8	0.7	4.0	4.5	2.9	1.8	0.17	0.49	50– 400
Copper(II) bichronate dihydrate	1.6	2.1	12.2	11.1	5.3	3.0	0.80	0.815	30–1000
Copper chloride	1.9	1.8	9.4	8.6	3.7	2.1	0.17	0.66	1–1000
Copper borotung-stenate	0.8	1.0	5.3	4.6	2.7	1.9	0.063	0.705	10– 500
							0.00031	1.54	500–1000
Potassium chromate and bichromate	1.3	1.9	9.0	7.7	3.5	2.0	0.011	1.25	20– 100
							0.142	0.67	100–1000
Bismuth (III) bichromate	1.5	1.5	7.0	7.2	3.4	1.8	0.24	0.521	50– 400
							0.031	0.878	400–1000
Ammonium bichromate	1.3	1.1	4.7	4.5	2.8	1.7	0.22	0.506	30– 200
							0.0015	1.315	400–1000
Lithium bichromate, zinc, and cesium chromates	1.7	1.7	8.4	7.3	3.7	2.1	0.142	0.65	50– 500
							0.025	0.93	500–1000
Barium chromate	1.1	1.2	6.8	6.5	3.7	2.1	0.053	0.79	50– 500
							0.017	0.983	500–1000
Cadmium chromate	1.5	1.5	7.6	6.1	3.2	1.8	0.17	0.63	50–1000
Lead chromate	1.9	1.5	9.0	8.6	4.3	2.3	0.165	0.64	50– 400
							0.031	0.923	400–1000
Cobalt chromate	—	1.4	7.4	6.6	3.1	1.8	0.08	0.74	100–1000
Chromic oxide	1.2	1.6	6.6	6.2	3.2	1.8	0.044	0.937	50– 100
							0.044	0.834	300–1000
Chromium chloride	1.2	1.0	4.9	4.9	2.5	1.6	0.19	0.506	30– 300
							0.0054	1.115	300–1000
Vanadium pentoxide	0.9	0.9	4.3	4.6	2.5	1.6	0.068	0.70	50– 100
							0.0095	1.04	200–1000
Cobalt nitrate, 6-hydrate	—	0.6	4.0	4.6	2.5	1.4	0.0095	1.02	100–1000
Potassium permanganate	0.4	0.7	4.1	4.6	2.3	1.5	0.05	0.693	20– 300
							0.007	1.068	300–1000
Iron	0.9	1.0	3.6	4.4	2.6	1.8	0.058	0.764	1– 140
							0.0006	1.44	300–1000
Ferric (III) oxide	0.8	1.1	6.2	5.3	2.7	1.6	0.034	0.85	20–1000
Ferric chloride	0.8	1.0	5.9	5.1	2.3	1.5	0.066	0.725	50–1000

combustion, while the arrows indicate pressures at which these rates are achieved. One should primarily note that this diagram limits the upper and lower boundaries of manifestation of catalytic effect of the 23 inorganic additives we studied.

It is necessary to note first of all, that if one omits the transition region of 160 to 300 atm from the examination, in which value ν is negative, then the obtained diagram can be divided into two regions: 50 to 150 atm, where one observes a weak increase in the catalytic effectiveness K with the increase in the initial rate of combustion; and 300 to 1000 atm, where K decreases with

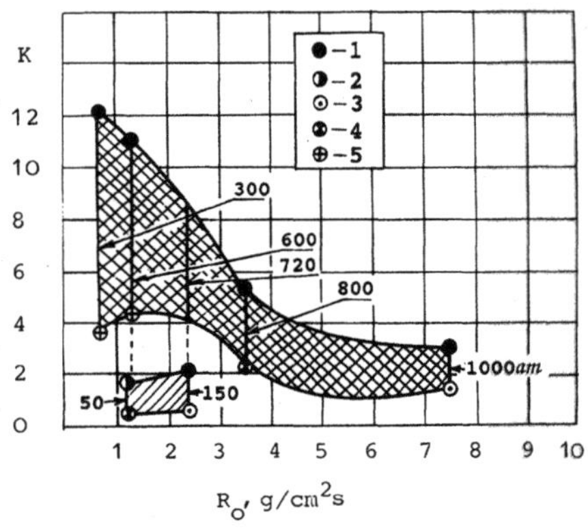

FIGURE 5.4. Relationship of the catalytic effectiveness of inorganic catalysts with the initial rate of combustion of ammonium perchlorate. (1) $CuCr_2O_7 \cdot 2\,H_2O$; (2) $Li_2Cr_2O_7$; (3) $Co(NO_3)_2 \cdot 6\,H_2O$; (4) $KMnO_4$; and (5) Fe. The numbers indicate the values of pressure (in atm).

the increase in R_0. If the equation of catalysis is expressed as $K = AR_0^n$, then the upper boundary of catalysis is described by the equation $K = +0.8\,R_0^{0.306}$, and the lower boundary by the equation $K = 0.15\,R_0^{0.425}$ in the range of the initial rates of combustion of 1.2 to 2.4 g/cm² s., in the pressure region of 50 to 150 atm. In the pressure region of 600–1000 atm, in the range of initial rates of combustion of 1.3, 7.5 g/cm² s the catalysis equation at the upper limit will be $K = 77\,R_0^{-0.76}$ and the lower boundary $K = 21.5\,R_0^{-0.63}$.

The upper boundary of catalytic effectiveness is limited by copper (II)-dichromate dihydrate, while the lower boundary is limited by potassium permanganate, cobalt nitrate, and iron (see Fig. 5.4) in both the first and second regions. The values of coefficient K for the most and least effective additives in the first region differ four times at 50 atm and 3.7 times at 150 atm. In the second region the most and least effective additives differ by 3.4 times at 300 atm and 2 times at 1000 atm.

It was considered that catalytic effects most sharply appear near the boundary conditions of combustion and for substances that cannot burn as well without catalysts or for slowly burning systems. Analysis of the results in Fig. 5.4 shows that the rate of combustion for ammonium perchlorate itself is not decisive in manifestation of effectiveness of the catalytic action of additives. In this case, the pressure at which the initial rate of combustion is reached plays an important role. Thus, at 50 and 600, 150 and 720 atm perchlorate burns at an identical rate (the vertical dash lines in the diagram). However, at 50 atm lithium dichromate increased the rate of combustion twofold, and at 600 atm copper (II) dichromate dihydrate increased it 11 times.

If one takes the same catalyst—copper (II) dichromate dihydrate—at 150 atm K is 2.1, and at 720 atm it is 8.2. This indicates that the magnitude of the initial rate of combustion is not important by itself, but rather those leading reactions as a result of whose occurrence it is achieved. In the pressure region of 50 to 150 atm, these reactions occur in the condensed phase; adding more effective catalysts—the dichromates of lithium, potassium, and copper to the ammonium perchlorate—led to a decrease in B and an increase in ν in the combustion equation, i.e., in this case the leading reaction shifted from the condensed to the gas phase. The results of experiments on the effect of density on the rate of combustion of ammonium perchlorate catalyzed by potassium dichromate can also serve as confirmation of this conclusion. As is evident from Fig. 5.1, the rate of combustion of catalyzed perchlorate did not depend on density, i.e., the reaction determining the rate of combustion occurred in the gas phase, while the reverse was the case for pure ammonium perchlorate.

On the contrary, in the high-pressure region the most effective additives reduced the proportion of reactions that occur in the gas phase. Thus copper (II) dichromate dihydrate reduced the value of ν 3 times (from 2.36 to 0.815).

One way to increase the rate of combustion with catalysts, particularly in the transitional region (160–500 atm) is acceleration of the decomposition of perchloric acid (Boggs et al. 1971) as well as that of its monohydrates and polyhydrates. The effect of catalytic additives on thermal decomposition of perchloric acid in vapor was studied by Solymosi et al. (1968). In this case the most effective catalyst was chromic oxide, particularly if zink oxide was added to it to enchance electrical conductivity.[1] As is evident from Table 5.3 the chromic oxide is less effective during combustion than compunds of hexavalent chromium.

The effect of metallorganic salts on characteristics of combustion of ammonium perchlorate are shown in Table 5.4. As is evident from Table 5.4, copper salicylate was the most effective of the studied salts of salycilic acid. Bismuth, magnesium, and mercury salicylates accelerated the combustion ammonium perchlorate only at pressures above 150 atm; in the region of low pressures they even reduced the rate of combustion.

Benzoic acid salts had precisely the same effect on the combustion of ammonium perchlorate as salicylic acid salts. Only sodium benzoate and lithium benzoate accelerated the combustion of perchlorate at a pressure of 500 atm. Fuchsin had a quite similar effect on the combustion of ammonium perchlorate.

Inasmuch as copper salicylate was the most effective of the studied organic catalysts, other copper-containing compounds were also studied: stabilin-9 (Smirnov 1969) copper oxiquinolate (oxinate), acetylacetonato-, bis-(ethylacetoacetonate)-, and bis-(3-nitroacetilacetonato)-copper (II). As is evident from the table the catalytic effect of stabilin-9 is equivalent to the effect of copper salicylate. Copper oxinate and other copper compounds are significantly more effective.

Finally, one should dwell on the effect of the organic salts of the aliphatic series, presented in Table 5.4. The data for ammonium perchlorate with the stearates of aluminium and zink show that aluminium stearate is less effective

[1] Combustion of ammonium perchlorate with various inorganic additives was studied in Boggs et al. (1988) within the pressure interval from 1 to 105 atm.

TABLE 5.4. The effect of metallorganic salts (5 percent by weight) on characteristics of combustion of ammonium perchlorate

Additive	K (atm) 50	150	300	600	800	1000	B	v	p (atm)
Copper oxinate	2.5	1.8	20.8	18.0	7.9	3.9	0.155	0.78	10–1000
Copper salicylate	1.4	1.5	10.9	12.0	5.9	3.5	0.110	0.68	10–1000
							0.019	1.05	100–1000
Copper benzoate	1.8	2.0	11.0	9.6	4.5	2.5	0.125	0.723	50–1000
Stabilin-9	1.2	1.4	10.1	12.0	5.8	3.6	0.142	0.607	1– 100
							0.019	1.05	100–1000
Copper (II) bis-	2.9	3.1	15.4	12.3	6.0	3.6	0.210	0.696	50– 400
(ethylacetoacetonate)							0.063	0.878	400–1000
Copper (II)	2.5	2.8	14.3	13.4	7.0	3.8	0.340	0.595	50– 400
bis(acetylacetonate)							0.041	0.945	400–1000
Copper (II) bis(3-	2.3	2.2	12.0	13.2	6.6	3.7	0.076	0.85	50–1000
nitroacetylacetonate)									
Cobalt (III) tris(3-	1.6	1.9	10.9	10.7	5.6	3.2	0.066	0.84	30– 900
nitroacetylaetonate)									
Cobalt (III) tris(3-	1.7	1.3	8.0	8.9	4.8	2.6	0.093	0.729	50– 400
nitroacetylacetonate)							0.0097	1.110	400–1000
Chromium (III) tris(3-	1.8	1.6	8.6	11.2	6.6	3.3	0.160	0.649	30– 200
nitroacetylacetonate)							0.0088	1.152	200–1000
Bismuth salicylate	0.5	0.7	3.6	6.4	4.3	2.3	0.0360	0.720	50– 200
							0.0002	1.670	300–1000
Magnesium salicylate,	0.7	0.8	4.6	5.5	3.3	2.0	0.0190	0.926	25– 300
hydrate							0.0017	1.318	200–1000
Mercury salicylate	0.7	0.5	3.7	5.2	3.0	1.9	0.116	0.515	10– 200
							0.0005	1.500	200–1000
Sodium salicylate	1.8	0.7	4.4	5.6	2.8	1.6	0.260	0.606	1– 40
							40	−0.786	40– 100
							0.0123	1.00	100–1000
Aluminium benzoate	0.9	1.8	10.3	9.1	4.3	2.4	0.043	0.88	50–1000
Lead benzoate	0.4	0.5	4.0	5.1	2.8	1.9	0.0085	1.05	50– 800
Bismuth benzoate	0.5	0.8	4.9	5.6	2.9	1.8	0.0085	1.045	50–1000
Sodium benzoate	1.3	0.6	4.1	4.5	3.3	1.1	0.100	0.832	10– 35
							plateau		40– 200
							0.0115	0.980	200–1000
Cadmium benzoate	0.6	0.8	5.2	5.7	3.0	1.8	0.0760	0.630	10– 200
							0.0098	1.056	200–1000
Lithium benzoate*	1.7	0.9	4.9	6.5	3.3	1.9	0.008	1.080	200–1000
Fuchsin	0.7	0.5	3.9	5.1	2.5	1.4	0.062	0.146	10– 200
							0.0047	1.135	200–1000
Aluminium stearate	1.2	1.3	8.6	10.8	4.8	2.4	0.0380	0.90	50–1000
Zinc stearate	0.9	1.2	7.7	9.1	5.0	3.1	0.048	0.826	30– 300
							0.0074	1.15	300–1000
tris- Pyrocatechin-	1.1	1.3	7.3	5.6	3.4	2.3	0.096	0.698	1– 500
ferric							0.0023	1.30	500–1000
Ferrocen	—	2.0	8.3	9.8	5.3	2.8	0.200	0.622	50– 600
							0.0029	1.283	600–1000

* Decrease of combustion rate with pressure and plateau on the curve took place before 200 atm.

up to 300 atm than benzoate, while in the region of pressure above 300 atm it is somewhat better relative to its effectiveness.

It is interesting to note that copper (II) dichromate dihydrate is more effective in the pressure range of 250 to 500 atm than some of the organic compounds examined earlier. This is probably associated with the better solubility of the inorganic salts in water, which have a significant effect on the combustion process in this range of pressures. The organic part of the molecule also exerts an extremely significant effect on the catalytic activity of the copper-containing compounds. Thus, it is evident from Table 5.4 that coefficient K can change from 1.2 to 3.0 at 50 atm and from 10 to 21 at 300 atm, depending upon the organic part of the molecule. However, the difference becomes smaller with the increase in pressure. In this case the difference in the catalytic effectiveness, for example, of the copper-containing organic compounds is not associated with the absolute amount of copper in the compound molecule. Thus, 5 percent copper salicylate by weight contains 0.96 grams of the metal, while the same amount of copper oxinate contains 0.83 grams. Nevertheless, the latter compound is significantly more effective as a catalyst. A similar picture is also observed for the sodium-containing salts, for example, at 50 atm $K = 1.3$ for sodium benzoate, $K = 1.8$ for sodim salicylate, and $K = 0.7$ for fuchsin.

One should pay attention to the complex picture of the catalytic effect of the additives studied earlier and in our work on the combustion of ammonium perchlorate observed during investigation of the process within a broad range of pressures. This is evidently associated with the complexity of the relationship $R_m(p)$ for ammonium perchlorate itself. The action of the most effective combustion catalysts (organic copper-containing salts, sodium salicylate, and benzoate) appeared, as is evident from Table 5.4, in a sharp increase in coefficient B in the combustion equation. Values of ν remained practically unchanged. These occurred in the region of low pressures (50–150 atm), when the reactions that determine the rate of combustion occur in the condensed phase. This can signify that the organic catalysts also increase the proportion of the reactions that occur in the condensed phase. In contrast, the effect of additives that reduced the rate of combustion in this region (bismuth, magnesium, mercury salicylates, lead, bismuth, cadmium benzoate, and fuchsin) was manifested in the fact that they reduced the values of B.

In the pressure region from 160–500 atm, where anomalies are observed on the curve $R_m(p)$ for pure ammonium perchlorate and where the law of combustion is characterized by extremely high values of B and negative values of ν (Glaskova 1963, 1970), all of these studied additives increased the rate of combustion. Finally, the effect of combustion catalysts appeared not only in an increase in the value of B, but also in a decrease in the power index ν in the region of pressure above 500 atm, when the reactions that determine the rate of combustion of pure perchlorate occur in the gas phase. These changes are particularly noticeable in the presence of the most effective catalysts.

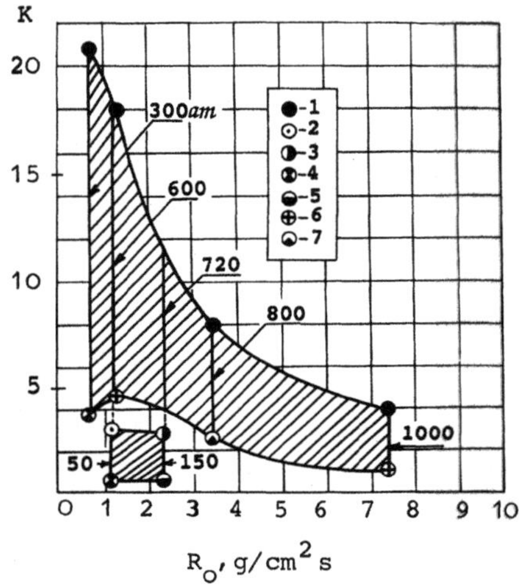

FIGURE 5.5. Diagram of dependence of catalytic efficiency of metallorganic salts on initial combustion rate of ammonium perchlorate. (1) copper oxinate; (2) bis-(ethylacetoacetonate) copper (II); (3) aceylacetonato-copper(II); (4) bismuth benzoate; (5) mercury salicylate; (6) sodium benzoate; and (7) fuchsin. Numbers on curves represent pressure (in atm).

One should note that catalytic effect of the metallorganic salts appeared during the combustion of ammonium perchlorate even at such high pressures as 500 to 1000 atm. Thus, at 1000 atm the rate of combustion of ammonium perchlorate in the presence of organic copper-containing salts exceeded the rate of combustion of pure perchlorate 3 to 4 times.

The paradoxical decrease in the rate of combustion of ammonium perchlorate in the presence of certain organic compounds (see Table 5.4), particularly in the low-pressure region, is probably associated with the fact that inasmuch as the ion of the given metal (for example, bismuth, mercury, magnesium,[2] or cadmium) has no catalytic action on the process, the effect of the organic part of the molecule, specifically its reducing properties, predominate (Glaskova and Popova 1967). Moreover, the participation of the metallic ion in an exchange reaction of the type described before (Glaskova 1969) is possible, as is retardation of combustion as a consequence of binding of perchloric acid, whose decomposition products are the oxidants for the perchlorate fuel elements.

Figure 5.5 gives the region of manifestation of the catalytic effect of 25 studied organic catalysts in a way similar to Fig. 5.4 for the inorganic catalysts. In the pressure range from 50 to 150 atm (the rate of combustion of pure ammonium perchlorate changed from 1.2 to 2.4 g/cm^2 s, the catalysis equation is the following: $K = 2.97\ R_0^{-0.92}$. The upper boundary of catalysis is described by the equation $K = 19\ R_0^{-0.144}$ in the region of pressures from 300 to 600 atm when R_0 changes from 0.7 to 1.3 g/cm^2 s and at still higher pressures $K = 22.7\ R_0^{-0.877}$ ($R_0 = 1.3 - 7.5$ g/cm^2 s, $p = 600$ to 1000 atm).

[2] During thermal decomposition of mixtures of ammonium perchlorate with magnesium perchlorate, the effect of the latter is equal to the catalytic effect of copper chromite and exceeds the effect of ferric oxide or charcoal (Acheson and Jacobs 1970).

Copper oxinate is on the upper boundary of manifestation of catalysis in the region of high pressures, and bismuth salicylate, sodium benzoate, and fichsin are on the lower one. Bismuth and mercury salicylates are on the lower boundary in the low-pressure region, while copper (II) bis-(ethylacetoacetonate) and copper (II) bis-(acethylacetonate) are on the upper boundary.

It should be emphasized that the catalytic effects are very great for the organic compounds. Thus, the rate of combustion of ammonium perchlorate with the most effective catalyst—copper oxinate—exceeded the rate of combustion of pure perchlorate 21 times at 300 atm. But even at 1000 atm (when, seemingly, reactions could occur more competely as a result of such a high pressure), copper oxinate increased the rate of combustion fourfold.

It is interesting to note that the lower boundary of catalysis during the combustion of perchlorate coincides for organic and inorganic catalysts (compare Figs. 5.4 and 5.5), for both the range of pressure from 50 to 150 atm, and 300 to 1000 atm. It is visibly evident from Figs. 5.4 and 5.5 how great the difference in the effectiveness of different additives is at precisely the same initial rate and pressure. Thus, adding iron or bismuth salicylate to perchlorate at 300 atm led to a fourfold increase in the rate of combustion, while adding copper oxinate led to a 20-fold increase (see Tables 5.3 and 5.4). At the same time sodium benzoate (on the lower limit) had no effect on the rate of combustion ($K = 1$); hence, nearly the same difference was preserved between the effect of the most and least active catalysts as at 300 atm.

Catalytic effects in the region of pressure from 50 to 150 atm are significantly less. It should be emphasized that the effectiveness of an additive not only depends upon the value of the initial rate of combustion itself, but also on the pressure at which this rate is achieved. Thus, the initial rates of combustion of ammonium perchlorate are identical at 50 and 600 atm. However, the maximum and minimum increase in the rate of combustion upon adding catalysts differs six times. $K_{max}/K_{min} = 4$ at 150 atm and 720 atm at precisely the same rate of combustion.

With respect to the character of the $K(R_0)$ relationship, for the upper boundary[3] of catalysis, a sharp drop in catalytic effectiveness with the increase in R_0 according to a power law $K = AR_0^n$ is characteristic. Table 5.5 gives the values of A and n in the cited equation, as well as the ranges of the rates and pressures in which it is satisfied.

The limits of manifestation of catalysis during the combustion of ammonium perchlorate with different additives were examined earlier. We shall now see how catalytic effectiveness depends on the initial rate of combustion for precisely the same catalyst during the combustion of explosives of different classes. Figure 5.6 shows such a relationship for different explosives (nitroguanidine (Glaskova 1971a), ammonium nitrate (Glaskova 1967), TNT,

[3] One could call it the upper limit, but we have deliberately not done this inasmuch as there is no absolute certainty that the catalytic effect of copper oxinate cannot be exceeded.

TABLE 5.5. Values of A and n in the catalysis equation $K = AR_0^n$ for ammonium perchlorate.

p(atm)	R_0(g/cm$^2\cdot$s)	A	n	Boundary of catalysis
		Inorganic Catalysts		
50–150	1.2–2.4	0.4	0.432	Lower
50–150	1.2–2.4	1.56	0.357	Upper
300–600	0.7–1.3	4.05	0.325	Lower
600–1000	1.3–7.5	5.2	−0.626	
300–600	0.7–1.3	11.8	−0.202	Upper
600–1000	1.3–7.5	13.6	−0.757	
		Organic Catalysts[a]		
50–150	1.2–2.4	2.97	−0.092	Upper
300–600	0.7–1.3	19.0	−0.144	
600–1000	1.3–7.5	22.7	−0.877	

[a] The lower boundary of catalysis coincides for the organic and inorganic catalysts.

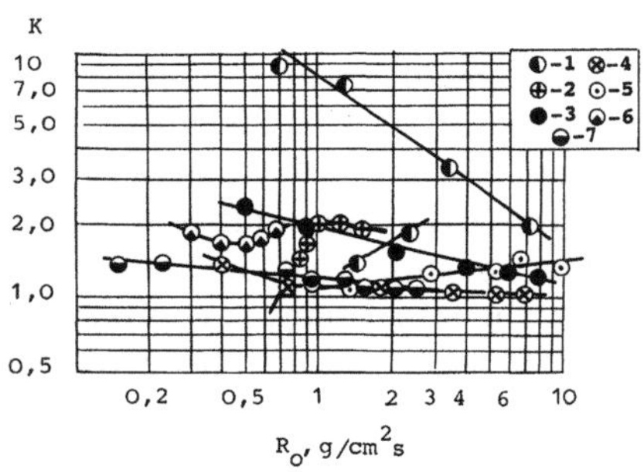

FIGURE 5.6. Relationship of the coefficient of catalysis with the initial rate of combustion for different explosives with 5 percent potassium dichromate by weight: (1) ammonium perchlorate; (2) ammonium nitrate; (3) picric acid; (4) trotyl; (5) tetryl; (6) dina; and (7) nitroguanidine.

tetryl, picric acid, and dina (Glaskova 1973, 1974) with 5 percent of potassium dichromate. Initial rate of combustion changed from 0.1 g/cm^2 s to 10 g/cm^2 s when pressure rose from 1 to 1000 atm. One should primarily note that, as for the cases examined for different catalysts, the value of K changes within broad limits for the same catalyst but during the combustion of different explosives.

Furthermore, the value of K can change four times precisely at the same initial rate of combustion. This indicates that it is not the magnitude of the initial rate of combustion of itself that is important in manifestation of catalytic effectiveness, but are those chemical reactions as a result of whose occurrence it is achieved. These reactions can be different for different ranges of pressures. This is why the same additive at the same value of R_0 can change its effectiveness 3 to 4 times with all other conditions being equal.

The shift of the leading reactions also determines the different course of the $K(R_0)$ curves for the different classes of explosives shown in Fig. 5.6. Thus, the

TABLE 5.6. Values of A and n in the catalysis equation $K = AR_0^n$ for different explosives (catalyst $-$5 percent $K_2Cr_2O_7$, pressure range 1–1000 atm).

Explosive	A	n	R_0(g/cm$^2\cdot$s)
Trotyl	1.0	−0.36	0.4 – 0.75
	1.1	−0.031	0.75– 7.0
Picric acid	1.95	−0.235	0.5 – 8.0
Tetryl	1.1	0.111	0.9 –10.0
Dina	1.45	−0.215	0.3 – 0.5
	2.3	0.46	0.5 – 0.7
Nitroguanidine	1.2	−0.1	0.15– 2.5
Ammonium perchlorate	1.15	0.565	1.2 – 2.4[a]
	8.2	−0.70	0.7 – 7.5[b]
Ammonium nitrate	2.0	1.90	0.8 – 1.0
	2.0	−0.12	1.0 – 1.6

[a] Pressure range 50–150 atm
[b] Pressure range 300–1000 atm.

effectiveness of potassium dichromate for picric acid decreases with an increase in the initial rate of combustion, while, on the contrary, it increases for tetryl. It passes through a maximum for ammonium nitrate, while it passes through a minimum for dina, and finally, for ammonium perchlorate (characterized by a complex $R_m(p)$ relationship), the effectiveness of the additive increases with increase in R_0 in the range of rates from 1.2 to 2.4 g/cm^2 s and decreases in the range of the latter of 0.7 to 7.5 g/cm^2 s. Table 5.6 gives the characteristics of the catalytic effectiveness of potassium dichromate for the studied explosives.

Certain General Aspects and Principles of Positive and Negative Catalysis

One should once again note the arbitrary nature of the concepts of "catalyst" and "inhibitor," not only during the combustion of condensed systems, but also in gaseous oxidation reactions. Thus, precisely the same substance, depending upon conditions, can either catalyze or inhibit a process. According to the data of Chamberlain and Hall (1973), certain surfactants were inhibitors of reactions of oxidation of coal by oxygen at low temperatures (up to 120°C), while at higher temperatures they did not influence the process. Iron sulphate, on the contrary, accelerated this reaction at temperatures up to 130°C, inhibited it in the region of temperatures from 130° to 170°C and once again accelerated it at temperatures above 170°C.

The double role of certain additives, for example, iodine and nitrogen oxide, which emerge in some cases as retardants, and in others as accelerators of chain reactions, was examined by Semenov (1958), while that for formaldehyde (in certain gaseous reactions of oxidation) was observed by Lewis and Elbe (1966). Therefore, when one is discussing catalysis, from the viewpoint of terminology it would be more correct to discuss not simply catalysis or inhibition, but positive and negative catalysis, and it would be more accurate to employ a single term—catalyst—with a definition added to it depending on the result of the additive on some particular reaction—positive or negative.

As an example of the relative concept of "catalyst" or "inhibitor," we shall recall the role of the alkali salts during the combustion of condensed systems, where they frequently emerge as catalysts (Glaskova 1952), and in chain reactions, where they are considered to be inhibitors. Thus, for example, the rate of recombination of radicals in flames of $H_2 + O_2 + N_2$, which contain additives of the alkali metals, increases (Cotton and Jenkins 1971). Lead oxide (Sharma and Bardwell 1965) is also an inhibitor of combustion during the oxidation of butane, aldehyde, and carbon monoxide at 310°C. Ketones, aldehydes, and peroxides rapidly oxidize on the surface of lead oxide and yield carbon dioxide gas and water (from our point of view this is also positive catalysis).

It is found in Salooja (1967) that lead oxides react with the hydroxyl derivatives of hydrocarbons, whose combustion they promote, and do not react with esters and hydrocarbons whose combustion they inhibit. We recall that the promoting effect of hydroxyl on the catalytic action of potassium dichromate was observed by us during the combustion of picric acid (Glaskova 1973, 1974), which was probably due to its promoting effect on the gas phase oxidation of CO. At precisely the same time, the organic compounds tetraethyl lead and pentacarbonyl iron strongly inhibit hexano-air flames (Lask and Wagner 1962). Moreover, they are the best antidetonators (Lewis and Elbe 1966).

The compounds of lead and iron are effective catalysts during the combustion of explosives. Thus, for example, lead chromate and chloride catalyzed the combustion of nitroguanidine and ammonium nitrate, while the organic compounds of iron are effective catalysts of the combustion of mixed powders based on ammonium perchlorate. At precisely the same time, lead oxide was an inhibitor during the thermal decomposition of ethylnitrate, while a copper surface accelerated decomposition (Ellis et al. 1955). A lack of correspondence on the effect of additives on thermal decomposition and combustion was also observed, for example, for potassium oxalate. Adding potassium oxalate to ammonium nitrate inhibited thermal decomposition and accelerated combustion (Glaskova 1976). But it is known (Birchall 1970; Boucart 1970; Demougin 1930–33; Glaskova et al. 1988; Prettre 1936) that potassium oxalate is one of the most effective inhibitors in chain reactions of oxidation.

The fact that the strongest catalytic effects appear for slowly burning systems and near the threshold conditions of combustion is the general principle of the effect of catalysts during the combustion of condensed systems. At the same time the inhibiting effect, on the contrary, is stronger for rapidly burning systems. A particularly visible illustration of this principle can be the resuls of experiments with mixtures based on ammonium perchlorate with fuels of varying chemical nature: an increase of 10 to 30 times in the rate of combustion upon adding catalysts was observed for the most slowly burning mixtures. Moreover, threshold pressure decreased by hundreds of atmospheres (Glaskova and Andreev 1972). A similar picture also existed for ammonium nitrate—adding small quantities of the salts of hexavalent chromium made it capable of combustion at atmospheric pressure, while in the pure form it did not even burn at 1000 atm.

It should also be noted that catalysts have an effect not only on the limits of combustion, but also on its stability. In this case, the stability of combustion decreases in the presence of catalysts, but upon adding inhibitors, it increases.

Thus, the combustion of poured dina occurred at accelerated regime at 100 atm (in tubes 6 to 7 mm in diameter) (Glaskova 1971). The addition of catalysts reduced this pressure to 60 atm, while that of inhibitors raised it to 350 atm (Glaskova 1969). An increase in the stability of combustion was also observed for hexogen (RDX) with inhibitors (Glaskova 1976).

It is interesting that the mechanism of action of even identical catalysts and compounds similar in chemical nature (for example, ammonia salts) is different. Thus, the halides and other compounds of the alkali metals—the most effective catalysts of the combustion of ammonium nitrate and mixtures based on it—inhibited the combustion of ammonium perchlorate. The mechanism of their initial effect in both cases is identical: they enter into exchange reaction with the primary products of dissociation of these salts. However, in the case of ammonium nitrate this reaction leads to the formation of a more effective oxidant and combustion accelerates. In the case of ammonium perchlorate, on the contrary, the more active oxidizing agent—perchloric acid—is bound, as a result of which the process of subsequent oxidation is inhibited. At the same time, there are also general features in the process of the catalytic effect of these salts: the catalytic effect on the combustion process for both salts is associated with the oxidation of ammonia. A general feature during catalyzed combustion of the studied systems is also the fact that the catalytic effects appear for most of them in a broad range of pressures—up to the studied limit of 1000 atm—and they are frequently quite great. Thus we once again recall that adding a catalyst for ammonium nitrate was equivalent to the effect of adding a fuel that raises the combustion temperature to 1500°.

Finally, one should also note still another general principle: as a rule, catalysts increase coefficient B in the combustion equation and decrease the power index ν for most studied systems. Inhibitors, on the contrary, reduce B and increase ν. These effects appeared most strongly in the region of low and moderately elevated pressures, which indicates that the reaction that determines the rate of combustion under these conditions occurs in the condensed phase. In fact, judging by the effect of density on the rate of combustion of catalyzed systems, the role of the catalyst consists of shifting the leading reaction from the condensed phase to the gas phase.

Thus, the rate of combustion did not depend upon density for ammonium perchlorate catalyzed by potassium dichromate (see Fig. 5.1) A similar picture was observed for ammonium nitrate, i.e., in both cases the reaction determining the rate of combustion (in the presence of catalyst) occurred in the gas phase.

We shall not dwell here on the mechanism of effect of metals of variable valence, inasmuch as it does not significantly differ from the mechanism of effect of catalysts on the conventional oxidative reactions. However, even here there are specifics: if oxides of metals of variable valence are used as catalysts in the gas oxidative reactions, then during the combustion of condensed systems they are significantly less effective and they occasionally even inhibit combustion. Various salts[4], particularly halides and salts of hexavalent chromium, are significantly more effective as catalysts.

The salts of hexavalent chromium were also effective in those cases when the organic salt with the given metal inhibited combustion. Thus, comparison of the data cited in Tables 5.3 and 5.4 shows that the benzoates of lead and cadmium inhibited the combustion of perchlorate in the low-pressure region, while the chromates of lead and cadmium catalyzed it. Here one probably has the promoting effect of chromium. At the same time, other metals had a sharply pronounced inhibiting action, for example, adding borotungstanate of copper to nitroguanidine led to inhibition of combustion; a similar weakening of the catalytic action was also observed upon adding this compound to ammonium perchlorate.

All of this again indicates the complexity of the phenomenon of catalysis during the combustion of condensed systems. In this case if one takes into account that a parallel is practically absent between the effect of the additives on the thermal decomposition and combustion, then it becomes obvious that the only current method of selecting positive combustion catalysts is the empirical one.

To wit, one should note that at present there is no complete scientific theory of selecting catalysts (Krylov 1967) even for chemical reactions that are simpler than the complicated combination of reactions that occurs during combustion. Thus, according to modern concepts, catalytic processes can be classified as two types: electron (oxidative-reducing, homolytic), and ion (acid-base, heterolytic). The catalysts for these are also, respectively, semiconductors or acids and bases (Krylov 1967). For example, the fact that both chlorides of alcali metals and salts of metals of variable valence are effective catalysts during the combustion of ammonium nitrate indicates that both types of reactions occur during combustion.

The problem of retarding chemical reactions during combustion is in significantly better shape as was shown by the author, a decrease in the rate of combustion of individual explosives (Glaskova 1968), ammonium nitrate and perchlorate, and mixtures based upon them, can be realized on the basis of

---

[4] Thus cooper oxide inhibited the combustion of nitroguanidine while copper chloride is one of the most effective catalysts.

examined hypotheses (Glaskova 1967, 1969). The circumstance that substances that bind the active intermediate products of primary decomposition or inhibit this primary stage are the strongest inhibitors is a general principle in this case. Thus ammonia salts that shift the equilibrium of dissociation to the left for ammonium nitrate and perchlorate are more effective than inhibitors that bind the products of decomposition of the formed acids.

More detailed data about mechanism of catalysts effect during combustion of condensed and gaseous systems can be found in Glaskova (1976).

References

Acheson, R.J., and Jacobs, P.W.M., 1970, *AIAA J.*, *8*, 1483.

Birchall, J.D., 1970, *Comb. and Flame*, *14*, 85.

Boggs, T.L., Petersen, E.E., Watt, D.M. Jr., 1972, *Comb. and Flame*, *19*, 131.

Boggs, T.L., Price, E.W., and Zurn, D.E., 1971, *Thirteenth Symposium (International) on Combustion*, The Combustion Institute, p. 995.

Boggs, T.L., and Zurn, D.E., 1972, *Comb. Sci. and Tech.*, *4*, 227.

Boggs, T.L., Zurn, D.E., Cordes, H.F., and Covino, J., 1988, *J. Propulsion*, *4*, 1, 27.

Boucart, J., 1960, *Explosifs*, 4, 127.

Chamberlain, E.A.C., and Hall, D.A., 1973, *The XVth International Conference of Scientific Research Institutes on Working Safety in the Mining Industry*, 1, Karlovy Vary, ČSSR, 173.

Cotton, D.H., and Jenkins, D.R., 1971, *Trans. Faraday Soc.*, *67*, 730.

Demougin, P., 1930–33, *Mem. Poudres*, *25*, 130.

Ellis, W.R., Smithe, B.M., and Trecharne, E.G., 1955, *5th Symposium (International) on Combustion*, Reinhold, 641.

Friedman, R., Nugent, R.G., Rumbel, K.E., and Scurlock, A.C., 1957, *Sixth Symposium (International) on Combustion*, Reinhold, p. 612.

Glaskova, A.P., 1952, "Candidate's Dissertation," *MKhTI in D.I.* Mendeleev, M.

Glaskova, A.P., 1963, *PMTF*, 5, 121.

Glaskova, A.P., 1967, *Explosifs*, 1, 5.

Glaskova, A.P., 1968a, *DAN SSSR*, *181*, 383.

Glaskova, A.P., 1968b, *FGV*, 4, 314.

Glaskova, A.P., 1969, *Comb. and Flame*, *13*, 55.

Glaskova, A.P., 1970, *Explosivstoffe*, 4, 89.

Glaskova, A.P., 1971a, *FGV*, 2, 211.

Glaskova, A.P., 1971, *FGV*, *7*, 1, 153.

Glaskova, A.P., 1973, *Explosivstoffe*, 4, 137.

Glaskova, A.P., 1974, *FGV*, *10*, 3, 323.

Glaskova, A.P., 1976, *Catalysis in the Combustion of Explosives*, Nauka, Moscow.

Glaskova, A.P., and Andreev, O.K., 1972, *The Collection "Combustion and Explosion,"* Nauka, Moscow, 78.

Glaskova, A.P., Karpov, V.P., and Phil, P.V., 1988, *Archiwum combustionis*, *8*, 2, 167.

Glaskova, A.P., and Popova, P.P., 1967, *DAN SSSR*, *177*, 1341.

Glaskova, A.P., and Tereshkin, I.A., 1961, *ZhFKh*, *35*, 1622.

Jacobs, P.W.M., and Russel-Jones, A., 1967, *Eleventh Symposium (International) on Combustion*, Combustion Institute, 457.

Krylov, O.V., 1967, *Catalysis by Non-metals*, Khimia, Leningrad.

Lask, G., and Wagner, H.G., 1962, *8th Symposium (International) on Combustion*, Williams and Wilkins, 432.
Lewis, B., and Elbe, G., 1966, *Combustion, flames and explosion in gases*, Moscow, Mir Press.
Prettre, M., 1936, *Mem. Poudres, 25*, 160; *J. Chem. Phys., 33*, 193.
Salooja, K.C., 1967, *Comb. and Flame, 11*, 511.
Semenov, N.N., 1958, *Certain Problems of Chemical Kinetics and Reaction Capacity*, Moscow, Published by AN SSSR.
Sharma, R.K., and Bardwell, J., 1965, *Comb. and Flame, 9*, 511.
Smirnov, L.N., 1969, *Candidate's dissertation*, Moscow., Inst. Iskusstvenogo Volokna.
Solymosi, F., Borcsok, S., and Lazar, E., 1968, *Comb. and Flame, 12*, 398.

6
Stability of Ignition Transients of Reactive Solid Mixtures

V.E. ZARKO

ABSTRACT: The problem of ignition stability arises in the case of the action of intense external heat stimuli when, resulting from the cutoff of solid substance heating, momentary ignition (gas flash) is followed by extinction. A physical pattern of solid propellant ignition is considered in detail and ignition criteria available in the literature are discussed. It is shown that this problem amounts to a problem on transient burning at the given arbitrary temperature distribution in the condensed phase. A brief review of published data on experimental and theoretical studies on ignition stability is offered. The comparison between theory and experiment is shown to prove qualitatively the efficiency of the phenomenological approach in the theory. However, the methods of mathematical simulation, as well as those of experimental studying of ignition phenomenon, especially at high heat fluxes, need to be improved.

Introduction

Analysis and mathematical description of ignition process involve terminological and methodical difficulties caused by dual content of the phenomenon: ignition is the initiation of fast exothermic reaction and, at the same time, the initial phase of combustion. The latter involves the concept of ignition stability, which implies the possibility of self-sustaining burning after cutting off an external energy stimulus. Note that until now the problem of ignition stability has not been sufficiently discussed in the literature on combustion; that seems to be due to experimental and theoretical difficulties caused by the complexity of the phenomenon.

Thus, for example, the simplest experiments reduce to measuring time delay at a certain intensity of an external heat flux and determining the boundary of the 50 percent extinction probability after cutoff of an external source. However, such experiments are not informative enough for an effective physical pattern of the phenomenon to be constructed. Actually, one needs reliable measurements of the heat flux really impinging on the initially inert and then reacting surface of a condensed substance. Besides, one needs information about the time evolutions of surface temperature and of sample weight or the intensity of gas release from the surface, especially after cutting off an external energy stimulus.

Analysis of such information allows the physical mechanism of the phenomenon to be formulated in more detail and this in turn requires an adequate mathematical description. It is also necessary to find appropriate methods for the problem solution. The oversimplifications of the early concepts (see, for example, Ya.B. Zeldovich's pioneer work (1942) now seem to be obvious. It was assumed that for stable ignition to occur it is necessary to warm up a solid propellant surface to the steady-state burning surface temperature provided that at the ignition instant the surface temperature gradient does not exceed the steady-state burning one. Actually, to solve the problem, one apparently should employ the methods of the unstable burning theory, although it must be noted that this theory is developed insufficiently and has no profound mathematical substantiation.

Herein, we have made an attempt to perform a critical review of available information on the experimental and theoretical studies of condensed substance ignition stability as well as to outline the ways for solving the problem. It should be mentioned that we will discuss the ignition of the substances capable of self-sustaining burning in inert gaseous media (explosives, propellants, etc.).

The Definition of Ignition Phenomenon

Terminological Aspects

Analyzing a number of ignition phenomenon definitions available in the literature, one may notice that their essence is the identification of the true ignition with the stable transition to burning. Thus, for example, in an early review Kulkarni et al. (1980) it was stated: "It is generally understood that ignition is incomplete if steady-state combustion does not follow the ignition event after removal of external energy stimulus." In a more recent review Hermance (1984) this statement has been formulated even more strictly: "Ignition process... involves a transient from a nonreactive state via thermochemical 'runaway,' followed by rapid transition to full-scale combustion. It is a transient process with a definite point of initiation but an endpoint that depends completely upon the definition. We must remember that something is not really ignited until it finally burns steadily."

The quoted definitions are rather practical and, in this sense, completely correct. However, it is easy to notice that they suffer from limitations since both experiment and qualitative analysis of the phenomenon show that, in many cases, the unstable ignition can be transformed into the stable one if the intensity of an external heat source is appropriately varied with time. Moreover, the question arises of how to interpret the ignition of the substances that are able to burn only under the action of an external heat supply. Pure ammonium perchlorate combustion at pressures below 20 atm occurring only under either external heat fluxes (about 5 to 10 cal/cm^2 s) or heating to above 200°C is a classical example of such a case. Another question is how to interpret the ignition by heat fluxes that are much more intensive than in a stationary combustion wave, when after sudden removal of the external stimulus the extinction process is obviously realized.

Thus, the necessity of a general definition of ignition phenomenon, which could be a base for an ignition stability definition, is obvious. At the same time, the concept of ignition moment should be defined more correctly and the question of ignition criteria should be discussed. To solve these problems, one must analyze the physical pattern of the phenomenon.

The Physical Pattern of Ignition

Early investigations of the physics of the phenomenon under consideration were based on experiments on ignition at low and moderate heat fluxes. According to experimental observations, in the initial phase of heating, solid propellants with controlling reactions in the condensed phase behave as inert materials. Then in a relatively short time period a self-accelerating exothermic reaction is developing and a local thermal explosion in the surface layer takes place. A similar picture seemingly takes place in the case when controlling reactions proceed on a heterogeneous surface. The picture of the process is more complex in the case when endothermic reactions only proceed in the condensed phase and exothermic reactions in the gas phase. However, thermochemical "runaway" in the condensed or gas phase after an initial induction period is a matter in common for all the cases; this moment is considered as ignition. In such descriptions of the process the characteristic features of ignition—sharply rising temperature, reaction rate, glow intensity, etc.—are realized practically simultaneously. Consequently, the ignition moment can equally well be determined using different ways of detection of physicochemical parameters.

However, it should be noted that such an idealized picture can differ essentially from reality. Moreover, the physical pattern of ignition can change with varying external conditions. As an example, let us consider the pattern of double-base solid propellant ignition under continuous irradiation. Figure 6.1 shows curves of the signals from a surface thermocouple, a photodiode, and a reactive force transducer on ignition in air by heat fluxes of 3 to 8 cal/cm^2 s of a propellant N doped with 1 percent carbon black (N + CB) and a propellant N doped with 1 percent PbO (N + PbO). The attenuation coefficient of Xenon lamp radiation is 450 cm^{-1} for the N + CB propellant and approximately 40 cm^{-1} for the N + PbO propellant. Visual observations and analysis of oscillograms show that upon radiative heating, an aerosol of

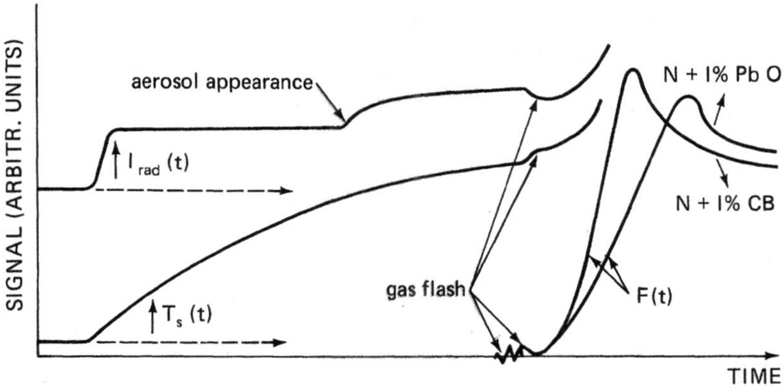

FIGURE 6.1. Temporal behavior of double-base propellant radiative ignition in air.

volatile components (nitroglicerin, dinitrotoluene) is formed above the propellant surface,[1] before a gas flash associated with the fast reaction of the aerosol takes place. On igniting the catalyzed propellant, the flash takes place at the initial stage of reactive force signal increase. In this case, a relatively slow increase in the reactive force signal is partially due to high transparency of the propellant and partially to local spot-type reactions in the surface layer.

It should be emphasized that these characteristic properties of the phenomenon are due to both propellant nature and external conditions. The nature of double-base propellants implies that exothermic conversions on the heated solid substance can take place in both the condensed and gas phases. The specific character of thermal radiation ignition exhibits itself in the fact that the products of solid propellant devolatilization and decomposition are mixed with surrounding cool gas, which leads to their condensation and slowing down of the gas-phase reaction rate.

The picture of the phenomenon is essentially modified if propellant is heated with hot gas. Heating with a quiescent hot gas, which can be realized, for example, with the help of metal foil placed above the propellant and being heated by electric current, is the limiting case. As shown by experiment Vilyunov and Zarko 1989), ignition time in this case is substantially reduced, approximately by a factor of 1.5 to 2, provided the experiment is carried out at atmospheric pressure. The cause of the ignition acceleration is obvious: solid propellant devolatilization and decomposition products are heated in the gas phase and an "accelerated" gas flash is realized. If in any way the gas-phase reactions are retarded, double-base propellant ignition occurs in approximately the same times as in the case of heating by a radiant flux. Effective retardation can be performed in various manners: by increasing inert gas pressure up to 30 to 40 atm or by blowing propellant surface with gas streams at a velocity above 5

[1] This fact is also confirmed by sampling liquid particles on a moving glass plate.

to 10 m/s. Besides, the "retardation" also takes place due to effective dilution by the diffusion of decomposition products, when igniting samples of very small cross sections. Thus, the ignition time for 0 = 1 mm propellant N samples in the quiescent hot gas turns out to be practically the same, at a certain pressure, as the time of radiative ignition for 0 = 10 mm samples.

The qualitative mechanisms under consideration are rather general for a wide set of condensed systems. In particular, they correspond completely to the case of ignition of the modified composite propellants that contain double-base propellants as binders. Strange as it may seem the ignition pattern for ordinary composite solid propellants with inert binders is the less detailed one. It should be noted that radiation energy absorption in the depth of the condensed phase plays an important part in the ignition of such propellants. Therefore, when analyzing experimental data, one should take into account the relationship between the characteristic depth of radiation attenuation and crystalline oxidizer grain size.

The physical pattern of ignition processes in the case of the action of high-intensity heat fluxes turns out to be much more complex compared to that described earlier since chemical reaction self-acceleration occurs simultaneously with the rapid heating and intense pyrolysis of the propellant. In this case, it is difficult to point out such evident ignition criteria as the fast self-acceleration of chemical reactions, of heat release, etc. Nevertheless, without any special substantiations, it is almost universally accepted that ignition takes place at the moment of gas flash appearance. It is, however, obvious that as different characteristics show different behaviors during the ignition processes, such criteria cannot be adopted as universal and unambiguous. In this connection it is appropriate to say a few words about the choice and use of adequate criteria of ignition.

It is well known that at present there are a great number of experimental and theoretical criteria of ignition (Kulkarni et al. 1980; Vilyunov and Zarko 1989). In particular, besides the detection of flame appearance by means of a photodiode or high-speed movie, experimenters can determine the instants of sharp changes in mass or reactive force, in thermocouple or ionizing probe signals, or find the boundary of the 50 percent probability of propellant extinction after removing the external thermal stimulus. On the other hand, the fact of reaching the critical values of a series of parameters, such as gas temperature, radiation intensity, the first or second derivative of surface or gas temperature, and rate of heat release due to internal chemical sources, have been taken by theorists as an ignition criterion.

It has been mentioned (Kumar and Hermance 1971; Vilyunov and Zarko 1989) that the use of arbitrary couples of the criteria in experiment and theory may lead to a discrepancy between experimental and theoretical data on ignition. At the same time, the question arises of how to interpret the cases when one of the criteria is not satisfied, when others are met. For example, on igniting double-base propellants at atmospheric pressures, flameless burning takes place, with the visible flame being completely absent. In ignition by powerful heat fluxes, on the contrary, a momentary flash in gas occurs and removal of an external heat source leads to the propellant extinction. What is the ignition phenomenon in these cases and how can we determine the instant of ignition? How can we determine the ignition stability, especially for ignition under a time-variable external heat stimulus? These questions show the necessity of developing general concepts and terminology that will allow us to compare objectively the results of different investigations and to plan further studies.

General Definitions

Let us define the ignition phenomenon in general terms as a transient process involving the local initiation of exothermic reactions in the condensed or gas phase of a gasified solid propellant under the action of a pulse or continuous external energy stimulus.

It is important to indicate the local character of the phenomenon in order to accentuate its difference from thermal explosion when a substance is heated homogeneously in the bulk. Treating ignition as a transient process gives a free hand to choose the criteria for the ignition instant determination. Here, however, it is important to specify for what purpose we need to determine the ignition instant. It should seemingly be recognized that searching for a universal ignition criterion is unreasonable since the process characteristics differ substantially for different condensed substances and conditions of heating. At the same time, the essence of an ignition criterion depends on the problems to be solved, among which one can distinguish the determination of global kinetic parameters of an exothermic reaction from data on ignition delays and simulation of ignited system transition to developed combustion.

To solve the former problem one needs an ignition criterion for which it is possible to determine correctly and reliably by experiment whether this criterion is satisfied or not. Using the criterion investigators obtain data on ignition delays and then, by means of solving the inverse chemical kinetics problem, find the global kinetic parameters of exothermic reactions. In doing so, it is highly important to ensure an adequacy of theoretical and experimental methods for ignition instant determination. Kinetic parameters found in such a way are employed in calculuations on fire and explosion prevention, on burning rates, and in some other fields.

Conventionally, on solving the first problem one uses the criteria based on the concept of explosive-type running of the process (reaching extremely high rates of temperature rise in the condensed or gas phases, flame appearance, pressure jump, etc.). It is evident that the formulation of such criteria was determined historically; in early investigations hardware for detecting the transient process characteristics were not developed sufficiently, the necessity of the exact correspondence between experimental and theoretical criteria of ignition was not always recognized.

Within the frames of the second problem, when we deal with calculation of the transient process from the initial nonreacting state of a solid propellant to combustion using a complete mathematical formulation of the problem, it is unnecessary to determine specifically the ignition instant and criterion. This corresponds to the detailed physical pattern of the phenomenon, according to which the process characteristics change continously with time. The necessity to indicate the instant of the "switch on" of a global exothermic reaction that ensures the substance's self-accelerated heating arises only in the case of an approximate formulation of the problem.

From the previously mentioned considerations it is seen that, in principle, the ignition criterion need not be formulated in the form of reaching some critical values of the process parameters. Provided the intense exothermic reaction is indeed initiated in the course of the process, the moments of reaching, for example, 50 or 90 percent levels of stationary rate of pyrolysis, reactive force, gas luminosity, can be taken as ignition instant. However, it is essentially important to derive similar characteristics in theoretical approaches. This will allow a direct comparison between theory and experiment. At the same time, it is clear that experiments on the global kinetic parameter determination should be carried out with low intensities of external heating since only in this case can one reliably distinguish the contribution of the substance's self-heating due to exothermic reactions.

Proceeding from the proposed definition, let us discuss the problem of ignition stability. In fact, the problem is about the stability[2] of transient burning at the given law of external heat flux time evolution. In its simplest form the question may be formulated as follows: does the transition to self-sustaining burning take place if the external heat supply is suddenly removed at once or at some later time, e.g., after gas flame appearance or other ignition indications? Such a definition is based on intuitive ideas about instantaneous ignition and ignores the evolution of the process with time. Hence, it does not take into account a number of details, such as the external stimulus cutoff rate, limiting time range of the "overexposure" of the external stimulus action, and time behavior of burning rate in transient process.

To answer these questions let us briefly consider the physics of transient burning during ignition. The analysis of solid propellant heating shows that transient phenomena in ignition are inevitable if at the ignition moment the temperature distribution in the solid phase of the propellant is different from that in a steady-state combustion wave. The same distribution can be obtained by reproducing the same laws of heating time evolution during the ignition process as in the stationary combustion wave. What are they?

First of all it should be noted that in the steady-state combustion wave the heat flux in the condensed phase depends on coordinate. This means that, as an approximation, at some point at a distance l from the burning surface the

---

[2] A special problem not considered here concerns the stability of self-sustaining combustion itself, which can be oscillatory both in the absence of external heat supply (auto-oscillations) and affected by the stationary heat supply.

heat flux varies with time:

$$q(t) = \begin{cases} q_s \cdot \exp\left(-\dfrac{\bar{r}l}{\alpha} + \dfrac{\bar{r}^2 t}{\alpha}\right) & 0 < t \leq \dfrac{l}{\bar{r}} \\ q_s \equiv \text{const} & t > \dfrac{l}{\bar{r}} \end{cases} \tag{1}$$

To rigorously reproduce heating conditions in the combustion wave is seen to be highly difficult. In particular, on igniting by a constant heat flux equal to that on the solid propellant burning surface the augmented warm-up in the initial stage is realized and, as a consequence, at the time instant of reaching the given surface temperature (e.g., corresponding to steady-state burning) in the condensed phase the heat storage turns out to be less than in the steady-state combustion wave. As is shown later, for heat accumulation the burning rate should be less than the stationary one under certain external conditions. That is what gives rise to transient burning with a decrease in burning rate followed by the transition to self-sustaining combustion (or extinction). Similarly, one can analyze the qualitative behavior of the burning rate for other $q(t)$ laws.

Consider now the problem on limiting "overexposures" of the heat flux in ignition. Note that such a problem may be transformed logically and rigorously only for a constant heat flux, with the time-dependence of heat flux cutoff being indicated. For time-variable heat flux the question about ignition stability should be formulated for each particular mode of flux change with time.

It is clear from general considerations that at "overexposures" long enough for steady-state external heating supported combustion to be established, upon cutting off the external stimulus the normal transient process from one stationary regime to another, namely, to self-sustaining burning, must take place. The result is dependent on the steepness of the back side of the heat pulse. It is evident that the ignition stability may be considered only for "overexposure" durations of the same order of magnitude as the consumption time of the heated material prior to ignition in the surface layer in the condensed phase and for characteristic times of external stimulus removal much less than the thermal relaxation time for the propellant condensed phase.

Experimental results

Experimental Technique

Various aspects of the experimental technique have been discussed in the literature. Mention should be made in this connection of a special issue of the *Progress in Astronautics and Aeronautics* (Zinn and Boggs 1978). However, it seems to be important to make some remarks concerning methods of heating and techniques of recording ignition characteristics.

In practice, ignition of solid propellants is known to be realized due to simultaneous convective, radiative, and conductive heating. Conventionally, for better understanding of the physics of the phenomenon, ignition is explored under the action of a certain heat transfer mode. Here it seems to be important to control carefully the correctness of the physical formulation of the problem and to estimate probable extension of known mechanisms to the case of the combined heat transfer.

The effect of propellant surface roughness on ignition delay should be mentioned. It was found in early works on ignition of double-base propellants heated by a reflected shock wave that the calculated surface temperature corresponding to ignition instant, as a rule, does not exceed 60 to 100°C, i.e., it is abnormally low. One of the first explanations of this fact was given by Kiselev et al. (1965).

They pointed out the possibility of overheating some local areas of the heated surface and verified this assumption by experiments with ideally smooth and artificially roughened propellant surfaces. A more detailed interpretation was given recently by Vorsteveld and Hermance (1987), who represented a rigorous mathematical consideration of the problem on heating a wedge by high heat flux. The effect of asperities has been found to manifest itself when their dimensions are less or comparable with the thickness of preheated ignition surface layer.

Flash visualization by means of high-speed movie or glow detection with a photodetector, as well as temperature monitoring by thermocouple or spectral techniques, are the conventional methods of ignition parameter detection. However, in each particular case one should analyze specific features and limitations of the registration methods used. In particular, it should be taken into account that the monitoring of glow is frequently integral (in the gas volume and radiation spectrum), while the thermocouple measurements are always local. Further, in measuring propellant surface temperature the reliability of the thermocouple measurements is highly dependent on the thermocouple—propellant surface heat contact perfection as well as on the difference in thermal-physical and optical characteristics of the thermocouple and propellant matter.

Analysis of experimental data on ignition must involve the measurement error for every recording method. At the same time one should take into account the errors due to the difference of the local and integral characteristics of the phenomenon. One of the causes of this difference—the inhomogeneity of propellant surface heating due to the surface roughness—was mentioned earlier. The inhomogeneity of reactions on the propellant surface due to local inhomogeneities of component distribution can be another reason of such a difference. As shown experimentally, the "hot spot" ignition effect can also be observed in experiments with homogeneous double-base propellants. The effect of the inhomogeneities of propellant surface heating and reaction on ignition stability has not hitherto been studied.

Ignition at a Moderate[3] Heat Supply

Mikheev was the first to perform a systematic experimental study of the stability of ignition by thermal radiation (Mikheev and Levashov 1973). He has carried out experiments using a graphite radiator with maximum heat fluxes of about 9 cal/cm^2 s. Double-base propellant ignition in nitrogen at atmospheric pressure was studied. It has been found that at a deradiation time of 0.03 sec the propellant N doped with 1 percent carbon black ignites steadily at any overexposure durations if the heat flux is less than 1 cal/cm^2 s. At heat fluxes exceeding 3 cal/cm^2 s extinction is observed even if a radiant flux is cut off immediately after ignition detected by flash in gas. Similarly, for the more transparent virgin propellant N the lowest limiting value of radiant flux, q_{min}, is 2 to 2.5 cal/cm^2 s and its upper limiting value, q_{max}, is over 9 cal/cm^2 s. It has been refined by later experiments that the upper limiting value is approximately 20 cal/cm^2 s. Experimental data on ignition stability are shown in Fig. 6.2. The critical values of radiant fluxes whose cutoff at stationary burning results in extinction have been determined. Within the spread in experimental data these values coincide with the lowest limiting values of radiant fluxes in ignition.

In further experiments performed with the author's participation, the more

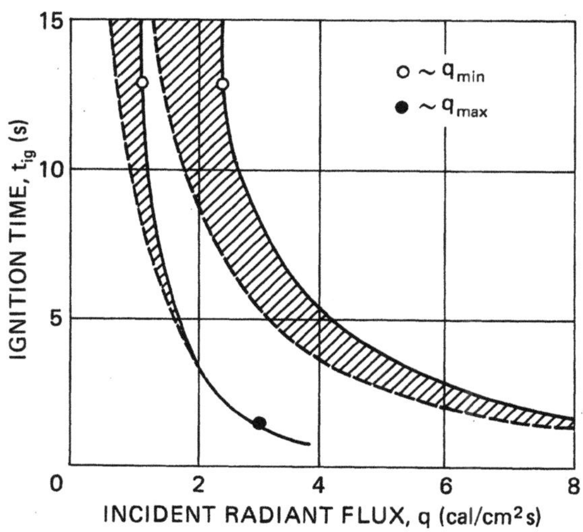

FIGURE 6.2. Stability map for the radiative ignition of double-base propellants. Dashed line-minimum ignition time, solid–upper limit for self-sustained ignition (Ref. 22).

[3] The term "moderate," in this case, means that the external heat flux does not exceed in value the heat flux on the surface of stationarily burning propellant in the condensed phase.

detailed physics of the phenomenon under study has been revealed. The continuous measurements of the recoil signal have shown that a drastic cutoff of the radiant flux from a Xenon lamp causes a two-step decrease in gas release intensity (or mass burning rate), first sharp and then more smooth. With radiant fluxes above 4 cal/cm^2 s and the deradiation time 0.03 sec, if the external stimulus is removed immediately after ignition (gas flash above propellant surface) the recoil signal is detected to decrease practically to zero, which corresponds to complete extinction on ignition in nitrogen. If the experiment is performed in the air, a temporary extinction is followed by a gradual repeat ignition with a time of transition to combustion on the order of 10 characteristic thermal relaxation times in the condensed phase under steady-state combustion. The high-speed movie and momentary photography of the transient process have shown that after sharp deradiation the double-base propellant surface gets covered with a dense net of gas bubbles and the initial phase of subsequent combustion is of a local character. This pattern is confirmed also by local thermocouple measurements of temperature in the gas and on the propellant surface. At the same time experimental data show the critical values of heat fluxes to be dependent on the optical characteristics of the solid propellant (or the spectral composition of radiation) and the form of flux variations with time.

Similar results have been obtained in Princeton, New Jersey, U.S.A. De Luca et al. 1976; Ohlemiller et al. 1973). Fig. 6.3 presents a generalized diagram of the experimental data on ignition illustrating different ways of determining the process characteristics. It should be emphasized that the position or even

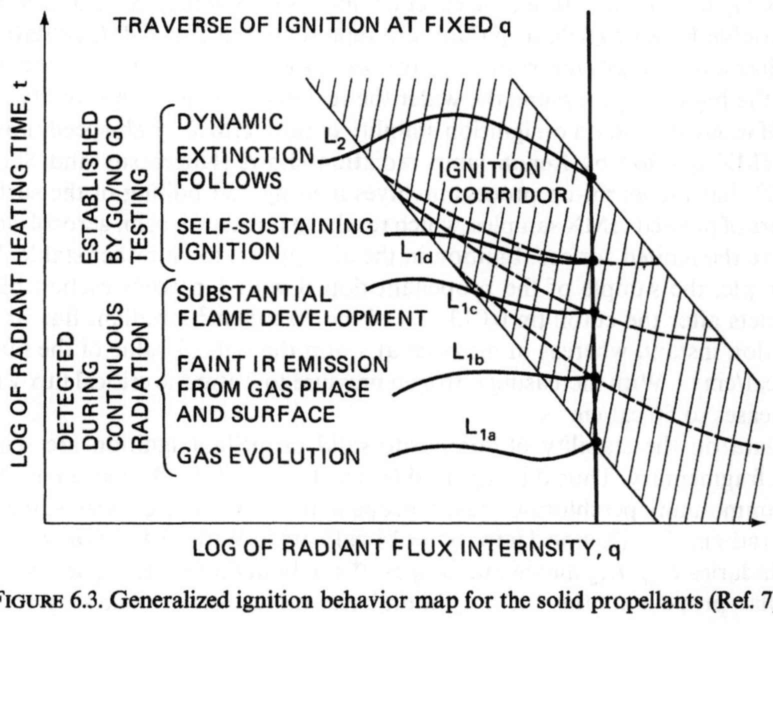

FIGURE 6.3. Generalized ignition behavior map for the solid propellants (Ref. 7).

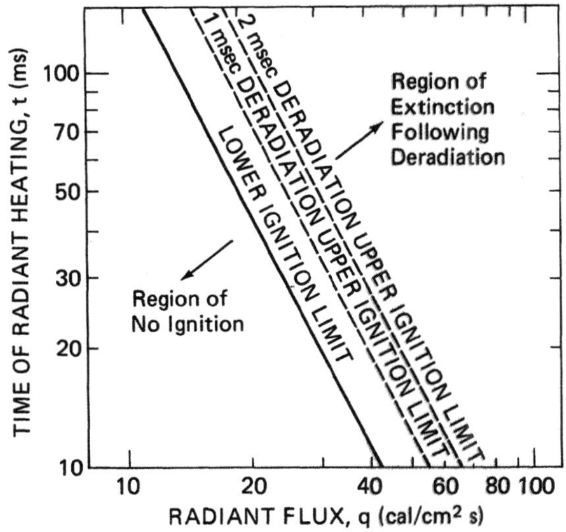

FIGURE 6.4. Effect of deradiation interval on measured upper and lower limits of double-base propellant, P = 1 atm (57.8% nitrocellulose, 40% nitroglycerin, 2.2% additives). (Ref. 26).

presence in the diagram of the curves L_{1a}–L_{1d} corresponding to chosen parameters of the phenomenon depend on both the sensitivity of detectors and the peculiarities of the solid propellant reaction mechanism under certain conditions. In particular, a luminescing flame is known to appear sometimes much later than gas release starts. In some cases it is not observed at all (e.g., on ignition of double-base propellants at low pressures). Quantitative data on the CO_2-laser deradiation time effect on admissible overexposure of heat flux in double-base propellant ignition are represented in Fig. 6.4. Qualitatively similar data on ignition stability have been obtained for increased pressure, i.e., the higher the pressure, the wider the region of the ignition stability.

Of interest are data on ignition stability of pure HMX. High-speed movies of HMX ignition by Xenon lamp radiation shows (Kuzentsov and Skorik 1977) that the warm-up process involves melting and boiling of the surface layers of pressed HMX samples, which results in formation of an aerosol cloud above the ignited surface. Ignition in the air appears to be rather stable. For example, the sample of the propellant doped with 1 percent carbon black extincts after the cutoff (in 0.013 sec) of the 70 cal/cm^2 s radiant flux at the ignition instant, whereas in nitrogen at 1 atm the critical value of the flux is 6.5 cal/cm^2 s. With increasing nitrogen pressure to 2 atm the critical flux value increases to 50 cal/cm^2 s.

Data on the stability of composite solid propellant ignition are scanty and fragmentary. Thus it is reported (Kuznetsov et al. 1974) that on ignition of ammonium perchlorate–based propellants there is a critical value for the radiant flux (Xenon lamp) $q^* = 14$ cal/cm^2 s. With fluxes below q^* the boundaries L_{1a}–L_{1d} merge and at $q > q^*$ the boundaries are separated. For similar propellants the L_{1a}–L_{1d} boundaries merge for the fluxes up to 100 cal/cm^2 s at the pressures above 5 atm (De Luca et al. 1976). Data on admissible ignition overexposures are lacking in these papers.

Ignition by Powerful Heat Fluxes

A considerable increase in heat supply rate may substantially modify the pattern of solid propellant ignition. Only for the substances with pronounced solid-phase reactions are the ignition patterns likely to be identical for every heat flux value. However, in practice propellants that contain melting or sublimating components are widely used, and the pattern of their ignition changes as the surface temperature becomes equal to or exceeds the value of phase transition temperature.

It has been reported (Strakovskii and Frolov 1980) that on high heat flux ignition of secondary explosives for which the calculated ignition temperature exceeds that of boiling, exothermic reactions in the gas develop with a significant delay. Gas flash is preceded by intensive gasification of the explosives, the fact of gas flash (bright luminosity) appearance depending on the presence of oxidizer in the environment. For instance, on ignition of tetril in nitrogen ($P = 1$ atm) the CO_2-laser radiation fluxes above 70 cal/cm^2 s fail to produce flame.

The specificity of the radiant heating determines the peculiarities of the semitransparent explosive ignition under extremely high radiation. Thus, the experiments on tetryl ignition by Nd-glass laser ($\lambda = 1.06$ μm) radiant fluxes of 10^3 to 10^4 cal/cm^2 s testify that the calculated surface temperature at the moment of glow appearance above the sample is below 30° to 60°C (Strakovskii 1985). For this case the glow is first of a local character. Based on this evidence the hypothesis has been formulated that with powerful radiant fluxes the ignition is due to the local warm-up of local absorbing inclusions (PbO particles, presumably) that are 10^{-3} cm and less in diameter. Data on ignition stability in such heat transfer conditions are lacking in the referenced papers.

Unfortunately, technical difficulties arising in the experiments involving a highly intense heat supply restrict the possibilities of making correct experiments on ignition stability. For example, a powerful radiant flux (Strakovskii 1985) was produced by a laser with rigorously restricted pulse duration. Only total pulse energy was controlled. No temporal or space distribution of beam energy was measured. The high-speed movie detected gas luminosity appearance near the surface, which was taken as ignition criterion. This information is insufficient and one should seek to perform a complete registration of the process parameters.

An attempt to realize such an approach has been made in the works on solid propellant ignition under fast pressure loading (Kumar et al. 1982; Yu et al. 1983). Propellant samples were placed in a closed end of a crack-like cavity and heated by the jets of hot gases from a main charge. The pressure in the cavity increased at the rate up to 10^6 atm/s to reach 300 to 500 atm. The ignition process was recorded with the help of a high-speed movie camera,

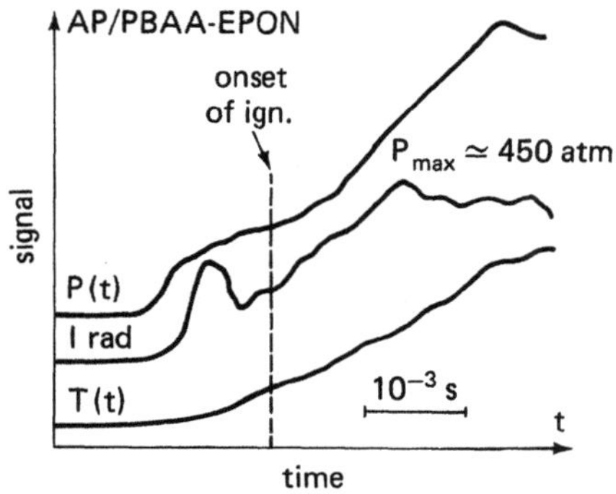

FIGURE 6.5. Time correlated pressure, light intensity, and thermocouple signals (Ref. 18).

thermocouple, and photodiode systems. Typical oscillograms of the ignition process of the ammonium perchlorate–based propellant are shown in Fig. 6.5. The instant of the onset of emission of bright light from the sample surface was considered the ignition instant. The high-speed movie data served as initial information, which was then refined from photodiode-detected luminosity data. Thus, reliable determination of the ignition instant seems to be highly difficult.

It is even more difficult to determine, in similar conditions, the ignition moment for nitramine–based solid propellants. In this case, the bright light appearance near the sample surface is hardly fixed by the motion picture film and photodiode signal oscillograms fail to show any characteristic point. Note that for the nitramine propellants this moment has been identified (Yu et al. 1989) not with ignition but with the onset of gas evolution. This difference in terminology is based on the fact that the ammonium perchlorate–based samples regressed entirely after ending a pressurization (time of its action was ≤ 50 ms) but the nitramine propellants, as a rule, extinguished (except for the cases of relatively low pressurization rates in the cavity—less than $2 \cdot 10^5$ atm/s). This terminology is sure to be conventional.

Note that difficulties in interpretation of such experiments are also related to changes in pressure in the course of heating: in all the cases the appearance of the luminous zone near the propellant surface was followed by an increase in pressure in the cavity and then by a decrease. Thus, it may be assumed that the extinction of the nitramine propellants is also due to transient combustion at drastic pressure decreases. This assumption agrees with the fact of correlation of the change in the curves of pressure in the cavity and the luminosity near the sample surface.

Ignition at Subatmospheric Pressures

Ignition at low pressures is rather specific due to the difficulties connected with the production of self-sustaining combustion in the gas phase. A series of interesting results have been obtained at Tokyo University (Harayama et

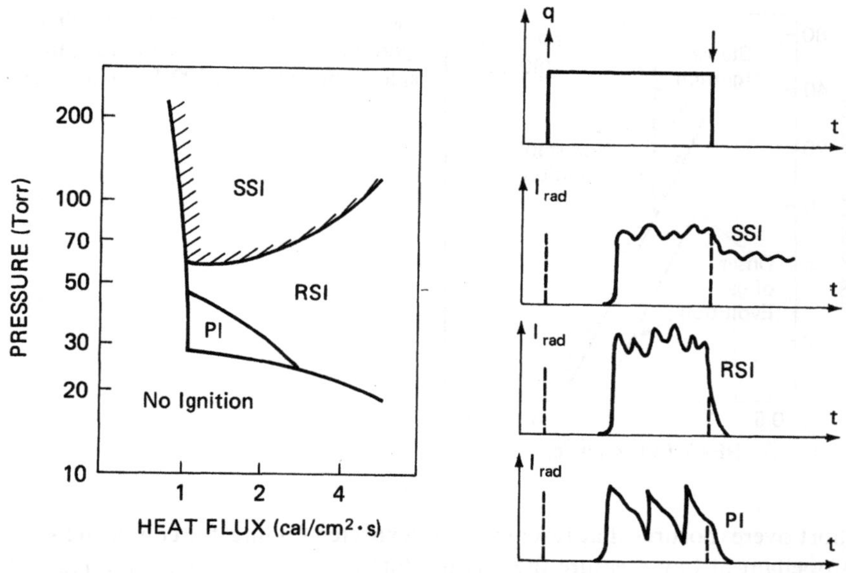

FIGURE 6.6. Modes of radiative ignition at different argon pressures for CTPB propellant (AP/binder 85/15). SSI—Self-Sustaining Ignition, RSI—Radiation-Sustaining Ignition, PI—Pulsating Ignition (Ref. 12).

al. 1983). Experiments were carried out under the pressure 1 to 500 Torr in argon. The samples of ammonium perchlorate (AP)–based propellant with carboxyl-terminated polybutadiene (CTPB) were heated with a CO_2 laser. The regimes of complete self-sustaining ignition (SSI) and nonstable ignition (RSI and PI) corresponding, respectively, to steady combustion and extinction after deradiation were revealed by varying the radiation intensity and medium pressure.

The map of ignition characteristics for basic propellant (without catalyst) is shown in Fig. 6.6. It is seen that in the region of nonstable ignition the pulsating ignition subregion (PI), which exhibits large brightness pulsations in the gas in spite of the constant irradiation of the sample surface, is distinguished. As shown in a later paper (Saito et al. 1985), this pattern may be changed substantially after adding catalysts to the propellant. In particular, addition of a few percent copper chromite significantly broadens the self-sustaining ignition region. In this case the subregion (PI) vanishes.

It is noteworthy that the definitions of ignition stability proposed (Harayama et al. 1983; Saito et al. 1985) are insufficiently correct regarding terminology. Fig. 6.6 testifies that the experiments were performed under very prolonged radiation overexposures. Consequently, the results characterize the stability of the transient burning under external heat supply. However, Harayama et al. (1983) have also reported results on the ignition stability under

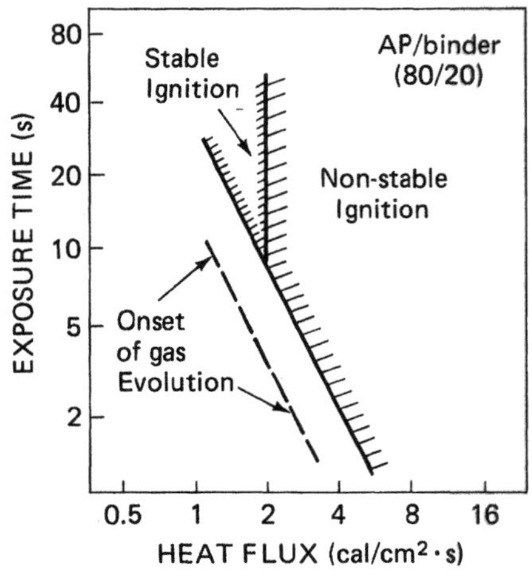

FIGURE 6.7. Stability map for the radiative ignition of CTPB propellant at subatmospheric pressure (Ar, 100 Torr) (Ref. 12).

short overexposures that revealed an unexpected finding: for composite solid propellant at low pressure there is no difference between q_{max} and q_{min} (for designations see Fig. 6.2) corresponding to the finitely small and finitely high overexposures of radiation. With the CO_2-laser deradiation time of 0.08 sec and the argon pressure of 100 Torr, the samples of basic composite solid propellant with a cross section of 3×3 mm^2 are steadily ignited at fluxes less that 1.9 cal/cm^2 s and any overexposures. With larger fluxes extinction is caused by the cutoff of the external stimulus at the moment of flame appearance (Fig. 6.7).

Theoretical Approaches

Formulation of the Problem

Experimental results show that transient stability is dependent on heat storage in solid propellant, external energy stimulus cutoff rate, pressure level, and physico-chemical parameters of the propellant including its transparency and kinetic characteristics. The question arises of how adequate the theory explaining the facts observed experimentally is.

One of the first attempts to theoretically describe this phenomenon was made in 1963 by Librovich (Zeldovich et al. 1975). To consider the problem on the solid propellant ignition by intense convective fluxes, he used a hypothesis (Zeldovich 1942) that stable transition to combustion is ensured if the surface temperature equals a given value of $\bar{T}_s$ ($\bar{T}_s$ is the steady-state burning surface temperature) and the temperature gradient on the surface is below a critical value. At the same time the assumptions have been made that the surface temperature of stationarily burning propellant under a given pressure is constant and there is endothermal gasification of the propellant. Solution of the problem of propellant heating followed by substance gasification yields the dependence of the ignition time on the magnitude of convective heat flux. A similar statement was used in the problem on radiative ignition stability (Khlevnoi 1971). These approaches are, however, insufficient for describing the transient dynamics. Neglecting the heat supply and temperature distribution in the condensed phase as well as the burning rate temperature sensitivity it is impossible to predict a priori either the transient duration or its stability.

Possibilities of the analytical description of nonsteady-state combustion are limited since this problem is substantially nonlinear. Hence, one should employ the methods of numerical simulation with subsequent generalization of results. The first attempt of such simulation was probably made by Vilyunov and Sidonskii (1965), who considered in succession the propellant heating by an external source and the transient after cutting off the external energy stimulus.

From the viewpoint of the analysis of the available results of numerical modeling it is important to know details of the formulation of a problem on transient combustion of solid propellant since calculation results are dependent on both assumption and choice of parameters. In all formulations the equation for energy with corresponding initial and boundary conditions is a necessary element. In addition, the complete formulation also includes the equations for reactant consumption, which makes the problem more complex.

Vilyunov and Sidonskii (1965) and Baklan et al. (1986, 1989) have made a global assumption on the adiabacity of the processes in the propellant condensed phase (heat flux from gas is zero) and took into account that the reaction in solid propellant is first order and functions up to a prescribed amount of substance depletion.

Baer and Ryan (1968) have formulated the problem without taking into account the substance depletion. However, they probably were the first to consider in detail time-dependent heat flux impinging the surface. When exothermic chemical reactions on the solid propellant surface are "switched on," their heat flux is summed up with the external one and the heat flux from the gas phase. The latter is assumed to be constant in time and equal to the flux in a steady-state combustion wave at a given pressure. The nonstationary burning rate is the function of total heat flux through the propellant surface.

In a number of studies during the last 15 years the transition to combustion problem is formulated within the framework of the Zeldovich-Novozhilov phenomenological approach. A characteristic feature and advantage of this approach is the use of empirical information about burning rate dependence on initial temperature that, in terms of the assumption of negligible relaxation time of the heat release zone in the gas and condensed phases (compared to the characteristic time of thermal relaxation in the condensed phase), provides an indirect account of the chemical kinetics in the combustion wave. Let us consider the problem formulation in which the change with time of the heat flux penetrating into the condensed phase is given in detail (Ohlemiller et al. 1973):

$$c\rho\left(\frac{\partial T}{\partial t} + r\varphi\right) = k\frac{\partial \varphi}{\partial x} \tag{2}$$

$$T(x,0) = T_{\text{in}}; \qquad \varphi \equiv \frac{\partial T}{\partial x} \to 0 \qquad \text{at } x \to -\infty \tag{3}$$

$$k\varphi|_{x=0} = \begin{cases} q(t) & 0 < t \leq t_0 \\ q(t) - r\rho L & t_0 < t \leq t_f \\ q(t) + k\varphi_s(T_s, P) & t_f < t \leq t_{\text{off}} \\ k\varphi_s(T_s, P) & t > t_{\text{off}}. \end{cases} \tag{4}$$

Here t_0 = gasification beginning time; t_f = flame formation time; t_{off} = radiant flux removing time; and L = heat of gasification.

To complete the problem the relation connecting nonstationary values $r(t)$ and $\varphi_s(t)$ should be introduced. To define this relation, let us consider the first integral of Eq. 2 at $\partial T/\partial t = 0$. We obtain:

$$c\rho\bar{r}(\bar{T}_s - T_{\text{in}}) = k\bar{\varphi}_s. \tag{5}$$

According to the phenomenological approach, Eq. 5 holds for the nonstationary case as well. However, instead of the true T_{in} value it must contain an apparent $T_{\text{in}}(r)$ value corresponding to the instantaneous burning rate value at a given instant of time. The $T_{\text{in}}(r)$ value is determined from the stationary expression $\bar{r} = \bar{r}(T_{\text{in}})$, e.g., if $\bar{r} = \text{const}_1 \cdot \exp \sigma_p \cdot T_{\text{in}}$, $T_{\text{in}}(r) = \ln(r/\text{const}_1)/\sigma_p$. Hence, to complete the problem (Eqs. 2–4) the nonstationary (Eq. 5) analogue is used:

$$c\rho r(t)[T_s(t) - T_{\text{in}}(r)] = k\varphi_s(t) \tag{6}$$

In addition, the relations of the pyrolysis law type are commonly used:

$$T_s(t) = E/2R \ln(\text{const}_2/r) \tag{7}$$

A set of equations (2–7) allows one to calculate the nonstationary burning rate with varying amplitude and duration of external thermal effect. It is impossible, however, to predict analytically the entire $r(t)$ pattern. Nevertheless, analysis of the problem reveals an important qualitative rule (Zarko and Kiskin 1980). Let us integrate (Eq. 2) over space taking into account Eq. 6:

$$c\rho\int_{-\infty}^{0}\frac{\partial T}{\partial t}dx = k\varphi_s - c\rho r(T_s - T_{\text{in}}) = c\rho r(t)[T_{\text{in}} - T_{\text{in}}(r)] = \frac{\partial H}{\partial t} \tag{8}$$

Eq. 8 suggests that $\dfrac{\partial H}{\partial t} > 0$ if $T_{\text{in}} > T_{\text{in}}(r)$ or with $r(t) < \bar{r}$ and vice versa. Thus, to accumulate the enthalpy in the combustion wave the propellant must burn at decreased, as compared to stationary, combustion rate. Alternatively, to decrease the enthalpy the burning rate must exceed its steady-state value.

A similar problem has been formulated (Assovskii and Zakirov 1987). It has been taken into account that at the initial stage of heating $r \equiv 0$ and in the condensed phase the volumetric reactions occur, which leads to the breakdown of thermal equilibrium. After ignition, which is determined as reaching a given intensity for chemical heat release, the nonstationary burning rate has been calculated in terms of the Zeldovich-Novozhilov approach with varying problem parameters.

This problem has recently been studied in more detail (Armstrong and Koszykowski 1988). The heterogeneous reaction on the surface of opaque solid propellant irradiated by a radiant flux pulse has been taken into account. The one-step chemical reaction in the gas phase has been considered. The solid propellant regression rate obeys the pyrolysis law.

The so-called flame model for solving the problem on the nonstationary solid propellant combustion is also available in the literature (Denison and Baum 1961). Its destinctive feature is the fact that Eq. 4 for heat feedback from the gas phase is found by solving a problem on steady-state combustion of solid propellant decomposition products. As shown by a number of works (e.g., Novozhilov 1973), this method yields results entirely coincident with those obtained by using the Zeldovich-Novozhilov approach.

Note that up to now theoretical approaches have treated mainly the models for homogeneous solid propellant combustion with a single global exothermic reaction. No theory is available for nonstationary combustion of real heterogeneous multicomponent mixtures. However, promising results have been achieved in developing ignition models for modern composite solid propellants. Thus, the problems have been considered (Glotov and Zarko 1981, 1984) on solid propellant ignition with independent and consecutive exo- and endothermic reactions on both the bulk and the heterogeneous surfaces. The same model for the ignition of perchlorate ammonium–based composite solid propellant has been proposed (Kumar et al. 1984). The authors have proposed a detailed account of chemical kinetics in the condensed and gas phases as well as on the burning surface. The structure heterogeneity has been taken into account in two-dimensional geometry. To model the ignition under the real conditions of a rocket motor the gas heating near the propellant surface due to fast pressure rise has been taken into account.

Calculation Results

The main aim of theoretical calculations is to obtain reliable correlations between experimentally determined parameters, e.g., ignition delay time dependence on heat flux. Note, however, that the final result may be achieved by using various assumptions that are sometimes opposite in meaning, which is quite a common situation in multiparameter problems. For example, the delay time reduction at the ignition of solid propellants at constant pressure in still hot oxygen-rich gas can readily be attributed to the acceleration of gas-phase reactions. However, the explanation in terms of the solid-phase

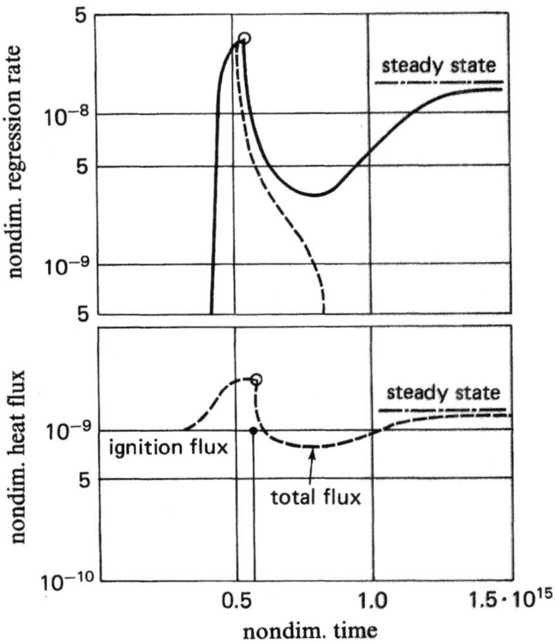

FIGURE 6.8. Temporal behavior of regression rate and heat fluxes during ignition by constant radiant flux. (Ref. 3).

reaction is shown (Vilyunov and Zarka, 1989) to be true if the heat transfer to the propellant due to effective increase in the coefficient of gas mixture thermal conductivity is taken into account.

Thus, it is rather important to obtain theoretical data on the time-dependent behavior of the process characteristics studied and compare them with experimental ones to establish a correlation between the phenomenon under study and its developed mathematical model.

One of the first theoretical results on the time-dependent behavior of the solid propellant burning rate in the course of ignition has been obtained by Baer and Ryan (1968). Fig. 6.8 illustrates the fact that the ignition by a constant flux equal to that on the steady-state burning propellant surface stimulates the transient process with a sharp initial maximum of the burning rate. To eliminate this burning rate peak the external heat flux must be decreased with time. The ignition delay time has been determined (Baer and Ryan 1968) according to the "go–no go" criterion, i.e., the ignition has been considered to occur if the cutoff of the external heat flux was followed by achievement of the stationary burning rate with the duration of the transient not exceeding the time of initial heating. Within the framework of the designations used in Fig. 6.3, only the boundary L_1 has been determined (Baer and Ryan 1968); the L_2 boundary corresponding to the extinction under overirradiation has not been revealed. The position of the L_1 boundary has been determined to

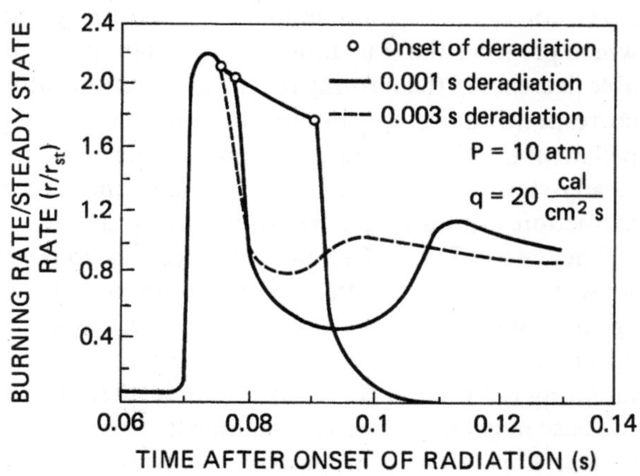

FIGURE 6.9. Calculated curves for the regression rate during radiative ignition of double-base propellant. (Ref. 26).

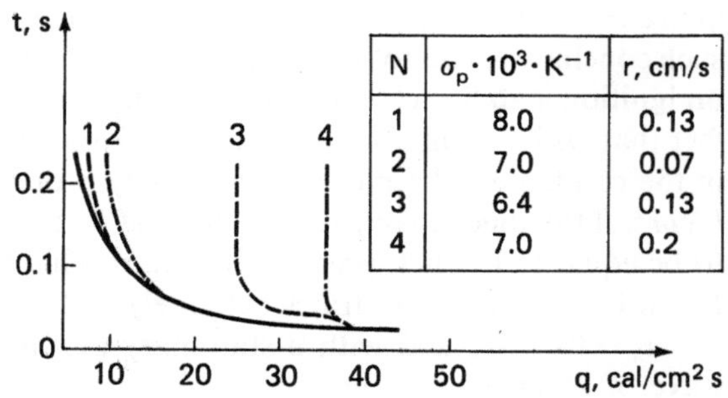

FIGURE 6.10. Effect of burning rate and $\sigma_p = \partial \ln r/\partial T$ in, values on the upper boundary of self-sustained ignition (solid line—lower ignition boundary for reference single-base propellant). $t_d = 0.02$ s; $t_h = 0.05$ s (curve I), 0.16 s (2), 0.05 s (3) and 0.02 s (4). (Ref. 2).

depend on heat flux cutoff rate. At low levels of heat fluxes the trapezoidal heat pulse needs a larger energy consumption for ignition than the rectangular one. However, at high flux levels the energy consumption on ignition by trapezoidal pulse are lower than on ignition by rectangular pulse.

A more thorough theoretical approach used by some authors (Armstrong and Koszykowski 1988; Assovskii and Zakirov 1987; Ohlemiller et al. 1973) attempts to model qualitatively many experimentally observed dependences including extinction at overexposure of the heat flux. Fig. 6.9 exemplifies the calculated behavior of transient burning rate during ignition of an opaque propellant by trapezoidal radiation pulse. It is shown that the decreased rate of the cutoff of the heat flux favors the stability of the transition to self-sustaining combustion whereas the "overexposure" leads to extinction.

Data on propellant transparency influence on ignition stability have been presented (Assovskii and Zakirov 1987). An experimentally consistent result has been obtained indicating that an increase in the transparency leads to an increase in ignition stability. The same paper reports some theoretical results on the influence of the burning rate temperature sensitivity and pressure level on ignition stability. These data are given in Fig. 6.10. The lower envelope curve corresponds to the minimum time of irradiation by trapezoidal pulse, which provides the transition to self-sustaining combustion without a noticeable decrease in the burning rate after radiation cutoff. The enumerated curves correspond to the limiting irradiation times. When they exceeded, the propellant extinction occurs. A burning rate decrease to less than 5 percent of the steady-state value under self-sustaining combustion serves as a criterion of extinction, although this criterion is arbitrary.

The data of Fig. 6.10 show that a decrease in the burning rate temperature sensitivity and an increase of the burning rate value lead to increase in ignition stability. Note, however, that variations of some of the problem parameters cause changes in others. Therefore one should be careful in formulating the general rules. For example, there is widespread opinion that an increase in pressure leads automatically to an increase in ignition stability. It is clear, however, that in the case of substantial change in combustion parameters with pressure, in particular if the burning rate temperature sensitivity increases with pressure, it may cause a decrease of ignition stability. Thus, such conclusions cannot be overgeneralized. In addition, when analyzing data on ignition stability with varying pressure, one should take into account the fact that with varying burning rate the characteristic time of heat relaxation in the condensed phase changes inversely proportionally to its square, $t_h = k/c\rho r^2$. If the time of the cutoff of the heat flux t_d remains the same, t_h is likely to be equal to t_d when pressure increases. This ratio is typical of curve 4 in Fig. 6.10, which demonstrates extremely high ignition stability. In this case, $t_d = t_h = 0.02$ sec and with a stationary flux in self-sustaining combustion wave of 25 cal/cm^2 s the ignition (and combustion) remains stable at any duration of the radiant flux overexposure up to 35 cal/cm^2 s. The same results have been obtained in the experiments (Ohlemiller et al. 1973) with double-base propellants at pressures above 34 atm when the time of radiation completion was 10^{-3} sec ($t_h \leq 1.2 \cdot 10^{-3}$ sec) and at pressures above 11 atm when $t_d = 10^{-2}$ sec ($t_h \leq 10^{-2}$ sec).

The specificity of radiative ignition of solid propellant with exothermic reactions proceeding on the propellant surface and in the gas phase has been demonstrated by the results of numerical calculation presented in Armstrong and Koszykowski (1988). It has been shown that at the initial state of heating the solid propellant surface temperature is higher than that of the surrounding gas. Then the self-accelerating gas-phase reaction stimulates a sharp rise of temperature in the gas and, together with heterogeneous heat release, provides the transition to self-sustaining combustion. Variation of kinetic parameters of both the reactions allows one to describe different types of dynamic behavior of burning rate.

These results of theoretical calculations are concerned with ignition at simplified conditions of heat transfer neglecting the interaction with condensed products of igniter combustion, the laws of pressure, and heat flux variations with time, i.e., the factors corresponding to the real process in rocket motors. It is, however, of importance that, so far, in the theoretical work on solid propellant ignition in rocket motors, no attempts have been made to take into account the peculiarities of nonstationary combustion during ignition (Caveny et al. 1979; Guttman 1984; Peretz et al. 1973; Ryzberg 1968). The main attraction is the solution of gas-dynamic problems; the stationary combustion of solid propellant was "switched on" just after the achievement of a given temperature on its surface. Further development of such approaches, particularly for small motors with a short period of main charge, combustion must be based on the detailed description of transients on achieving combustion regime.

Conclusions

Now let us make some final conclusions.

1. The question of ignition stability, important from the viewpoint of designing various generators of gas and plasma, as well as their reliability and safety in operation, is thus far not sufficiently covered in the scientific literature. From the viewpoint of modern knowledge this phenomenon must be interpreted on the basis of the nonstationary combustion theory. As for the optimal ignition conditions, they should be defined more correctly. In general it is impossible to achieve the optimum for several process characteristics simultaneously. Therefore, only the local extrema may be considered in detail, e.g., the minimum of the ignition delay time or the minimum of heat amount transferrred to the propellant from an external source prior to ignition.
2. The theories of nonstationary solid propellant combustion formulated to solve the problem of ignition stability are developed insufficiently. First, this is because, from a mathematical point of view, there was the assumption of quasistationary processes in the gas phase. Only recently papers have appeared developing more comprehensively the theory of transient burning of condensed substances with finite relaxation times in the condensed and gas phase reaction zones (Novozhilov 1988a, 1988b). Note, however, that on ignition the reaction zone in the gas is very often more extended than in stationary combustion. This may be easily confirmed by examining numerical results of the problem simulation (Armstrong and Koszykowski 1988). Hence, the widely used assumption of quasi-steady-state conditions for gas-phase processes with respect to the solid-phase ones is rather ambiguous. Another difficulty is in the adequate description of real heterogeneous systems. Efficient means should be elaborated to take into account in the problem formulation the effects of non-onedimensional propellant structure and different reactivity of the components.
3. The development of new theoretical approaches must be based on fundamental experimental results. To this end the modern techniques of recordng time-dependent process parameters, such as regression rate, weight losses, propellant recoil, and concentration of particles and components, must be

developed and adopted. These measurements should be performed under the conditions of controlled heat transfer to the surface of the ignited propellant and experimental results should be compared with corresponding theoretical approaches.

Nomenclature

c = specific heat, cal/Kg.
E = activation energy in pyrolysis law, cal/g-mol.
F = reactive force, g.
H = specific enthalpy, cal/cm^2.
k = thermal conductivity, cal/cm · K · s.
L = heat of gasification before flame formation, cal/g.
P = pressure, atm.
q = radient flux, cal/cm^2 · s.
R = universal gas constant, 1.98 cal/g-mol · K.
r = burning rate, cm/s.
T = temperature, K.
t = time, s.
t_d = characteristic time for heat flux cutoff, sec.
t_h = characteristic time for condensed phase thermal relaxation, s.
x = distance, cm.
α = thermal diffusivity, cm^2/s.
λ = wavelength of radiation, μm.
ρ = density, g/cm^3.
σ_p = burning rate temperature sensitivity at constant pressure, K^{-1}.
$\varphi = \frac{\partial T}{\partial x}$ temperature gradient, K/cm.

Subscripts

s = surface.
in = initial.
rad = radiation.

Superscripts

- = steady-state conditions.
* = critical value.

References

Armstrong, R.C., and Koszykowski, M.L., 1988, "A Theoretical and Numerical Study of Radiative Ignition and Deradiative Extinction in Solid Propellants," *Combustion and Flame*, *72*, 13–26.

Assovskii, I.G., and Zakirov, Z.G., 1987, "On the Ignition of Gasified Solid Propellant by Heat Pulse," *Khim. Fiz., 6*, 1583–1589 (in Russian).

Baer, A.D., and Ryan, N.W., (1968) "An Approximate but Complete Model for the Ignition Response of Solid Propellants," *AIAA J., 6*, 872–877.

Baklan, S.I., Vilyunov, V.N., and Dik, I.G., 1986, "On Transition to Combustion of a Condensed Matter under the Action of the Light Flux Pulse," *Combustion, Explosion, and Shock Waves, 22*, 6.

Baklan, S.I., Vilyunov, V.N., and Dik, I.G., 1989, "On Question of the Stability Criterion for the Condensed Substance Ignition," *Combustion, Explosion, and Shock Waves, 25*, 1.

Caveny, L.H., Kuo, K.K., and Shackelfol, B.W., 1979, "Thrust and Ignition Transients of the Space Shuttle Solid Rocket Motor," *AIAA Paper*, pp. 79–1137.

De Luca, L., Caveny, L.H., Chlemiller, T.J., and Summerfield, M., 1976, "Radiative Ignition of Double-Base Propellants: I. Some Formulation Effects," *AIAA J., 14*, 940–946.

Denison, M.R., and Baum, E., 1961, "A Simplified Model of Unstable Burning in Solid Propellants," *ARS J., 31*, 1112–1122.

Glotov, O.G., and Zarko, V.E., 1981, "Numerical Modeling of Ignition in a Condensed Substance with Independent Endo- and Exothermal Reactions," *Combustion, Explosion, and Shock Waves, 20*, 359–365.

Glotov, O.G., and Zarko, V.E., 1981, "Numerical Simulation of Sodium Nitrate and Magnesium Mixture Ignition by Constant Heat Flux," In *Problems of Technological Combustion*, Institute of Chemical Physics Division, Chernogolovka, 1, 83–86 (in Russian).

Guttman, D.N., 1984, "Ignition Transient Modeling for the Space Shuttle Solid Rocket Boosters," *AIAA Paper*, N 1359, 4 pp.

Harayama, M., Saito, T., and Iwama, A., 1983, "Ignition of Composite Solid Propellant at Subatmospheric Pressures," *Combustion and Flame, 52*, 81–89.

Hermance, C.E., 1984, "Solid-Propellant Ignition Theories and Experiments," In K.K. Kuo and M. Summerfield (Eds.), *Fundamentals of Solid Propellant Combustion, Progress in Astronautics and Aeronautics, 90*, 239–304.

Khlevnoi, S.S., 1971, "Explosives Extinction after Cutting off the Radiant Flux," *Combustion, Explosion, and Shock Waves, 7*, 150.

Kiselev, E.E., Margolin, A.D., and Pokhil, P.F., 1965, "On Powder Ignition under Shock Wave," *Problemy Goreniya i Vzryva, 1*, 83–84 (in Russian).

Kulkarni, A.K., Kumar, M., and Kuo, K.K., 1980, "Review of Solid Propellant Ignition Studies," *AIAA Paper*, N. 1210.

Kumar, R.K., and Hermance, C.E., 1971, "Ignition of Homogeneous Solid Propellant under Shock Tube Conditions: Further Theoretical Development," *AIAA J., 9*, 1615–1620.

Kumar, M., Wills, J.E., Kulkarni, A.K., and Kuo, K.K., 1982, "Ignition of Composite Propellants in a Stagnation Region under Rapid Pressure Loading," *19th Symposium (International) on Combustion*, The Combustion Institute, Pittsburg, Pa., 757–767.

Kumar, M., Wills, J.E., Kulkarni, A.K., and Kuo K.K., 1984, "A Comprehensive Model for AP Based Composite Propellant Ignition," *AIAA J., 22* 526–534.

Kuznetsov, V.T., Marusin, V.P., and Skorik, A.I., 1974, "On Ignition Mechanism for Heterogeneous Systems," *Combustion, Explosion, and Shock Waves, 10*, 456–458.

Kuznetsov, V.T., and Skorik, A.I., 1977, "HMX Ignition by Radiant Fluxes," *Combustion, Explosion, and Shock Waves, 13*, 228–230.

Mikheev, V.F., and Levashov, Yu. V., 1973, "Experimental Study of Critical Conditions for Powder Ignition and Combustion," *Combustion, Explosion, and Shock Waves, 9*, 438–441.

Novozhilov, 1973, *Nonstationary Combustion of Solid Propellants*, Nauka, Moscow, 176 pp. (in Russian).

Novozhilov, B.V., 1988a, "Effect of Gas Flame Characteristic Time on the Combustion Stability of Volatilized Condensed Systems," *Khim. Fiz.*, 7, 388–396.

Novozhilov, B.V., 1988b, "Theory of Nonstationary Combustion of Condensed Systems Including the Finite Relaxation Times," *Khim. Fiz.*, 7, 674–687.

Ohlemiller, T.J., Caveny, L.H., De Luca, L., and Summerfield, M., 1973, "Dynamic Effects on Ignitibility Limits of Solid Propellants Subjective to Radiative Heating," *14th Symposium (International) on Combustion*, The Combustion Institute, Pittsburg, Pa., 1297–1307.

Peretz, A., Caveny, L.H., Kuo, K.K., and Summerfield, M., 1973, "The Starting Transient of Solid-Propellant Rocket Motors with High Internal Gas Velocities," *AIAA J., 11*, 1719–1727.

Ryzberg, B.A., 1968, "Physical Background and Mathematical Model of Flame Spreading over Solid Propellant Surface During Ignition Process," *Combustion, Explosion, and Shock Waves, 4*.

Saito, T., Shimoda, M., Yamaya, T., and Iwama, A., 1985, "Effects of Some Additives on the Ignition Response of Composite Solid Propellants Subjected to CO_2 Laser Heating at Subatmospheric Pressures," *Proc. 16th Int. ICT Jahrestagung 1985*, Fraunhofer-Institute für Propellants and Explosives, Karlsruhe, 12/1–12/14.

Strakovskii, L.G., 1985, "Hot Spot Mechanism of Ignition under Monochromatic Radiation for Several Secondary Explosives," *Combustion, Explosion, and Shock Waves, 21*, 38–41.

Strakovskii, L.G., and Frolov, E.I., 1980, "Peculiarities of Transparent Secondary Explosives Ignition by Monochromatic Radiation," *Combustion, Explosion, and Shock Waves, 16*, 598–604.

Vilyunov, V.N., and Sidonskii, O.B., 1965, "On Ignition of Condensed Systems by Thermal Radiation," *Fizika Goreniya i Vzryva, 1*, 39–43 (in Russian).

Vilyunov, V.N., and Zarko, V.E., 1989, *Ignition of Solids*, Elsevier, Amsterdam, 441 pp.

Vorsteveld, L.G., and Hermance, C.E., 1987, "Effects of Geometry on Ignition of a Reactive Solid: Square Corner," *AIAA J., 25*, 622–624.

Yu, S., Hsien, W.H., and Kuo, K.K., 1983, "Ignition of Nitramine Propellants under Rapid Pressurization," *AIAA Paper*, N 1194, 11 pp.

Zarko, V.E., and Kiskin, A.B., 1980, "Numerical Simulation of the Nonstationary Solid Propellant Burning under Irradiation," *Combustion, Explosion, and Shock Waves, 16*, 650–653.

Zeldovich, Ya.B., 1942, "On the Combustion Theory of Powders and of the Explosives," *Zh. Eksper. i Teor. Fiz., 12*, 498–510 (in Russian).

Zeldovich, Ya.B., Leipunsky, O.I., and Librovich, V.B., 1975, *Theory of Powder Nonstationary Combustion*, Nauka, Moscow, 132 pp.

Zinn, B., and Boggs, T.L., (Eds.), 1978, Experimental Diagnostics in Combustion of Solids," *Progress in Astronautics and Aeronautics, 63*, 173–187.

7
Combustion Modeling and Stability of Double-Base Solid Rocket Propellants

L. De Luca and L. Galfetti

Abstract: Modeling and stability of solid rocket propellant combustion waves, featuring spacewise thick flames and characteristic gas phase time depending on pressure as well as temperature, are discussed. Within the usual framework of monodimensional deflagration and quasi-steady gas phase, this comprehensive approach is applied to double-base compositions by implementing an extended macrokinetics for the fizz-zone and a two-step distributed pyrolysis for the condensed phase. The proposed general model includes other models commonly accepted in the literature as particular cases and yields reasonable results over a large pressure range. Both the thickness of the flame and the temperature dependence of the characteristic gas phase time conspire against combustion stability but, being well understood, their influences can be fully settled. Exothermic thermal degradation of the condensed phase also strongly conspires against combustion stability, mainly in the super-rate burning region. It is found that distributed heat release in the high temperature degradation layer helps the static stability of super-rate burning, but worsens dynamic stability properties; in turn, this is improved by heat release distributed in the low-temperature degradation layer. Overall, properly partitioning the heat release removed from the surface concentrated layer to volumetrically distributed degradation layers sensibly augments intrinsic combustion stability. The results obtained may be sensitive to the data cited here. Nevertheless, the two-step consecutive pyrolysis mechanism deserves attention for all processes in which nitrate ester decomposition is relevant and in any case is a prerequisite for super-rate burning modeling.

Background

Modeling of solid rocket propellant transient flames by this research group has been evolving for years starting from the classical KTSS model (thin, uniform, and anchored flame structure) originally proposed by the Princeton group about 20 years ago (Krier et al. 1968). Within the framework of overall monodimensionality in space, quasi-steadiness of the gas phase in time, and thermal nature of the combustion model, transient burning of several classes of solid propellants is reasonably well numerically simulated and analytically predicted. The theoretical treatment is based on fundamental principles only; among its features is the fact that, over a wide range of operating conditions, the model incorporates and/or reproduces the most detailed experimental information available. In particular, the relevant steady-state burning properties are not predicted, but experimentally evaluated and embedded in the governing set of equations with no modification whatsoever.

Double-base (DB) solid propellants, whether catalyzed or not, clearly manifest a multizone flame structure over a large pressure range. However, for most current compositions burning at pressures below, say, 150 atm, the dark-zone effectively filters away the heat feedback to the burning surface from the luminous zone (if no erosive burning occurs). On the other hand, for most current compositions, overall monodimensionality in space requires a minimum operating pressure of, say, 2 atm. In addition, the particular but important class of catalyzed DB manifests the peculiar effect of super-rate burning, usually in a narrow range near the low end of the previously defined pressure interval, consisting of a spectacular increase of burning rate with ballistic exponent largely bigger than 1. This implies that, within wide pressure limits, monodimensional modeling of DB propellants is permissible and, moreover, DB flames require only the fizz-zone to be quasi-steady. However, peculiar problems have to be expected in the narrow pressure range over which super-rate occurs.

The purpose of this line of research is to model the transient combustion of DB propellants and test the related (intrinsic) stability properties. The results reported in this chapter are based on a succession of oral presentations offered at different meetings as well as some publications not easily circulated in the pertinent scientific community. A critical review of the merits and associated constraints of thin flame approaches was conducted in De Luca (1984) within a unified mathematical formalism; among other things, inherent limitations of the so-called linearized heat feedback laws were underlined (p. 695). The approximate but nonlinear analytical stability methods developed by this research group for thin flames was extended to spacewise thick flames in De Luca et al. (1986a), where a variety of experimental results was also presented for a catalyzed DB propellant. The limited validity of both spatially thin and thick flames to model transient burning, even including variable thermal properties in the condensed phase, was recognized in Bruno et al. (1986). The importance of the characteristic gas phase time was discussed on physical ground in De Luca et al. (1986b), where the general formulation of a widely applicable transient flame model (called $\alpha\beta\gamma$) was given as well, with particular reference to nonmetallized AP-based composite propellants. A first application to the specific configuration of a catalyzed DB flame was made in De Luca et al. (1987), yielding the conclusion that theoretical and experimental results are in good agreement in the pressure range above that of super-rate burning. An attempt to probe the super-rate pressure interval by the same approach, enforcing a pyrolysis submodel either totally concentrated at the burning surface or volumetrically distributed in the condensed phase, did not produce satisfactory results (Grimaldi et al. 1987). All of the previous studies were conducted for a propellant strand burning in an open vessel; computations of AP-based composite propellants burning in a confined geometry, by the $\alpha\beta\gamma$ flame model and including pressure coupling, were reported in Galfetti et al. (1988). A detailed critical review of the state of the art in transient flame modeling of solid rocket propellants was recently offered by De Luca (1990).

The specific objective of this chapter is to implement a pyrolysis submodel consisting of two-step consecutive processes volumetrically distributed in the condensed phase. The preliminary work reported in Grimaldi et al. (1987) on a one-step distributed pyrolysis was not conclusive. On the other hand, the need of a two-step distributed pyrolysis is strongly advocated in literature and quite motivated on a physical basis (Cohen and Holmes 1983; Fifer 1984; Zenin 1983, 1990); extensive numerical work dealing with one-step distributed pyrolysis again was not decisive (Birk 1983; Bizot et al. 1985). The results obtained herein are based on the accurate quantitative information provided in particular by the excellent experimental work of Zenin (1990).

It is important to underline that a burning propellant requires at least the following four basic steady-state dependences to be identified:

1. Experimental burning rate versus pressure, $\bar{r}_b = r_b(\bar{p})$;
2. Experimental surface temperature versus pressure, $\bar{T}_s = T_s(\bar{p})$;
3. Experimental surface heat release versus pressure, $\bar{Q}_s = Q_s(\bar{p})$;
4. Experimental or computed flame temperature, $\bar{T}_f = T_f(\bar{p})$.

Should any of these pieces of information be missing, then appropriate assumptions have to be made, which would, however, transform the problem under scrutiny into an exercise of limited use. It is stressed that experimental knowledge of steady behavior is a prerequisite to solve unsteady problems or predict intrinsic stability. As to the flame temperature, in general, results from standard thermochemical codes are adequate as long as the operating pressure is enough larger than the corresponding pressure deflagration limit (PDL). Additional pieces of information, needed to fully identify a burning sample, strictly depend on the propellant nature and require specialized treatments; in this instance they are discussed later.

The plan of presentation is the following. Main assumptions and equations are grouped together in the next section, emphasizing the peculiar features of DB burning in the gas phase and showing how to take them into account. An application of the proposed overall modeling effort is discussed in the following section, with reference to a catalyzed DB composition whose ballistic laws were extensively tested experimentally. The specific effects of a volumetrically distributed pyrolysis, consisting of two consecutive overall steps with different thermochemical features, on dynamic combustion and burning stability of the tested DB are discussed in the next section. Conclusions and suggestions for future work are reported in the final section.

Assumptions and Problem Statement

Consider a strand of propellant burning in a vessel at uniform pressure, possibly subjected to a radiant flux originating exclusively from a continuous-wave external source. Assume a monodimensional processes, no velocity coupling, no radiation scattering, no photochemistry, and irreversible gasification. Define a Cartesian x-axis with its origin anchored at the burning surface and positive in the gas phase direction; see Fig. 7.1. Nondimensional quantities are obtained by taking as reference values those (maybe only nominal) associated with the conductive thermal wave in the condensed phase at 68 atm, 300 K, and under adiabatic operations.

The nondimensional energy equation in the condensed phase ($X \leq 0, \tau \geq 0$) can be written as

$$C_c(\theta)\left[\frac{\partial\theta}{\partial\tau} + R\frac{\partial\theta}{\partial X}\right] = \frac{\partial}{\partial X}\left[K_c(\theta)\frac{\partial\theta}{\partial X}\right] + F_0 f(X, \lambda) + H_{c_{\ell}c} \tag{1}$$

$$\theta(X, \tau = 0) = \text{some assigned } \theta(X) \tag{2}$$

$$\theta(X - -\infty, \tau) = 0 \tag{3}$$

$$\left(K_c\frac{\partial\theta}{\partial X}\right)_{c,s} = \left(K_g\frac{\partial\theta}{\partial X}\right)_{g,s} + RH_s \tag{4}$$

where $C_c(\theta)$ and $K_c(\theta)$ are, respectively, assigned temperature-dependent specific heat and thermal conductivity in the condensed phase; K_g is an assigned pressure-dependent thermal conductivity in the gas phase; $f(X, \lambda)$ is an assigned function depending on the optical properties of the condensed phase and nature of the external radiation source emitting with known intensity F_0

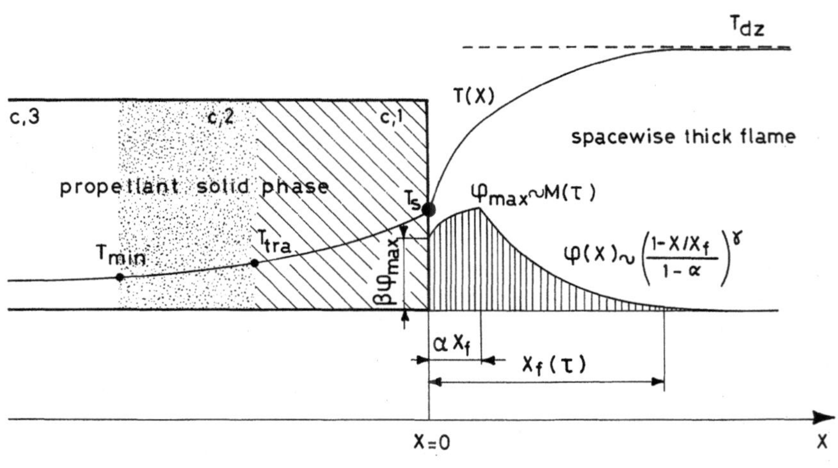

FIGURE 7.1. Sketch of the implemented combustion wave structure showing spacewise thick flame in the gas phase and two-step distributed pyrolysis in the condensed phase.

at wavelength λ. The boundary condition of Eq. 3 assumes that the propellant strand is in thermal equilibrium with the environment maintained at reference temperature. In general, except where explicitly indicated, all quantities are deduced from experimental data (collected in our laboratories or elsewhere).

As to the pyrolysis law, essentially the standard Arrhenius formalism is implemented for the concentrated surface gasification:

$$R(\Upsilon_s, P) = P^{n_s} \exp\left[-E_s\left(\frac{1}{\Upsilon_s} - 1\right)\right] \tag{5}$$

with minor modifications in the low (near ambient) temperature to avoid the classical cold boundary difficulty of the Arrhenius expression (see De Luca 1984, p. 692); these modifications have no practical effect in this work. The temperature dependence of the (net) heat release concentrated at the burning surface is computed as

$$H_s(T_s, P) = \frac{Q_s(\bar{p}) + \langle c_g\rangle \cdot (\bar{T}_s - T_s) - \int_{T_s}^{\bar{T}_s} c_c(T)\, dT}{Q_{\text{ref}}} \tag{6}$$

where the steady-state dependence $Q_s(\bar{p})$ is experimentally deduced and $\langle c_g(T)\rangle$ is an assigned pressure-dependent average specific heat of the gas phase mixture.

The group $H_c\varepsilon_c$, for a two-step consecutive pyrolysis reaction distributed in the condensed phase between the surface temperature and some minimum temperature below which no thermal degradation is assumed to occur, is computed as

$$H_c\varepsilon_c \equiv H_{c,1}\varepsilon_{c,1} = H_{c,1}A_{c,1}\exp(-E_{c,1}/\Upsilon) \qquad \Upsilon_s \le \Upsilon < \Upsilon_{\text{tra}} \tag{7}$$

$$H_c\varepsilon_c \equiv H_{c,2}\varepsilon_{c,2} = H_{c,2}A_{c,2}\exp(-E_{c,2}/\Upsilon) \qquad \Upsilon_{\text{tra}} < \Upsilon \le \Upsilon_{\min} \tag{8}$$

where γ_{tra} is the switching temperature between the two pyrolysis steps. At any rate, under steady-state conditions, the following balance must be satisfied by all energy sources:

$$\bar{H}_g(P) + \bar{H}_s(P) + \sum \bar{H}_c(P) = \frac{\int_{\text{Tref}}^{\bar{T}_s} c_c(T)\, dT + \langle c_g\rangle \cdot (\bar{T}_f - \bar{T}_s)}{Q_{\text{ref}}} \tag{9}$$

which bounds the sum $\bar{H}_x(P) = \bar{H}_s(P) + \sum \bar{H}_c(P)$ of the total (concentrated and distributed) heat release in the condensed phase. All relevant quantities in these expressions will be discussed later.

For *all* quasi-steady gas phase flames of a thermal nature, the nondimensional heat feedback from the gas phase to the burning surface is

$$q_{g,s}(P, R) = \int_{0^+}^{X_f} H_g\varepsilon_g \frac{\rho_g}{\rho_c} \exp\left(-\frac{C_g}{K_g} RX\right) dX, \tag{10}$$

where the usual assumption is made that

$$(\partial\theta/\partial X)_{g,s} \gg (\partial\theta/\partial X)_f \exp(-R^2\langle\tau_g'\rangle).$$

The quantity $\langle\tau_g'\rangle$ is a characteristic gas phase time parameter conveniently defined as

$$\langle\tau_g'\rangle \equiv \langle\tau_g\rangle \frac{C_g}{K_g} \frac{\rho_c}{\langle\rho_g\rangle}, \tag{11}$$

where $\langle\tau_g\rangle$ is the residence time in the gas phase. Resorting to the quasi-steady mass conservation across the burning surface, one finds

$$\langle\tau_g\rangle \equiv \frac{X_f}{\langle U\rangle} = \frac{X_f}{R} \frac{\langle\rho_g\rangle}{\rho_c} \tag{12}$$

and

$$\langle\tau_g'\rangle = \frac{C_g}{K_g} \frac{X_f}{R}. \tag{13}$$

The formal integration of Eq. 10 holds true for any integrable expression of the heat-release rate distribution $H_g \varepsilon_g \rho_g/\rho_c$. Detailed and systematic work carried out in several international (and independent) laboratories clearly shows that the heat-release rate is neither uniform (Krier et al. 1968) nor sharp (Culick 1968) in space, as commonly assumed. Rather, a comet-like structure as in Fig. 7.1 is suggested, featuring an elongated flame zone with heat-release rate very intense near the burning surface (the comet head) and slowly weakening far from the burning surface (the comet tail); details are fully discussed in De Luca (1990). The $\alpha\beta\gamma$ flame model simply recognizes this space structure, which does not necessarily exclude the previous simpler configurations accepted in the literature. By enforcing the $\alpha\beta\gamma$ flame model, the following *unified* expression, valid for all the currently available transient flame models of thermal nature, is found for the transient heat feedback (De Luca 1990):

$$q_{g,s}(P, R) = \frac{H_g}{X_f(\tau)} \frac{K_g}{C_g} \frac{\gamma + 1}{1 + \alpha \dfrac{\beta(\gamma + 1) + \gamma - 1}{2}} F(\alpha, \beta, \gamma; R^2\langle\tau_g'\rangle) \tag{14}$$

where

$$\begin{aligned} F(\alpha, \beta, \gamma, R^2\langle\tau_g'\rangle) = {} & \beta + \frac{1 - \beta}{\alpha R^2\langle\tau_g'\rangle} - \frac{(-1)^\gamma \gamma!}{(1 - \alpha)^\gamma (R^2\langle\tau_g'\rangle)\gamma} \exp(-R^2\langle\tau_g'\rangle) \\ & + \left[\sum_{i=1}^{\gamma} \frac{(-1)^i \gamma!}{(\gamma - i)!(1 - \alpha)^1 (R^2\langle\tau_g'\rangle)^i} - \frac{1 - \beta}{\alpha R^2\langle\tau_g'\rangle}\right] \\ & \times \exp(-\alpha R^2\langle\tau_g'\rangle). \end{aligned} \tag{15}$$

Notice that for $\alpha = 0$ (which necessarily implies $\beta = 1$) and $\gamma = 0$, the $\alpha\beta\gamma$ approach exactly recovers the results obtained by the KTSS type of transient model (thin, uniform, and anchored flame). For $\alpha = 0$ and $\gamma > 0$, the flame thickness associated with $\alpha\beta\gamma$ approach is $\gamma + 1$ times larger than the thickness associated with the KTSS type of flame. The $\alpha\beta\gamma$ approach yields a transient heat feedback law, which can be written in both "linear" (the group $R^2\langle\tau_g'\rangle$ assumed much larger than 1) and nonlinear versions, just as in the case of the KTSS type of flame. The linearized version is directly obtained from Eq. 14 as:

$$q_{g,s}(P,R) = \frac{H_g}{X_f(\tau)}\frac{K_g}{C_g}\frac{\beta(\gamma+1)}{1+\alpha\dfrac{\beta(\gamma+1)+\gamma-1}{2}}, \tag{16}$$

which immediately recovers the special case of KTSS, as usually enforced in the literature, for $\alpha \simeq 0$, $\beta \simeq 1$, and $\gamma = 0$.

As to the characteristic time (or time parameter) of the gas phase, proper allowance must in general be made for both pressure and temperature dependences, especially for transient operations. In general, this requires an appropriate submodel, which actually is the most difficult task; in broad terms it is assumed that:

$$\langle\tau_g'(P,R)\rangle = f(P)g(R; E_g\ldots) \tag{17}$$

where the function $f(P)$, depending on pressure only, is evaluated under steady operations but, in the spirit of gas phase quasi-steadiness, is assumed valid under transient conditions as well. The function $g(R, E_g\ldots)$ depends primarily, but not solely, on temperature (notice that, in the spirit of gas phase quasi-steadiness, reference is made to the variable R, the instantaneous burning rate); it has to be specifically modeled for each class of solid propellant based on the prevailing physical mechanisms. For DB propellants, the following approximate but convenient expression was deduced by manipulating the formal macrokinetics determined in the pioneering work by Zenin (1983, 1990):

$$\langle\tau_g'(P,R)\rangle = f(P)\left\{\exp\left[\frac{\tilde{E}_g}{\Re T_s}\right] + \exp\left[Z(P) + \frac{\tilde{E}_g}{\Re T_{dz}}\right]\right\}. \tag{18}$$

Zenin experimentally shows that the volumetric heat-release rate distribution in space cannot be described by the standard power laws adopted in the gas phase combustion chemical kinetics, but rather an exponential law is deduced by a multiparameter fitting procedure:

$$\varphi(X) \propto \exp[-Z(P)\eta(X)]\exp\left[\frac{-\tilde{E}_g}{\Re T(X)}\right] \tag{19}$$

where both the fizz-zone activation energy and the function $Z(P)$ are slightly pressure-dependent; $\eta(X)$ is the reaction progress $\eta(X) = [T(X) - T_s]/[T_{dz} - T_s]$. It is the factor $\exp[-Z(P)\eta(X)]$ that makes the thermokinetics of DB fizz-zone markedly different from that obtainable by any of the standard approaches in this technical area.

Under steady-state operations, one necessarily finds for the characteristic gas phase time parameter:

$$\langle \tau_g'(P,R) \rangle = \frac{(\gamma + 1)H_g}{\bar{R}^2(\int_0^{\bar{\theta}_s} C_c(\theta)\,d\theta - \bar{H}_s - \sum \bar{H}_c)} F(\alpha = 0, \beta = 1; \gamma, R^2\langle \tau_g' \rangle) \quad (20)$$

for the common configuration of $\alpha \simeq 0$ and $\beta \simeq 1$; in general, the solution of Eq. 20 requires an iterative loop that is usually quickly convergent. Again, for $\gamma = 0$, Eq. 20 recovers the corresponding KTSS characteristic time parameter (see De Luca, 1984, p. 694 where $C_c(\theta) \equiv 1$ and $\sum H_c \equiv 0$ were also assumed). Notice that under steady conditions only the flame thickness in space (through the parameter γ) plays a role; possible temperature effects will take place only during transient operations. This is fully in the spirit of the enforced quasi-steady gas phase approach based on embodied experimental steady-state properties. By matching the steady-state constraint of Eq. 20 with the definition of Eq. 18 (or Eq. 17, in general), the function $f(P)$ is immediately evaluated once the activation energy in the fizz-zone has been selected and $Z(P)$ deduced from the experimentally reconstructed curves of the heat-release rate distribution in the fizz-zone. Typical results are given in the next section.

Finally, the heart of the overall $\alpha\beta\gamma$ approach for modeling spacewise thick transient flames is the following. First, $X_f(\bar{P})$ is experimentally obtained under steady operations from microthermocouple measurements; from this, $\gamma(P)$ is evaluated by a best fitting procedure (see next section) and $\langle \tau_g'(\bar{P}) \rangle$ is computed through Eq. 13 (or, equivalently, $\langle \tau_g(\bar{P}) \rangle$ is computed through Eq. 12]. Then, under nonsteady operations, the instantaneous $\langle \tau_g'(P,R) \rangle$ [or, equivalently, $\langle \tau_g(P,R) \rangle$) is computed through a proper submodel by just introducing the instantaneous values of the relevant quantities. The transient flame thickness $X_f(\tau)$ at this stage is computed backward through Eq. 13 (or Eq. 12). It is obvious that the simple mass balance across the burning surface, in terms of either Eq. 13 or Eq. 12, is the pivotal point for the whole procedure.

All this implies that the space structure of the volumetric heat-release rate is first experimentally deduced, for the specific propellant under test, and then combined with a realistic dynamic behavior of the associated flame region. Therefore, once properly calibrated through Eq. 18 (see next section), the $\alpha\beta\gamma$ flame model should in principle approximate reasonably well the complex macroscopic patterns of real flame kinetics.

Application to a Catalyzed Double-Base Propellant

Several DB compositions, catalyzed and not, were provided by a national manufacturer and tested during an experimental project carried out over some years at CNPM laboratories (De Luca et al. 1986a, 1987; Grimaldi et al. 1987). For a matter of space, application of the theoretical guidelines presented in the previous section are restricted to a catalyzed DB propellant denoted as DB-5. In this section, in view of its experimental character and for easy comparisons with other data sets from the literature, the use of dimensional quantities is preferred. The steady-state burning rate, measured up to 250 atm by different nonintrusive techniques, features a super-rate burning in the pressure interval from about 4 to 9 atm and a plateau burning near 200 atm De Luca et al. 1986a). By embedding microthermocouples, surface and dark-zone temperatures were measured, under steady-state operations, in the pressure range from subatmospheric to about 40 atm (De Luca et al. 1986; 1987). The accepted best fitting laws, for dimensional temperatures measured in K, are the following:

One power expression can suffice for the surface temperature:

$$T_s(\bar{p}) = 581.2\bar{p}^{0.06} \qquad 0.14 \le \bar{p} \le 38\ \text{atm};$$

Three branches were found suitable for the dark-zone temperature:

$$T_{dz}(\bar{p}) = 1232.0\bar{p}^{0.15} \qquad 0.14 \le \bar{p} \le 2\ \text{atm},$$

$$T_{dz}(\bar{p}) = 1252.2\bar{p}^{0.13} \qquad 2 \le \bar{p} \le 5\ \text{atm},$$

$$T_{dz}(\bar{p}) = 1225.3\bar{p}^{0.01} \qquad 5 \le \bar{p} \le 38\ \text{atm}.$$

The surface pyrolysis law used in this work is obtained by the Arrhenius plot shown in Fig. 7.2 (taken from De Luca et al. (1987), but first proposed in Dondé (1987), resulting from the *combination* of the previous experimental information on steady burning rate and surface temperature. The Arrhenius plot of Fig. 7.2 clearly points to the existence of three different branches for the surface activation energy, the middle one corresponding to the super-rate region; in addition, several numerical checks (see De Luca et al. 1987; Grimaldi

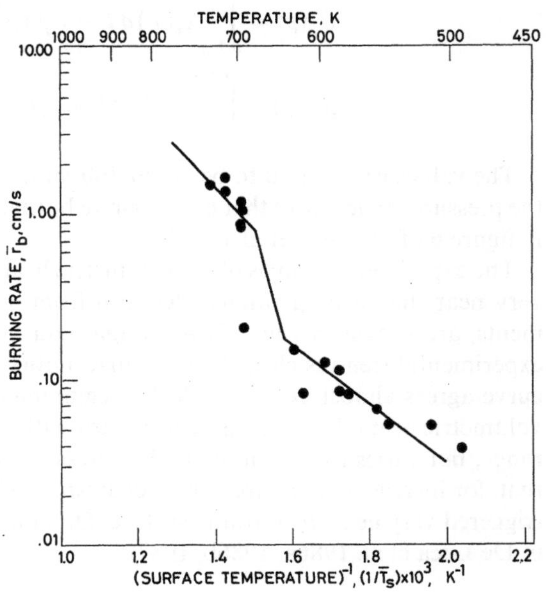

FIGURE 7.2. Experimental Arrhenius plot for burning surface gasification suggesting the existence of three branches taken from Dondé (1987).

et al. 1987, for example) and experimental evidence convinced the authors that no pressure dependence is involved at least during steady burning ($n_s \equiv 0$). The deduced values of the surface activation energy (in cal/mole) and the associated ballistic exponents are:

$$\tilde{E}_{s,1} = 11081.0 \qquad n_1 = 0.496 \qquad \text{for } 0.14 \le \bar{p} \le 4.04\,\text{atm}$$

$$\tilde{E}_{s,2} = 38274.2 \qquad n_2 = 1.889 \qquad \text{for } 4.04 \le \bar{p} \le 8.99\,\text{atm}$$

$$\tilde{E}_{s,3} = 8146.4 \qquad n_3 = 0.456 \qquad \text{for } 8.99 \le \bar{p} \le 38\,\text{atm}.$$

From the same steady-state temperature profiles, the thermal wave thickness in both the condensed phase and fizz zone could be measured. The collected experimental values, reported in De Luca et al. (1987) for both condensed phase and fizz zone, feature a slope discontinuity near 1 atm for the condensed phase thickness and near 2 atm for the fizz-zone thickness; this confirms that the whole theoretical treatment breaks down below 2 atm. By matching experimental versus computed steady-state temperature profiles, thermal diffusivity at the cold end of the burning propellant was deduced to be $\alpha_c(T_{\text{ref}}) = 0.0015\,\text{cm}^2/\text{s}$, with the density $\rho_c = 1.637\,\text{g/cm}^3$. As to the fizz zone, the average specific heat was taken as linearly increasing with pressure $\langle c_g(T) \rangle = 0.32 + 0.0006 \cdot p$, while thermal conductivity was assumed constant: $k_g = 0.0002\,\text{cal/cm} \cdot \text{s} \cdot \text{K}$. The effects of this initial data set for the fizz-zone properties will be discussed at the end of the next section.

The total heat release in the condensed phase $\bar{Q}_x = \bar{Q}_s + \sum \bar{Q}_c$ and fizz zone $\bar{Q}_{fz}$ were deduced from the experimental steady-state temperature gradients at the gas phase side of the burning surface, being in general by definition:

$$\bar{Q}_x(p) = \int_{T_{\text{ref}}}^{\bar{T}_s} c_c(T)\,dT - k_g(T_s)\frac{(dT/dx)_{g,s}}{\rho_c \bar{r}_b} \tag{21}$$

$$\bar{Q}_{fz}(p) = \int_{\bar{T}_s}^{\bar{T}_{fz}} c_g(T)\,dT + k_g(T_s)\frac{(dT/dx)_{g,s}}{\rho_c \bar{r}_b} \tag{22}$$

The values were found to be about 100 cal/g and 300 cal/g, respectively, in the pressure range above that of super-rate burning; detailed plots are reported in figure 6 of De Luca et al. (1987).

The experimental values of the volumetric heat-release rate in the gas phase very near the burning surface, deduced from microthermocouple measurements, are shown in Fig. 7.3. Although data handling is very delicate, the experimental trend is clear. In particular, notice that the break of the fitting curve agrees almost exactly with the beginning of the super-rate region; the volumetric heat release rate is nearly linear with pressure in the lower pressure range, but varies as $p^{1.9}$ in the higher pressure range. This seems to suggest that, for increasing pressure, the mechanisms yielding super-rate burning are triggered very near the burning surface. Other experimental results are given in De Luca et al. 1986a, 1986b, 1987).

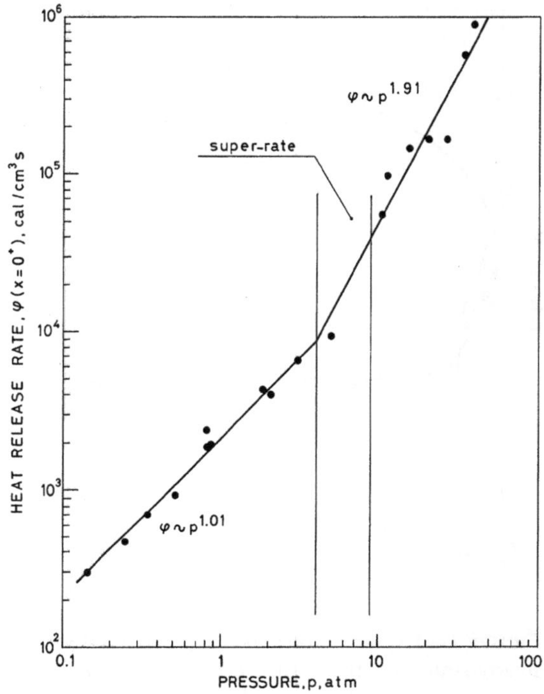

FIGURE 7.3. Experimental volumetric heat-release rate in the gas phase near the burning surface taken from De Luca et al. (1987).

At this point, it is possible to infer from the experimental results the functions $\alpha(p)$, $\beta(p)$, and $\gamma(p)$ required to implement the $\alpha\beta\gamma$ transient flame model. This is performed by closely matching, under steady-state operations, the experimental and theoretical quantities necessary to satisfy the surface boundary condition (Eq. 4) of the condensed phase energy equation; among other parameters, this equation includes the flame thickness. After some algebraic work, the following convenient steady version of Eq. 4 is suggested for the case under test:

$$\frac{\overline{x}_f}{\dfrac{k_g}{c_g \rho_c \overline{r}_b} + \dfrac{\overline{T}_{fz} - \overline{T}_s}{(d\overline{T}/dx)_{g,s}}} = \frac{\gamma + 1}{1 + \alpha \dfrac{\beta(\gamma + 1) + \gamma - 1}{2}} F(\alpha, \beta, \gamma; c_g/k_g \rho_c \overline{r}_b \overline{x}_f) \quad (23)$$

where, under any condition, $\rho_c r_b^2 \langle t_g' \rangle = c_g/k_g \rho_c r_b x_f$ by virtue of Eq. 13 (rewritten in dimensional terms). Since the detailed behavior of the combustion wave near the burning surface is too difficult to ascertain, the simplifying assumption of $\alpha(p) \cong 0$ and $\beta(p) \cong 1$ is invoked. Then the function $\gamma(p)$ is essentially evaluated from the experimental flame thickness values $x_f(\overline{p})$ under steady operating conditions. By enforcing the quasi-steady behavior for the fizz zone (see De Luca et al. (1986a), for example), the same dependences

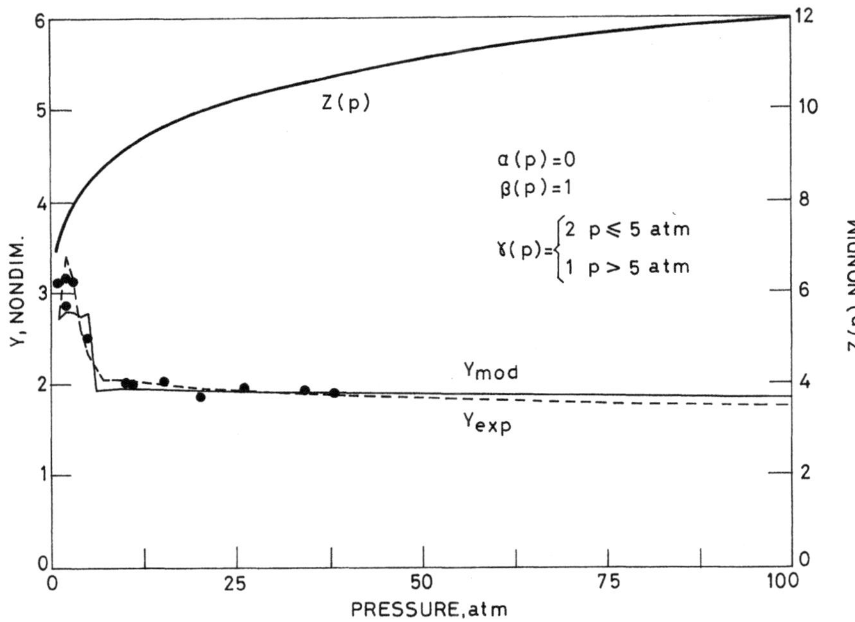

FIGURE 7.4. Experimental and computed nondimensional function associated with the steady-state energy balance across the burning surface (see Eq. 27) and deduced dependence of the fizz-zone elongation parameter $Z(p)$ on pressure.

of $\alpha(p)$, $\beta(p)$, and $\gamma(p)$ are taken to be valid under transient conditions as well. The results shown in Fig. 7.4 were obtained by putting $\gamma(p) = 2$ below 5 atm but $= 1$ above 5 atm for the "model part" of Eq. 23 (the right-hand side). The behavior of the "experimental part" of Eq. 23 (the left-hand side) in the low pressure range looks wild, due also to the lack of a well-developed fizz zone and the simultaneous occurrence of a super-rate region. By patient smoothing of the experimental data set, it was made sure that modeling and experimental quantities are congruent; Eqs. 21–23 and 9 should neither conflict each other nor contradict measurements, and this can only be obtained through repeated fitting attempts.

The final agreement reached between model and experiments is very reasonable in the pressure range above the super-rate interval up to 100 atm; for higher pressures no computation was effected. A better matching between the experimentally evaluated and theoretically evaluated parts of the steady boundary condition could perhaps be effected, but it seems a futile exercise in view of the forcibly approximate knowledge of so many ingredients.

One can now realistically describe the heat-release space distribution in the fizz zone of the tested DB composition; the full implementation of the $\alpha\beta\gamma$ transient flame model requires to properly define the gas phase time parameter as well (see Eq. 18). The function $Z(P)$ was obtained from the experimental values of the volumetric heat-release rate, reported in Fig. 7.3, by properly manipulating Eq. 18 for the limiting condition of $X \to 0^+$; details are discussed in De Luca et al. (1987). For the specific case of DB-5 propellant, the results shown in Fig. 4 were obtained for the indicated data set (to be discussed in the next section). The missing function $f(P)$ was computed by matching Eq. 18 with Eq. 20 under steady-state operations.

For the fizz zone of a common Soviet DB composition burning in the pressure range from 20 to 80 atm, with p measured in atm and T in K, Zenin, (1983, 1990) found:

$$\tilde{E}_g(p) = 11{,}300 + 135p$$

$$\tilde{E}_s = 9900$$

$$Z(p) = 7.15 - 0.000675p$$

$$c_g(T) = 0.42 - 55/T$$

$$k_g(T) = (0.167 \cdot \sqrt{T} - 2.67) \cdot 10^{-4}$$

Notice that the fizz-zone activation energy is an input parameter in this chapter and therefore the Zenin expression was accepted with no modification. Additionally, it is reassuring that the surface activation energy and function $Z(p)$, evaluated as described for the tested DB-5 propellant, are close to the corresponding quantities determined by Zenin in the pertinent pressure range. The average specific heat and thermal conductivity in this chapter are input parameters given versus pressure; the more detailed temperature-dependent information given by Zenin will be implemented in future work.

A Two-Step Volumetrically Distributed Pyrolysis

Thermal degradation of DB starts in the condensed phase at relatively low temperatures; this is recognized by several authors (Birk 1983; Bizot et al. 1985; Cohen and Holmes 1983; Fifer 1984; Zenin 1983, 1990). By far the most detailed information is supplied by Zenin (1990), who proposes two well-defined steps: a first degradation in the temperature range up to 523 K with activation energy at 48,000 cal/mole, followed by a further degradation in the temperature range from 523 K up to surface temperature with activation energy of 21,000 cal/mole. The first step corresponds to the well-known splitting of NO_2 from O—NO_2 bonds in nitrate esters (Fifer 1984; Zenin 1990); the second step, discussed in Fifer (1984) and clearly recognized in the experiments of Cohen and Holmes (1983), is seen in Zenin (1990) as due to the "highly exotermic reactions of oxidation of the semiproducts of the propellant decomposition triggered by the NO_2" liberated in the low-temperature degradation. The preexponential factors in this work are deduced by enforcing a normalization condition for each of the two processes:

$$A_{c,1} = \frac{\bar{R}(P)}{\displaystyle\int_{X_{tra}}^{0^-} \exp\left(-\frac{E_{c,1}}{\bar{\Upsilon}(X)}\right) dX} \tag{24}$$

$$A_{c,2} = \frac{\bar{R}(P)}{\displaystyle\int_{X_{min}}^{X_{tra}} \exp\left(-\frac{E_{c,2}}{\bar{\Upsilon}(X)}\right) dX} \tag{25}$$

An iterative procedure simultaneously processing these two integrals was implemented, starting by a fictitious temperature profile corresponding to a chemically inert condensed phase. A cutoff temperature was enforced for chemical activity in the condensed phase; this was arbitrarily assumed to be $T_{min} = 300$ K, but its value has no perceivable relevance. The number of iterative loops depends on the convergence requirement; several tens of iterations are needed to yield preexponential factors within a tolerance of 10^{-6}, which, in turn, is needed to obtain accurate burning stability predictions; the requested computer time is negligible. Typically, values of the order of $A_{c,1} \approx 10^{21}$ and $A_{c,2} \approx 10^7$ are found; these are reasonably close to the values suggested by Zenin (1990). The order of reaction is assumed to be zero for both pyrolysis steps. The energetics of each step is discussed now.

Predictions of intrinsic burning stability are conveniently effected by means of "bifurcation diagrams," obtained by the approximate but nonlinear stability analysis discussed in De Luca (1984) for spacewise thin flames and extended to spacewise thick flames in De Luca et al. (1986a, 1986b). Bifurcation diagrams are produced by plotting versus pressure the surface temperature of the characteristic roots of the perturbed energy equation in the condensed phase, for a given parameter of the problem (typically, heat release herein). They were called bifurcation diagrams because most often the plot reveals an abrupt transition of the steady-state burning rate from time-invariant to self-sustained oscillatory to zero steady-state reacting solution. For a matter of space, the rationale behind the approach cannot be repeated here, but explanations on how to read the bifurcation diagrams will be given where appropriate.

The initial data set, used to test the overall model and assess departures with respect to previous work, is that discussed in the previous section (based on De Luca et al. 1987; Grimaldi et al. 1987). Consider the bifurcation diagram of Fig. 7.5, obtained by enforcing a pyrolysis submodel fully concentrated at the burning surface with $Q_s = 100$ percent (baseline), 90 percent, 80 percent, and 70 percent of the experimental Q_x value. At each operating pressure, root A corresponds to the steady-state reacting solution, root B to the dynamic extinction limit, root C to the trivial nonreacting solution, and all other roots to further dynamic instability effects to be discussed. It can be seen that enforcing the full experimental value of Q_s yields static burning instability at and around the super-rate region; this would imply that steady-state time-

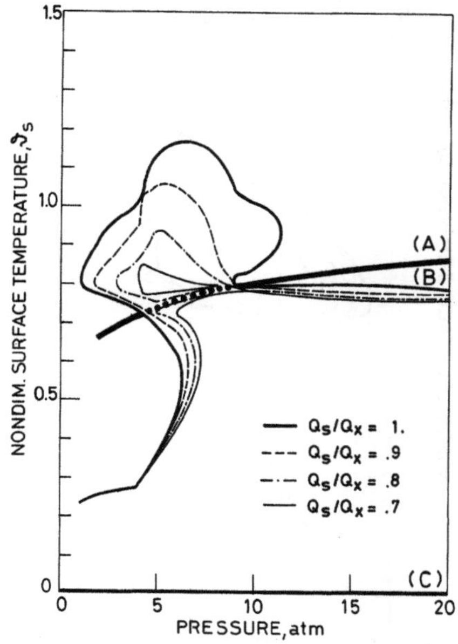

FIGURE 7.5. Bifurcation diagram showing strong dependence of burning stability feature on surface heat release.

invariant burning cannot be observed in the pressure interval between about 4 and 9 atm. This is not verified experimentally. Therefore, neither the model nor the enforced data set is adequate. Although the emphasis herein is on how to refine the pyrolysis submodel, it is appropriate to underline that changes of some of the input quantities, even within experimental uncertainties, are enough to drastically change the whole picture. In this regard, the surface heat release Q_s is a most delicate parameter; if arbitrarily reduced to 90 percent then 80 percent and finally 70 percent, the curves of Fig. 7.5 show first a decrease and then a total disappearance of static burning instability by a full detachment of the upper lobe (an island of upper dynamic instability is created whose practical effects are felt only during severe burning transients). The succession of bifurcation diagrams vividly illustrates the complex but interesting topological behavior of the nonlinear combustion model; similar trends were observed in many instances. It should, however, be clear that the whole data set has to keep its internal congruence and, therefore, other quantities have to be properly modified when an input parameter is changed (cf. the previous section). The objective of this preliminary test was to define a baseline bifurcation plot with a reference Q_s value (the experimental one) and stress that narrower tolerance bands have to be requested to the experimentalists, even though this is impervious to realize.

Now, a pyrolysis submodel consisting of two consecutive steps volumetrically distributed in the condensed phase will be considered. The bifurcation diagram of Fig. 7.6 refers to a reaction pattern by which the total condensed phase heat release Q_x is partitioned between surface-concentrated and high-

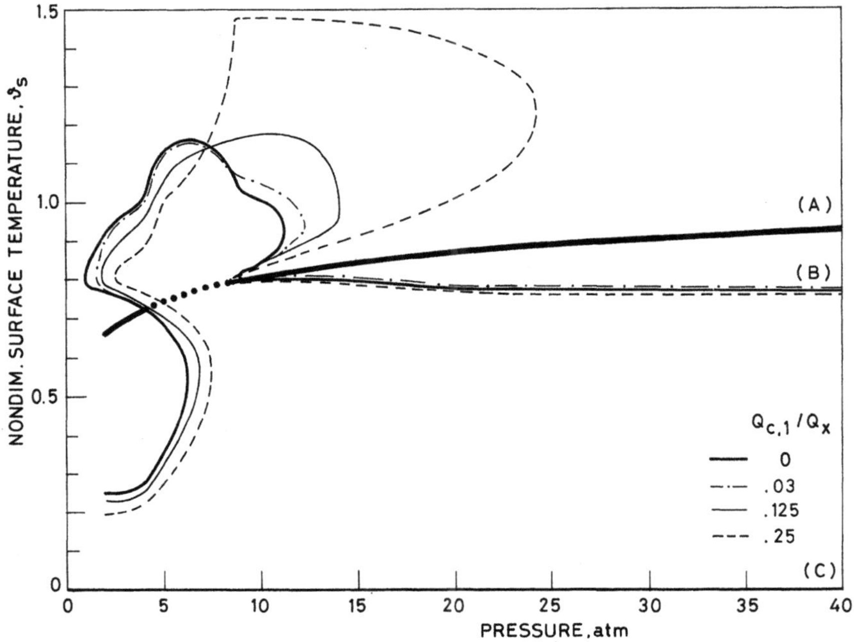

FIGURE 7.6. Bifurcation diagram showing that distributed pyrolysis in the high-temperature range favors static stability but also upper dynamic instability.

temperature distributed degradation; the low-temperature distributed degradation being energetically neutral. It is seen that removing some of the heat release from the burning surface to the hot layer near the burning surface improves static burning stability (in that the dashed portion of root A decreases), but also sensibly favors upper dynamic instability (in that the lobe above root A magnifies, especially in the large pressure range). See the succession of bifurcation diagrams for $Q_{c,1} = 0, 0.03, 0.125$, and $0.25\ Q_x$; notice that this effect could already be observed by implementing one-step distributed pyrolysis submodels (figure 7.6 of De Luca et al. (1987)). On the other hand, the same quantity of heat release removed from the burning surface to either the high-temperature or low-temperature distributed degradation shows a trend toward worse or better dynamic stability properties, respectively, if delivered in the high- or low-temperature range (see Fig. 7.7). This result might be not convincing since the tested fraction of heat release is small; yet, another conclusive example of this effect is shown in the next figure.

These first (purely analytical) results suggest that combining volumetrically distributed exothermicity, between high- and low-temperature degradation layers of the condensed phase, helps burning stability. A synergistic effect can be seen, by which the lack of static burning stability is alleviated by distributed heat release $Q_{c,1}$ in the high-temperature degration layer, while the lack of

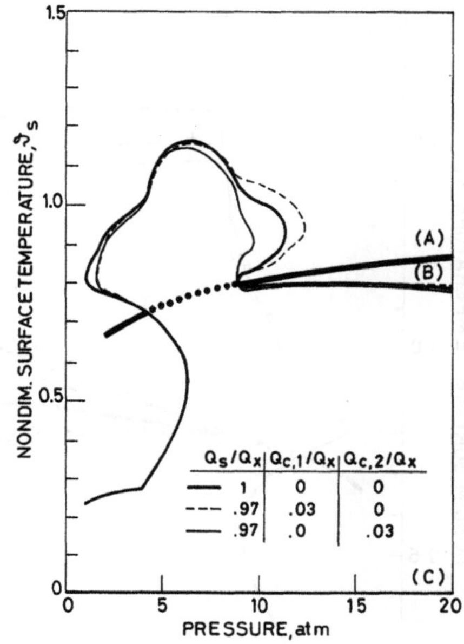

FIGURE 7.7. Bifurcation diagram showing that high- or low-temperature distributed pyrolysis, respectively, worsens or improves upper dynamic instability.

upper dynamic burning stability (i.e., above steady-state burning) is alleviated by distributed heat release $Q_{c,2}$ in the low-temperature degradation layer. This is confirmed by the results shown in Fig. 7.8, where the fraction of heat release removed from the burning surface is partly deposited in the high-temperature distributed degradation layer and partly in the low-temperature distributed degradation layer. It is observed that static burning stability is not appreciably affected, while upper dynamic instability decreases for the low-temperature degradation process going from endothermic to neutral to exothermic.

It is further observed that the bifurcation diagram for $Q_{c,1} = 0.25\ Q_x$ and $Q_{c,2} = 0.02\ Q_x$ predicts less burning instability than the bifurcation diagrams obtained with $Q_{c,1} = 0.27\ Q_x$ or $0.25\ Q_x$ being $Q_{c,2}$ neutral; this confirms the results shown in Fig. 7.7.

The reader is urged to bear in mind the hidden role being played in this discussion by the substantial difference existing between the activation energies associated with the two volumetric degradation processes. Large activation energies depress both static and dynamic burning stability, but to an extent strongly dependent on the accompanying heat release (De Luca et al. 1984). The synergism between the two pyrolysis steps allows the combustion model to digest as activation energy for the triggering degradation process the big value of 48,000 cal/mole otherwise impossible to handle, while the moderate value of 21,000 cal/mole for the sustaining degradation process allows a sensible fraction of exothermicity to be removed from surface-concentrated to volumetrically distributed.

After making sure that the overall combustion model is meaningful and

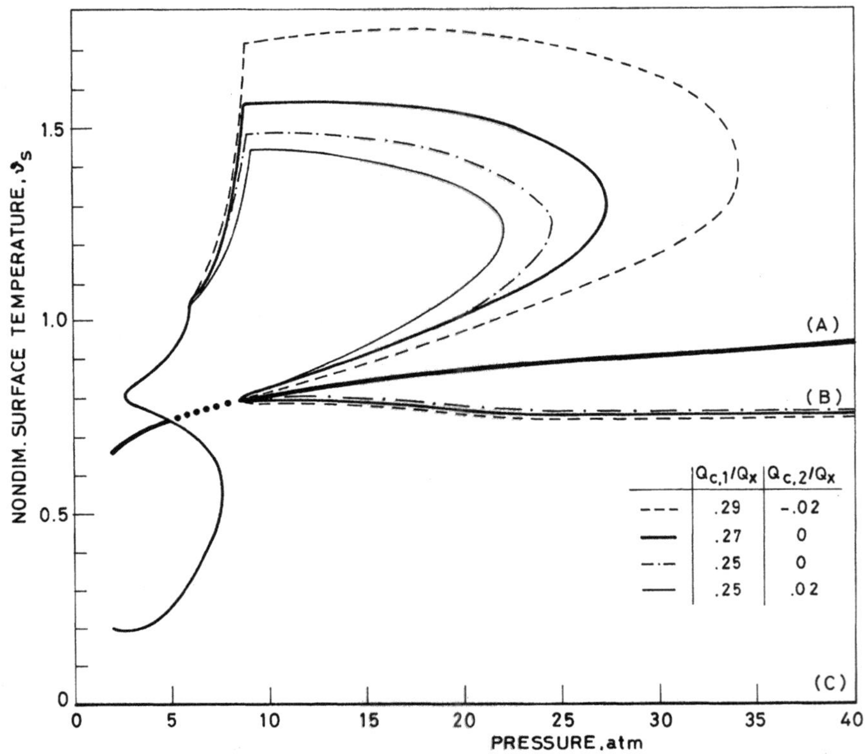

FIGURE 7.8. Bifurcation diagram showing a favorable synergism from the combined effects of high- and low-temperature distributed pyrolysis.

assessing relative trends and bounds, attempts were made to refine the enforced data set by resorting to the most detailed information available. The bifurcation diagram of Fig. 7.9 was obtained with $Q_{c,1} = 0.25\ Q_x$ and $Q_{c,2} = 0.02\ Q_x$. The two plots refer to different fitting laws of the average fizz-zone specific heat, both increasing with pressure: the lower values were taken from De Luca et al. (1987) and Grimabli et al. (1987), the higher values $\langle c_g(T) \rangle = 0.36 + 0.0002 \cdot p$ from adapting the expression suggested in Zenin (1990) implying a large value of the specific heat in the low-pressure range; see details in the previous section. Of course, the comments made earlier on this kind of exercise are still valid; the data set had to be recomputed, but only minor changes were found. The final result is that both static combustion instability associated with the super-rate burning and upper dynamic combustion instability vanish due to the larger fraction of heat released in the fizz zone rather than condensed phase (either surface or volume).

Several tests of this kind were effected, but are not reported here due to space. In particular, by assigning to thermal conductivity a slight pressure

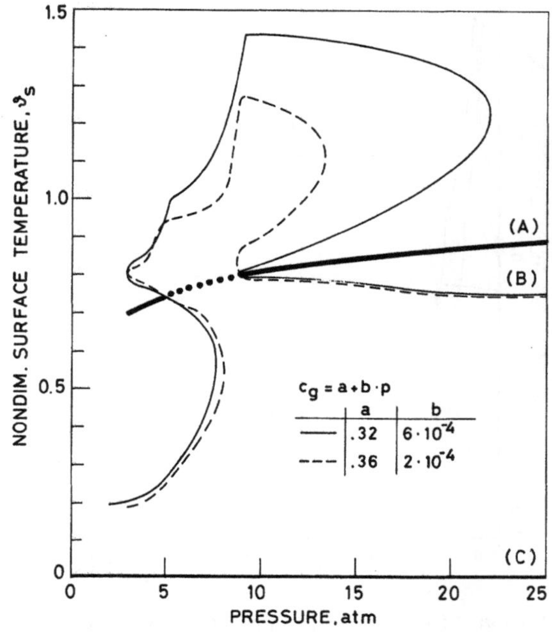

FIGURE 7.9. Bifurcation diagram showing that increasing fizz-zone specific heat decreases upper dynamic instability.

increase as suggested by Zenin (1990), the upper dynamic combustion instability decreases; this effect is sensibly emphasized by a minor increase of the thermal conductivity level from 2 to $2.25 \cdot 10^{-4}$ cal/cm·s·K. This could suggest that the super-rate burning is associated with a larger heat feedback from the fizz-zone region adjacent to burning surface. Anyway, the trend of combustion instability to vanish is again due to the larger fraction of heat released in the fizz zone rather than condensed phase (either surface or volume) when the fizz-zone thermal conductivity increases.

Numerical integration of the full set of governing equations (see the second section) proved the validity of the approximate but nonlinear analytical predictions. The results shown in Fig. 7.10 are numerically computed transient burning histories corresponding to time-invariant steady-state solution (3 and 25 atm) and self-sustained oscillatory steady-state solution (5 and 6 atm). The four burning histories were computed by enforcing slight pressurization transients, from some initial to the wanted final operating condition, with the initial data set of the previous section being $Q_{c,1} = Q_{c,2} = 0$ (no distributed pyrolysis, cf. baseline of Figs. 7.5–7). The corresponding transient burning histories with the same data set but $Q_{c,1} = 0.25$ and $Q_{c,2} = 0.02$ (two-step distributed pyrolysis, cf. Fig. 7.8) are reported in Fig. 7.11, pointing the different oscillatory behavior between concentrated and distributed pyrolysis processes. Of course, time-invariant steady-state solutions manifest differences between the two pyrolysis submodels only in the details of the combustion wave structure. All of these numerical runs are tests of static burning stability, whose outcomes

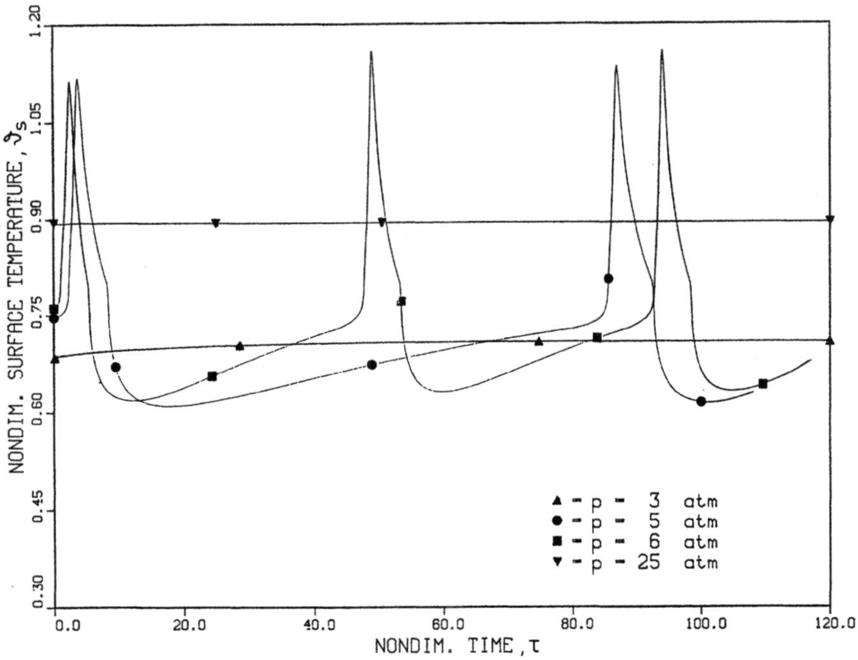

FIGURE 7.10. Numerical computations of transient surface temperature confirming the analytical expectations of the bifurcation diagrams of Figs. 5–7 for no distributed pyrolysis.

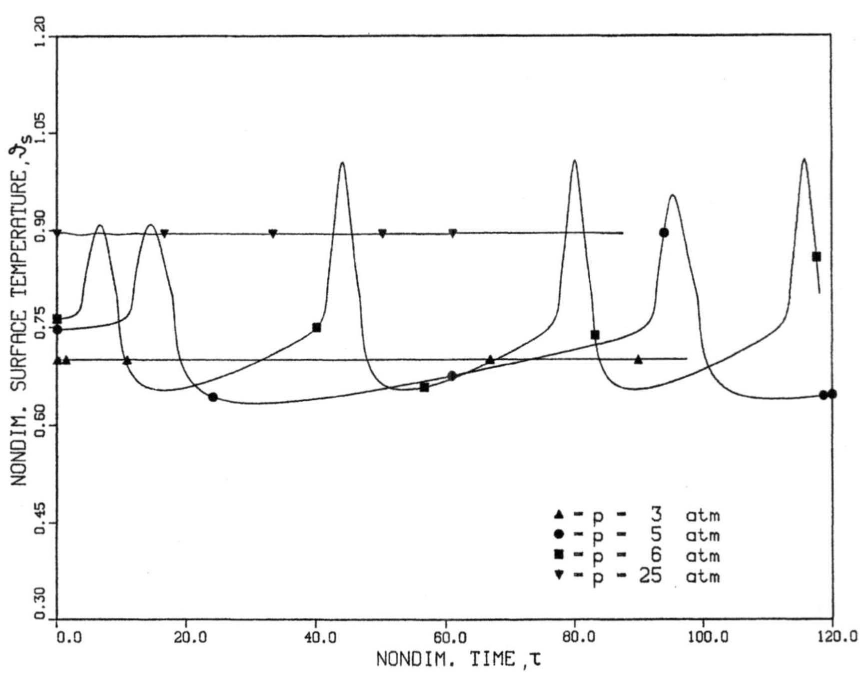

FIGURE 7.11. Numerical computations of transient surface temperature confirming the analytical expectations of the bifurcation diagrams of Fig. 8 for two-step distributed pyrolysis.

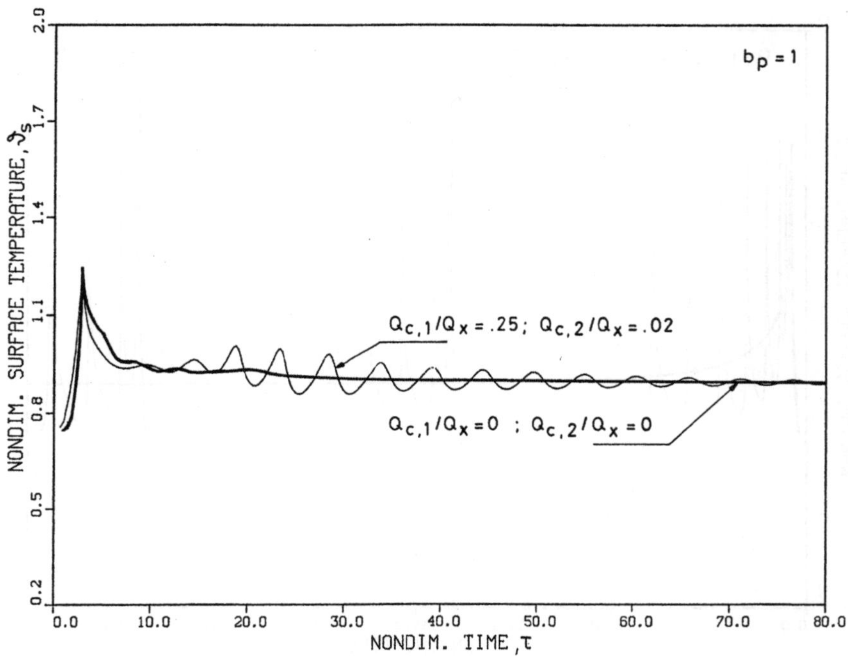

FIGURE 7.12. Numerical computations of transient surface temperature during a pressurization test, from 5 to 25 atm with $b_p = 1$, showing sensible effects of upper dynamic instability for two-step distributed pyrolysis with respect to no distributed pyrolysis.

agree with the approximate but nonlinear analytical expectations illustrated by the bifurcation diagrams of Figs. 7.5–9.

Computed transient burning histories following a forced pressurization test, a linear increase of pressure from 5 to 25 atm, are shown in Figs. 7.12–13 for different pyrolysis submodels. For concentrated pyrolysis, the bifurcation diagrams of Figs. 7.5–7 predict no upper dynamic instability and the transient surface temperature shows an overdamped behavior following the initial peak of surface temperature due to transient combustion; the peak is sharper for faster pressurization rate (cf. Fig. 7.12 for $b_p = 1$ against Fig. 7.13 for $b_p = 500$). For two-step distributed pyrolysis, the bifurcation diagram of Fig. 7.8 ($Q_{c,1} = 0.25$ and $Q_{c,2} = 0.02$) predicts upper dynamic instability: the computed transient surface temperature indeed shows a perceivable oscillatory behavior for moderate pressurization rate (Fig. 7.12 for $b_p = 1$) yielding a complex oscillatory pattern (consisting of attempts to extinguish followed by "reignition") for fast pressurization rate (Fig. 7.13 for $b_p = 500$). Overall, comparing the numerical response of the two pyrolysis submodels to the same pressurization test reveals that exothermicity distributed in the high-temperature degradation layer triggers dynamic instability in terms of oscillatory burning, possibly

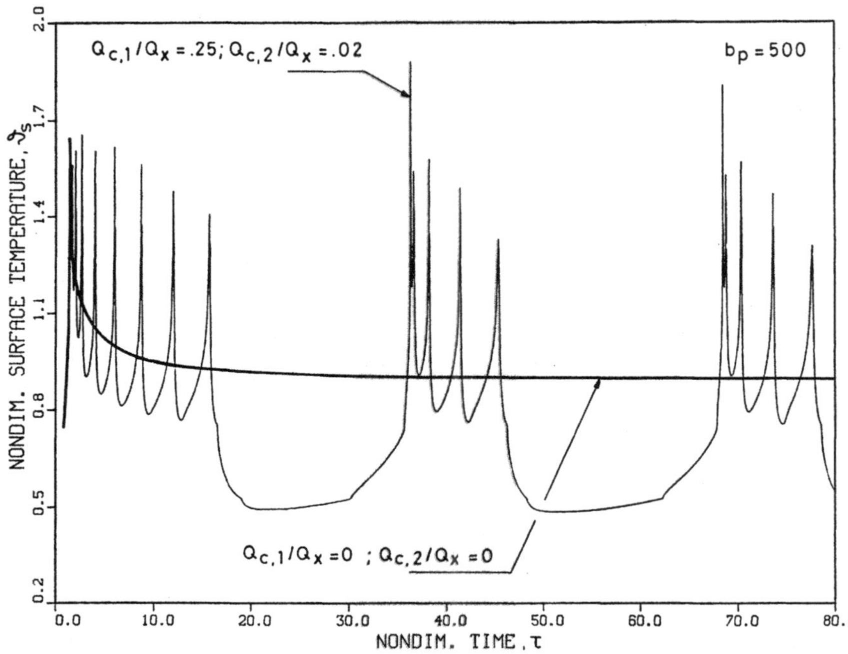

FIGURE 7.13. Numerical computations of transient surface temperature during a pressurization test, from 5 to 25 atm with $b_p = 500$, showing sensible effects of upper dynamic instability for two-step distributed pyrolysis with respect to no distribution pyrolysis.

yielding dynamic extinction, while exothermicity distributed in the low-temperature degradation layer promotes "reignition."

Conclusions and Future Work

The two-step consecutive overall reactions, implemented to model volumetrically distributed pyrolysis in the condensed phase of double-base propellants, proved meaningful and certainly deserve further attention. The specific mechanism enforced is based on the experimental work by Zenin (1990); possible important details concerning the reaction order and special effects related to the super-rate burning region have yet to be investigated. However, the results of this first detailed application on two-step distributed pyrolysis are quite encouraging. On the other hand, the inadequacy of surface-concentrated pyrolysis and puzzling problems encountered by enforcing one-step distributed pyrolysis were already discussed (Grimaldi et al. 1987). At any rate, two-step distributed pyrolysis is a necessary, although not sufficient, ingredient to understand super-rate burning.

The relevant result of this investigation is that properly partitioning volumetrically distributed exothermicity, between high- and low-temperature degradation layers in the condensed phase, helps all aspects of intrinsic burning stability. The lack of static burning stability in the super-rate burning region is alleviated by distributed heat release in the high-temperature aegradation layer, while the lack of upper dynamic burning stability is alleviated by distributed heat release in the low-temperature degradation layer. This synergistic effect, being of great interest, is under careful study for deeper understanding. However, this work only points a good direction and puts bounds on what to do and what to avoid; within current limitations, it is a sensitivity test for the overall transient combustion model.

The need for more accurate data defining the involved thermochemistry is a tough challenge for the experimenters, which has to be met; dynamic burning and combustion stability are just too sensitive to condensed phase heat release. However, in spite of the need for some refinements of the already accurate data set used in this work, it is believed that a two-step distributed pyrolysis mechanism is a must for the modeling of nitrate esters propellants. This conclusion is inescapable from the independent but convergent experimental work of Cohen and Holmes (1983), Fifer (1984), and Zenin (1990); the fundamental physics and chemistry of the thermal degradation process just conspire against any oversimplification.

Future work will focus on the intricacies of super-rate burning: how much can be deduced from first principles and how much might require ad-hoc modeling must still be ascertained. Some results of this investigation seem, however, to suggest that most, if not all, of the super-rate transient burning features can be interpreted in terms of the classical burning theories, once proper attention has been given to its complicated but experimentally detectable steady-state structure.

Acknowledgments. The authors wish to express their sincere gratitude to Drs. A.A. Zenin and M.S. Miller for helpful comments, to Mr. G. Colombo for skillful data reduction, and to an unknown referee for careful review of the manuscript. The propellant used for experimental tests was courteously provided by SNIA BPD. Financial support from MPI 40 percent, MPI 60 percent, and CNPM/CNR are gratefully acknowledged.

Nomenclature

A = nondimensional preexponential factor.
b_p = nondimensional pressurization rate.
c = specific heat, cal/g·K.
C = c/c_{ref}, nondimensional specific heat.
d = thickness, cm.
$\tilde{E}$ = activation energy, cal/mol.

$E_{(\ )}$	$= \tilde{E}/\Re/T_{(\ ),\mathrm{ref}}$, nondimensional activation energy.
F_0	$= I_0/\tilde{q}_{\mathrm{ref}}$, nondimensional external radiant flux.
H	$= Q/Q_{\mathrm{ref}}$, nondimensional heat release.
I_0	= external radiant flux intensity, $\mathrm{cal/cm^2 \cdot s}$.
k	= thermal conductivity, $\mathrm{cal/cm \cdot s \cdot K}$.
K	$= k/k_{\mathrm{ref}}$, nondimensional thermal conductivity.
M	= maximum value of nondimensional chemical reaction rate.
n	= ballistic exponent, defined by the law $\bar{r}_b = ap^n$.
n_s	= pressure exponent in the pyrolysis law.
p	= pressure, atm.
p_{ref}	= 68 atm, reference pressure.
P	$= p/p_{\mathrm{ref}}$, nondimensional pressure.
q	$= \tilde{q}/\tilde{q}_{\mathrm{ref}}$, nondimensional energy flux.
$\tilde{q}$	= energy flux intensity, $\mathrm{cal/cm^2\ s}$.
$\tilde{q}$	$= \rho_c c_c r_{b,\mathrm{ref}}(T_{s,\mathrm{ref}} - T_{\mathrm{ref}})$, reference energy flux, $\mathrm{cal/cm^2 \cdot s}$.
Q	= heat release, cal/g (positive exothermic).
Q_{ref}	$= c_{\mathrm{ref}}(T_{s,\mathrm{ref}} - T_{\mathrm{ref}})$, reference heat release, cal/g.
Q_x	= total condensed-phase heat release (surface + volumetric), cal/g.
r_b	= burning rate, cm/s.
$r_{b,\mathrm{ref}}$	$= r_b(p_{\mathrm{ref}})$, reference burning rate, cm/s.
$\bar{\bar{r}}_\lambda$	= average optical reflectivity of the burning surface, percent.
R	$= r_b/r_{b,\mathrm{ref}}$, nondimensional burning rate.
$\Re$	= universal gas constant; 1.987 $\mathrm{cal/mol \cdot K}$ or 82.1 $\mathrm{atm\ cm^3/mol \cdot K}$.
t	= time coordinate, s.
T	= temperature, K.
T_{ref}	= 300 K, reference temperature.
$T_{s,\mathrm{ref}}$	$= T_s(p_{\mathrm{ref}})$, reference surface temperature, K.
u	= gas velocity, cm/s.
U	$= u/r_{b,\mathrm{ref}}$, nondimensional gas velocity.
w	= power of pyrolysis law.
$\bar{\bar{w}}$	= average molecular mass of gas mixture, g/mol.
x	= space coordinate, cm.
X	$= x/(\alpha_{\mathrm{ref}}/r_{b,\mathrm{ref}})$, nondimensional space coordinate.
Y_{exp}	= function defined by the left-hand side of Eq. 23.
Y_{mod}	= function defined by the right-hand side of Eq. 23.
Z	= parameter of elongated flame kinetics.

Greek Symbols.

α	= thermal diffusivity, $\mathrm{cm^2/s}$; also parameter of a transient flame model.
β	= parameter of a transient flame model.
γ	= parameter of a transient flame model.
ε	= nondimensional reaction rate.
$\bar{\bar{\varepsilon}}_\lambda$	= average optical emissivity of the burning surface, percent.
θ	$= (T - T_{\mathrm{ref}})/(T_{s,\mathrm{ref}} - T_{\mathrm{ref}})$, nondimensional temperature.
$\Upsilon_{(\ )}$	$= T_{(\ )}/T_{(\ ),\mathrm{ref}}$, nondimensional temperature.

ρ = density, g/cm^3.
ΣH_c = total heat volumetrically released in the condensed phase, nondimensional.
ΣQ_c = total heat volumetrically released in the condensed phase, cal/g.
τ = $t/(\alpha_{ref}/r^2_{b,ref})$, nondimensional time coordinate.
$\langle \tau' \rangle$ = $\tau(C_g/K_g)(\rho_c/\langle \rho_g \rangle)$, nondimensional characteristic time parameter.
φ = $Q_g\, \rho_g\, \tilde{\varepsilon}_g$, heat-release rate per unit volume, cal/cm$^3 \cdot$ s.

Subscripts and Superscripts.
c = condensed-phase.
c, s = burning surface, condensed-phase side.
dz = dark zone.
f = flame; also final.
fz = fizz zone.
g = gas phase.
g, s = burning surface, gas-phase side.
i = ith term; also initial.
min = minimum.
ref = reference.
s = burning surface.
tra = transition.
x = total heat release in the condensed phase (surface + volumetric).
λ = spectral.
— = steady state.
~ = dimensional value.
= = average over chemical composition.
1 = first step (either high-pressure or high-temperature).
2 = second step (either middle-pressure or middle-temperature).
3 = third step (either low-pressure or low-temperature).
$-\infty$ = far upstream.
$\langle\ \rangle$ = space average.

Abbreviations.
AP = Ammonium Perchlorate (NH_4ClO_4).
DB = Double-Base.
KTSS = Krier-T'ien-Sirignano-Summerfield.
MTS = Merkle-Turk-Summerfield.

References

Birk, A., 1983, "Ignition Dynamics of Fully Reactive Propellants in Stagnation Flow," *AIAA J.*, *21*, 4, April.

Bizot, A., Ferreira, J.G., and Lengellé, G., 1985, "Modelization of the Ignition Process of Homogeneous Propellants," *AIAA Paper No. 85–1178*, AIAA/SAE/ASME 21st Joint Propulsion Conference, July.

Bruno, C., Dondé, R., Riva, G., and De Luca, L., 1986, "Computed Nonlinear Transient

Burning of Solid Propellants with Variable Thermal Properties," *17th International ICT Conference*, Karlsruhe, Germany, 25–27 June 1986, Proceedings, paper 73.

Cohen, A. and Holmes, H.E., 1983, "Convective Ignition of Double-Base Propellants," *XIX International Symposium on Combustion*, 691–699.

Culick, F.E.C., 1968, "A Review of Calculations for Unsteady Burning of a Solid Propellant," *AIAA J.*, *6*, 12, 2241–2255.

De Luca, L., 1984, "Extinction Theories and Experiments," In K.K. Kuo and M. Summerfield (Eds.), *Fundamentals of Solid Propellant Combustion*, Progress in Astronautics and Aeronautics, 90, 661–732.

De Luca, L., 1990, "A Critical Review of Solid Rocket Propellant Transient Flame Models," invited lecture for the 3rd International Seminar on Flame Structure (ISFS-89), Alma Ata, U.S.S.R. 18–22 Sept. 1989, *Pure and Applied Chemistry.*

De Luca, L., Dondé, R., and Riva, G., 1984, "Effects of Distributed Condensed Phase Reactions on Heterogeneous Deflagration Waves," *2d International Specialists' Meeting of the Combustion Institute on Oxidation*, Budapest, Hungary, 18–22 August 1982, Oxidation Communications, *6*, 1–4, 185–198.

De Luca, L., et al., 1986, "Burning Stability of Double-Base Propellants," *The Propulsion and Energetics Panel of AGARD*, 66th (A) Specialists' Meeting on Smokeless Propellants, Florence, Italy, 12–13 Sept. 1985, Conference Proceedings No. 391, paper 10.

De Luca, et al., 1986b, "Modeling of Spacewise Thick Flames," *IAF Paper No. 86–196*, presented at the 36th International Astronautical Federation Congress, Innsbruck, Austria, 4–11 October.

De Luca, L., et al., 1987, "Transient Modeling of Spacewise Thick Flames for Rocket Propulsion," presented at the Symposium on Commercial Opportunities in Space, Taiwan, China, 19–24 April.

Dondé, R., 1987, "Combustione Eterogenea in Regime Non Stazionario," *Doctoral Thesis*, Dipartimento di Energetica, Politecnico di Milano, Milano, Italy, April.

Fifer, R.A., 1984, "Chemistry of Nitrate Ester and Nitramine Propellants," in K.K. Kuo and M. Summerfield (Eds.), *Fundamentals of Solid Propellant Combustion*, AIAA Progress in Astronautics and Aeronautics, 90, 177–237.

Galfetti, L., Turrini, F., and De Luca, L., 1988, "Modeling of Transient Combustion in Solid Rocket Motors," *XVI International Symposium on Space Technology and Science*, Sapporo, Japan, 22–27 May 1988, Proceedings, 217–227.

Grimaldi, C., et al., 1987, "Modeling of Catalyzed Double-Base Transient Flames," presented at the 9th Congresso Nazionale AIDAA, Palermo, Italy, 26–29 October.

Krier, H., T'ien, J.S., Sirignano, W.A., and Summerfield, M., 1968, "Nonsteady Burning Phenomena of Solid Propellants: Theory and Experiments," *AIAA J.*, *6*, 2, 278–285.

Zenin, A.A., 1983, "Universal Dependence for Heat Liberation in the K-Phase and Gas Macrokinetics in Ballistic Powder Combustion," *Explosion, Combustion and Shock Waves*, *19*, 4, 444–446.

Zenin, A.A., 1990, "Thermophysics of Stable Combustion Waves of Solid Propellants," accepted for publication in *Nonsteady Burning and Combustion Stability of Solid Propellants*, a forthcoming volume of the series AIAA Progress in Astronautics and Aeronautics.

8
Combustion Instabilities and Rayleigh's Criterion

F.E.C. CULICK

ABSTRACT: In 1878, Lord Rayleigh formulated his criterion to explain several examples of acoustic waves excited and maintained by heat addition. It is a qualitative explanation successfully capturing the essence of the phenomena but not providing a basis for quantitative predictions. The widespread appeal of Rayleigh's criterion merits placing this important result on a more rigorous basis. To do so requires careful formulation grounded in the theory of small amplitude motions in a compressible fluid. In this chapter, we review the construction of an approximate analysis and establish the equivalence of Rayleigh's criterion and the condition for linear stability of small amplitude motions. Thus Rayleigh's criterion is formulated explicitly in the context of an analysis applicable to any combustion chamber. Some results are discussed for both linear and nonlinear motions. Recent experimental results discussed by others suggest that the criterion may offer a practical means for investigating the causes of instabilities in propulsion systems.

Introduction

It's my genuine pleasure to contribute to this volume commemorating Professor Casci's retirement after many years of dedicated service to his institution, to Italy, and indeed to the international affairs of aerospace engineering. I first met Professor Casci when I became a member of the U.S. delegation to the Propulsion and Energetics Panel of the Advisory Group for Aerospace Research and Development (AGARD). At that time one of the items of our business involved planning a meeting on problems of combustion instabilities in solid propellant rockets. Professor Casci was program chairman of the meeting that was held, very successfully, in Oslo, Norway. It is therefore particularly appropriate that I address here some fundamental aspects of combustion instabilities.

In the broad context of fluid mechanics, the phenomenon called "combustion instability" is a special—though common—form of unsteady motion excited and sustained by heat addition in a compressible fluid. Strictly the term is misleading because almost always an instability of combustion is not an issue. What is observed is an unstable motion caused by the conversion of heat released in combustion processes to mechanical energy of fluid motions. That transfer of energy occurs because the combustion is sensitive to fluctuations of the flow variables, notably the pressure and velocity. Thus under suitable circumstances a small disturbance is linearly unstable. To an observer, the system appears to be self-excited, the unstable motion growing without the action of any external agent.

Unsteady motions are always present in a combustion chamber as random fluctuations or noise. Noisy motions cause structural vibrations over a broad frequency range, usually requiring only routine qualification of equipment. The term "combustion instability" refers to organized vibratory motions having well-defined frequencies typically close to the resonant frequencies predicted by classical acoustics for the same geometry without combustion and flow. These oscillations rarely directly affect the steady performance of the system (i.e., thrust or power output). Serious problems arise because only a small portion of the available combustion energy is sufficient to support motions producing unacceptably large structural vibrations. In extreme cases, internal surface heat transfer rates may be amplified tenfold or more, causing excessive erosion of chamber walls.

Combustion instabilities in stationary powerplants (e.g., Putnam 1971) and in combustors designed to oscillate (Reynst 1961; Zinn 1986) have long been a practical concern. However, the most severe examples arise in combustion chambers designed for propulsion systems: gas turbine combustors, afterburners, ramjets, liquid rockets, and solid rockets. The chief reasons that instabilities are commonly encountered (in fact they must be expected in the development of a new system) are the high volumetric density of energy release and the relatively weak attenuating processes.

Throughout the long recent history of combustion instabilities (roughly 50 years) there has been a tendency to emphasize perhaps too strongly the obvious practical differences between systems. In reality, the similarities and common characteristics may dominate. Combustion chambers intended for different systems differ chiefly in two respects: geometry and the kind of reactants used. It is therefore possible to capture virtually all of the behavior within one theoretical framework. This point of view can be theoretically developed to the extent that analysis of a particular system requires chiefly modeling of the special processes, numerical computation of the normal modes and frequencies for the given geometry, and possibly some ancillary laboratory experiments to provide input data that cannot be calculated with tolerable uncertainty. With that information, one is positioned to compute and assess linear stability of the system as well as estimate certain aspects of nonlinear behavior. Culick (1988, 1989) and Culick and Yang (1990) have given recent reviews elaborating this point of view .

More than a century before combustion instabilities were a problem, there was already much interest in the general problem of acoustic waves generated by heat addition. Musical or singing flames were first studied in the late 18th century. Within a few years excitation of acoustic modes by a flame supported in a vertical tube had been discovered. Both Lord Rayleigh (1878b) and Tyndall (1897) have given detailed discussions of several phenomena involving the interactions of flames or hot surfaces and sound waves. Later, pulsating combustion was proposed as a basis for jet propulsion, by Lorin in 1908; see Zinn (1986) for a lengthy review of the subject of pulsating combustion in various systems.

As a result of his investigations of sound waves sustained by heat addition, Rayleigh (1878a; 1878b, Vol. II, p. 226) stated his famous criterion:

> If heat be periodically communicated to, and abstracted from, a mass of air vibrating (for example) in a cylinder bounded by a piston, the effect produced will depend upon the phase of the vibration at which the transfer of heat takes place. If heat be given to the air at the moment of greatest condensation, or be taken from it at the moment of greatest rarefaction, the vibration is encouraged. On the other hand, if heat be given at the moment of greatest rarefaction, or abstracted at the moment of greatest condensation, the vibration is discouraged.

This principle has probably been more widely invoked than any other in attempts to explain the occurrence of unsteady motions supported by heat addition, including combustion instabilities. Rayleigh never expressed his criterion in a form suited to quantitative tests. Only in the past 30 years have others done so. The most active user of the criterion has been Putnam (1971) who has applied a linearized form to a wide variety of practical problems arising in stationary powerplants and heating systems. Chu (1956) and Zinn (1986) have given derivations of the criterion, also for small-amplitude motions governed by the equations of linear acoustics with heat sources. The approach followed here begins in much the same fashion as Zinn's analysis but applies to nonlinear motions and places Rayleigh's criterion in the general context of the stability of acoustic waves.

In a short note (Culick 1987), I showed one way to interpret and use Rayleigh's criterion for nonlinear motions. It is important to establish results of that sort in order to be able to use the principle under the conditions commonly prevailing when combustion instabilities are observed. Subsequently, while preparing a review on combustion instabilities (Culick 1988), I realized that the criterion is obviously related to the usual formal result characterizing the stability of small amplitude motions. The invitation to contribute to Professor Casci's commemorative volume is a happy opportunity to merge and elaborate upon those ideas so fundamental to the behavior of disturbances in a combustion chamber. To provide the context for discussion of Rayleigh's criterion, the following three sections are a brief review of an approximate analysis constructed to accommodate combustion instabilities in any propulsion system.

Formulation

In order not to restrict the results to infinitesimally small amplitude motions in a particular geometry, it is essential to begin with the general equations of motion. Many systems contain not only a gas mixture, but also condensed phases. It is convenient to simplify matters formally by assuming that the actual medium may be approximated as a two-phase mixture, a single average gas and a condensed phase represented locally by an average particle. The conservation equations are then written for the medium comprising the two phases. After some rearrangement following the approach discussed by Marble (1963), we can write the governing equations in the form

$$\frac{\partial \rho}{\partial t} + \vec{u}\cdot\nabla\rho = \mathscr{W} \tag{1}$$

$$\rho\frac{\partial \vec{u}}{\partial t} + \rho\vec{u}\cdot\nabla\vec{u} = -\nabla p + \vec{\mathscr{F}} \tag{2}$$

$$\frac{\partial p}{\partial t} + \bar{\gamma}p\nabla\cdot\vec{u} = -\vec{u}\cdot\nabla p + \mathscr{P} \tag{3}$$

where $\vec{u}$ and p are the velocity and pressure of the gas but ρ and $\bar{\gamma}$ are mass-averaged values,

$$\rho = \rho_g + \rho_l = \rho_g(1 + C_m) \tag{4}$$

$$\bar{\gamma} = \frac{(C_p + C_m C_l)/((1 + C_m)}{(C_v + C_m C_l)/(1 + C_m)} = \frac{C_p + C_m C_l}{C_v + C_m C_l} \tag{5}$$

where C_l is the specific heat of the condensed material and $C_m = \rho_l/\rho_g$ is the ratio of the mass of condensed material to the mass of gas in a unit volume of chamber. Equation 3 is the energy equation written for the pressure and involves the assumption that the equation of state for the mixture is

$$p = \rho\bar{R}T \tag{6}$$

and $\bar{R} = (1 + C_m)^{-1}[(C_p + C_m C_l) - (C_v + C_m C_l)] = \bar{C}_p - \bar{C}_v$ is the mass-averaged gas constant.

If viscous effects are neglected, the source terms in Eqs. (1–3) are

$$\mathscr{W} = -\rho\nabla\cdot\vec{u} - \nabla\cdot(\rho_l\delta\vec{u}_l) \tag{7}$$

$$\vec{\mathscr{F}} = \delta\vec{F}_l + \delta\vec{u}_l w_l \tag{8}$$

$$\mathscr{P} = \frac{\bar{R}}{\bar{\bar{C}}_v}\left[Q + \delta Q_l + \delta\vec{u}_l\cdot\vec{\mathscr{F}}_l + \left\{(e_l - e) + \frac{1}{2}(\delta\vec{u}_l)^2\right\}w_l - \bar{C}_v T\nabla\cdot(\rho_l\delta\vec{u}_l)\right]. \tag{9}$$

Here $w_l = -\partial\rho_l/\partial t - \nabla\cdot(\rho_l u_l)$ stands for the rate of conversion (kg/m$^3\cdot$s) of condensed material to gas; $\delta\vec{u}_l = \vec{u}_l - \vec{u}$, $\delta T_l = T_l - T$ are local velocity and temperature differences between the liquid and gas phases and by definition,

$$\delta \vec{F}_l = -\rho_l\left[\frac{\partial \delta \vec{u}_l}{\partial t} + \delta \vec{u}_l \cdot \nabla \delta \vec{u}_l + \delta \vec{u}_l \cdot \nabla \vec{u} + \vec{u} \cdot \nabla \delta \vec{u}_l\right] \tag{10a}$$

$$\delta Q_l = -\rho_l C_l\left[\frac{\partial \delta T_l}{\partial t} + \delta \vec{u}_l \cdot \nabla T_l + \delta \vec{u}_l \cdot \nabla T + \vec{u} \cdot \nabla \delta T_l\right]. \tag{10b}$$

The interactions between the gas and condensed phases are computed by specifying the force $\vec{F}_l$ of interaction and the heat transfer, and solving the equations

$$\vec{F}_l = -\rho_l\left[\frac{\partial \vec{u}_l}{\partial t} + \vec{u}_l \cdot \nabla \vec{u}_l\right] \tag{11a}$$

$$Q_l = -\rho_l C_l\left[\frac{\partial T_l}{\partial t} + \vec{u}_l \cdot \nabla T_l\right]. \tag{11b}$$

Although some approximations not discussed here are implied (see Marble (1963) for further details) this formulation is quite general. For example, nonlinear interactions $\vec{F}_l$ and Q_l are accommodated and the condensed phase need not be uniformly distributed in the chamber. In fact no analysis of combustion instabilities has avoided further approximations, many of which we shall use here.

Approximate Analysis: Spatial and Time Averaging

The essential idea is that the motions called combustion instabilities appear, on the basis of experimental results, to be closely related to classical acoustic modes of a closed chamber. Thus the influences of combustion, the mean flow, etc., that distinguish the instabilities are treated as perturbations of the classical results. Given the obviously vigorous activity in a combustion chamber—nonuniform flow, turbulence, possibly flow separation, nonuniform burning, often large temperature gradients, heat transfer, and viscous processes—it seems at first acquaintance surprising that well-defined pressure waves should exist under such conditions.

However, the work by Chu and Kovasznay (1957) provides at least part of the explanation. Rayleigh (1878b, Vol. II, pp. 315 ff), following earlier work by Stokes and Kirchkhoff, had shown that pressure disturbances propagate independently as acoustic waves; fluctuations of density and temperature may participate in other unsteady motions. Chu and Kovasznay constructed a systematic procedure to show that small freely propagating disturbances may be synthesized (not uniquely, but that ambiguity is immaterial here) of three kinds of waves: acoustics waves carrying pressure fluctuations but no entropy; viscous waves having no pressure perturbations; and entropy waves also causing no pressure changes. (See Doak (1973) and references in that paper.)

In the linear approximation, these three kinds of waves propagate independently: the acoustic waves of course propagate at the speed of sound while viscous and entropy waves move with the mean flow speed.

Consequently, we may expect that in lowest order, small-amplitude pressure waves will propagate sensibly unaffected by other fluctuations in the flow and therefore in a chamber they may evolve into acoustic modes. Evidently that result is observed in combustion instabilities, providing, as often is the case, the Mach number of the average flow is not too large. As we shall see shortly, acoustic modes are distorted by the mean flow (e.g., refraction must occur) and by other processes, but the effects are commonly small. Useful results for combustion instabilities can be obtained without actually computing those distortions.

With that theoretical basis, and in view of the behavior commonly observed, we are justified in seeking a representation dominated by classical wave behavior but exhibiting the average flow and combustion processes as perturbations. Write all dependent variables as sums of mean and fluctuating parts, $p = \bar{p} + p'$, etc., and assume for simplicity that the average values are independent of time. To second order in the fluctuations, eqs. 2 and 3 become

$$\bar{\rho}\frac{\partial \vec{u}}{\partial t} + \nabla p' = -\bar{\rho}(\vec{\bar{u}}\cdot\nabla\vec{u}' + \vec{u}\cdot\nabla\vec{\bar{u}}) - \bar{\rho}(\vec{u}'\cdot\nabla\vec{u}') - \rho'\frac{\partial \vec{u}'}{\partial t} + \mathscr{F}' \tag{12}$$

$$\frac{\partial p'}{\partial t} + \bar{\gamma}\bar{p}\nabla\cdot\vec{u}' = -\vec{\bar{u}}\cdot\nabla p' - \bar{\gamma}p'\nabla\cdot\vec{\bar{u}} - \vec{u}'\cdot\nabla p' - \gamma p'\nabla\cdot\vec{u}' + \mathscr{P}' \tag{13}$$

Terms of order $|\vec{\bar{u}}|^2|\vec{u}'|$ and $|\vec{\bar{u}}|\,|\vec{u}'|^2$ are ignored, consistent with the assumption that the mean pressure is constant. The system of equations is completed by the perturbed forms in this second-order analysis, the equation of state,

$$p' = \bar{R}(\rho'\bar{T} + \bar{\rho}T') \tag{14}$$

and of the continuity equation (1). However, at least to second order, the assumption of isentropic behavior seems valid; that relates the density to the pressure, and the continuity equation is not required. Now differentiate Eq. 13 with respect to time and substitute Eq. 12 for $\partial\vec{u}'/\partial t$ to give the desired nonlinear wave equation:

$$\nabla^2 p' - \frac{1}{\bar{a}^2}\frac{\partial^2 p'}{\partial t^2} = h \tag{15}$$

where $\bar{a}^2 = \bar{\gamma}\bar{R}\bar{T}$ and to second order in the fluctuations,

$$\begin{aligned} h = & -\bar{\rho}\nabla\cdot(\vec{\bar{u}}\cdot\nabla\vec{u}' + \vec{u}'\cdot\nabla\vec{\bar{u}}) + \frac{1}{\bar{a}^2}\vec{\bar{u}}\cdot\nabla\frac{\partial p'}{\partial t} + \frac{\bar{\gamma}}{\bar{a}^2}\frac{\partial p'}{\partial t}\nabla\cdot\vec{\bar{u}} \\ & - \nabla\cdot\left(\bar{\rho}\vec{u}'\cdot\nabla\vec{u}' + \rho'\frac{\partial\vec{u}'}{\partial t}\right) + \frac{1}{\bar{a}^2}\frac{\partial}{\partial t}(\vec{u}'\cdot\nabla p') + \frac{\bar{\gamma}}{\bar{a}^2}\frac{\partial}{\partial t}(p'\nabla\cdot\vec{u}') \\ & + \nabla\cdot\vec{\mathscr{F}}' - \frac{1}{\bar{a}^2}\frac{\partial\mathscr{P}'}{\partial t}. \end{aligned} \tag{16}$$

Take the scalar product of the outward normal vector with Eq. 12 to find the boundary condition on the gradient of p':

$$\hat{n}\cdot\nabla p' = -f \tag{17}$$

and

$$f = \bar{\rho}\frac{\partial \vec{u}'}{\partial t}\cdot\hat{n} + \bar{\rho}(\vec{\bar{u}}\cdot\nabla\vec{u}' + \vec{u}'\cdot\nabla\vec{\bar{u}})\cdot\hat{n}$$

$$+ \bar{\rho}(\vec{u}'\cdot\nabla\vec{u}')\cdot\hat{n} + \rho'\frac{\partial \vec{u}'}{\partial t}\cdot\hat{n} - \vec{\mathscr{F}}'\cdot\hat{n}. \tag{18}$$

The values of the quantities on the right-hand side are set according to local characteristics of the boundary.

A procedure for spatial and time averaging has been discussed in several places (e.g., Culick 1988; Yang and Culick 1990); it is adequate here simply to quote the results. The pressure and velocity fields are written as syntheses of the normal modes ψ_n for the chamber, with time-varying amplitudes $\eta_n(t)$:

$$p'(\vec{r}_1 t) = \bar{p}\sum_{n=1}^{\infty}\eta_n(t)\psi_n(\vec{r}) \tag{19a}$$

$$\vec{u}'(\vec{r}_1 t) = \sum_{n=1}^{\infty}\frac{\dot{\eta}_n(t)}{\bar{\gamma}k_n^2}\nabla\psi_n(\vec{r}). \tag{19b}$$

Eq. 19a is a form of the conventional beginning of Galerkin's method (see Zinn and Powell 1971 and Culick (1976) for applications to combustion instabilities). The representation of Eq. 19b is precisely consistent with Eq. 19a in the limit of small-amplitude isentropic motions. Although there is no rigorous justification, numerical calculations (most recently by Culick and Yang (1989)) have shown that this assumed form for the velocity fluctuation works well for pressure fluctuations at least as large as 10 percent of the mean value.

Afer substitution of Eqs. 19a and b in Eq. 15, spatial averaging leads eventually to the set of ordinary differential equations for the η_n:

$$\frac{d^2\eta_n}{dt^2} + \omega_n^2\eta_n = F_n \tag{20}$$

where ω_n is the frequency of the nth mode and the force is

$$F_n = -\frac{\bar{a}^2}{\bar{p}E_n^2}\left\{\int h\psi_n\,dV + \oiint f\psi_n\,dS\right\} \tag{21}$$

where

$$E_n^2 = \int \psi_n^2\,dV. \tag{22}$$

The functions h and f contain p' and $\vec{u}'$, which are to be replaced by Eqs. 19a and b. Hence F_n depends nonlinearly on the amplitudes η_m, and Eq. 20 represents a collection of coupled nonlinear oscillators; one oscillator is associated with each classical mode.

The functions $\psi_n(\vec{r})$ depend on the geometry of the chamber and on the boundary conditions. At a rigid wall, $\hat{n}\cdot\nabla\psi_n = 0$, approximately true as well at the entrance to a choked exhaust nozzle if the Mach number at the entrance is small. Otherwise, adjustments must be made, possibly cumbersome in practice, but doable in principle.

After the explicit forms of h and f are determined (i.e., $\mathscr{W}'$, $\vec{\mathscr{F}}'$, and $\mathscr{P}'$ are specified), F_n is known and Eq. 20 can be solved numerically. Much more interesting results are obtained if the equations are time-averaged, producing a set of coupled first-order equations, two for each mode. Write $\eta_n(t)$ as an oscillation having slowly varying amplitude and phase,

$$\eta_n(t) = r_n(t)\sin(\omega_n t + \phi_n(t) = A_n(t)\sin\omega_n t + B_n(t)\cos\omega_n t \tag{23}$$

where

$$r_n = (A_n^2 + B_n^2)^{1/2}$$

where

$$\tan\phi_n = B_n/A_n. \tag{24}$$

After averaging over some interval τ characteristic of the slow variations, $A_n(t)$ and $B_n(t)$ satisfy the equations

$$\frac{dA_n}{dt} = \frac{1}{\omega_n\tau}\int_t^{t+\tau} F_n(t')\cos\omega_n t'\,dt' \tag{25a}$$

$$\frac{dB_n}{dt} = \frac{-1}{\omega_n\tau}\int_t^{t+\tau} F_n(t')\sin\omega_n t'\,dt'. \tag{25b}$$

Because the single function $\eta_n(t)$ has been replaced by the two functions $A_n(t)$ and $B_n(t)$ in the assumed form (Eq. 23), we have the freedom to impose one further constraint. As usual in the method of time-averaging, that is chosen so that the derivative of $\eta_n(t)$ has the same form as that for a simple harmonic oscillator for which A_n and B_n are constant,

$$\dot{\eta}_n = \omega_n[A_n\cos\omega_n t - B_n\sin\omega_n t]. \tag{26}$$

Thus we require $\dot{A}_n\sin\omega_n t + \dot{B}_n\cos\omega_n t = 0$.

High-frequency contributions to F_n in (Eqs. 25a and b) average to zero (approximately) leaving simplified right-hand sides. The details, and therefore the structure of the right-hand sides, depend on the problem considered. Results obtained to date have shown an important difference between the cases of longitudinal modes and transverse modes. For the first, in which the frequencies are integral multiples of the fundamental, the right-hand sides of Eqs. 25a and b do not depend explicitly on time. But for transverse modes, the equations are nonautonomous with some of the terms containing modulation factors. The presence of that time dependence has important consequences for the behavior in limit cycles (Yang and Culick 1989).

Linear Stability

If only second-order terms in h and f are retained and only the contributions from the gas dynamics are accounted for, F_n has the form (Culick 1976):

$$F_n = -\sum_{i=i}^{\infty} [D_{ni}\dot{\eta}_i + E_{ni}\eta_i] - \sum_{i=1}^{\infty}\sum_{j=1}^{\infty} [A_{nij}\dot{\eta}_i\dot{\eta}_j + B_{nij}\eta_i\eta_j]. \tag{27}$$

Note that products $\dot{\eta}_i\eta_j$ are missing. That is a result that has significant consequences for nonlinear behavior but will not be discussed here. All influences of linear processes are contained in the coefficients D_{ni}, and E_{ni}. For longitudinal modes when the natural frequencies are integral multiples of the fundamental, $\omega_n = n\omega_1$, coupling is absent: $D_{ni} = E_{ni} = 0$ for $i \neq n$ (Culick 1976). Generally, linear coupling is present, arising in particular from interactions between the mean flow and the acoustic field. The strengths of the couplings depend on the frequencies of the modes; for example, with the method discussed here we can reproduce the familiar phenomenon of beating between weakly coupled modes having nearly equal frequencies. To keep the present discussion simple, we ignore linear coupling and the linear part of F_n is

$$F_n = -D_{nn}\dot{\eta}_n - E_{nn}\eta_n. \tag{28}$$

Substitution in Eqs. 25a and b with τ equal to the period of the nth mode, $\tau = 2\pi/\omega_n$, gives

$$\frac{dA_n}{dt} = -\frac{1}{2}D_{nn}A_n - \frac{1}{2}\frac{E_{nn}}{\omega_n}B_n \tag{29a}$$

$$\frac{dB_n}{dt} = -\frac{1}{2}D_{nn}B_n + \frac{1}{2}\frac{E_{nn}}{\omega_n}A_n \tag{29b}$$

Multiply the first by A_n, the second by B_n, and add the equations to find

$$\frac{dr_n}{dt} = -\frac{D_{nn}}{2}r_n$$

with solution

$$r_n(t) = r_{n0}e^{-1/2D_{nn}t}. \tag{30}$$

From Eq. 23 with substitution of Eqs. 29a and b we have the equation for the phase,

$$\frac{d\phi_n}{dt} = \frac{1}{2}\frac{E_{nn}}{\omega_n}$$

and

$$\phi_n = \frac{1}{2}\frac{E_{nn}}{\omega_n}t + \phi_{n0}. \tag{31}$$

Hence the amplitude $\eta(t)$, Eq. 23 is

$$\eta_n(t) = r_{n0} e^{\alpha_n t} \cos[(\omega_n - \theta_n)t + \phi_{n0}] \tag{32}$$

where

$$\alpha_n = -\frac{1}{2} D_{nn} \tag{33a}$$

$$\theta_n = -\frac{1}{2} \frac{E_{nn}}{\omega_n} \tag{33b}$$

are the growth constant and frequency shift of the nth mode. Equation 32 is obtained as the solution to Eq. 20 if we assume $\alpha_n^2 \ll \omega_n^2$, which is normally the case and is consistent with the assumptions on which this theory is based.

Equations 33a and b are the main results of the theory of linear stability: if $\alpha_n > 0$ then the amplitude of the nth mode grows without limit if there is an initial disturbance r_{n0}. The values of α_n and θ_n depend on the physical processes according to the definitions of F_n (Eq. 20). They can be calculated by noting that h and f will have the forms

$$h = h_{1n}\dot{\eta}_n + h_{2n}\eta_n \tag{34a}$$

$$f = f_{1n}\dot{\eta}_n + f_{2n}\eta_n. \tag{34b}$$

Substitution in the definition Eq. 23 and equating the coefficients with corresponding terms in Eq. 28 gives

$$\alpha_n = -\frac{\bar{a}^2}{2\bar{p}E_n^2}\left\{\int h_{1n}\psi_n\, dV + \oiint f_{1n}\psi_n\, dS\right\} \tag{35a}$$

$$\theta_n = -\frac{\bar{a}^2}{2\bar{p}E_n^2\omega_n}\left\{\int h_{2n}\psi_n\, dV + \oiint f_{2n}\psi_n\, dS\right\}. \tag{35b}$$

Alternatively, it is often useful to express the formulas in complex form. Only linear terms are retained in h and f (Eq. 22) and we may assume exponential time-dependence,

$$\eta_n = \hat{\eta}_n e^{i\bar{a}kt}, \qquad h = \hat{h} e^{i\bar{a}kt}, \qquad f = \hat{f} e^{i\bar{a}kt}$$

where k is the complex wavenumber,

$$k = \frac{1}{\bar{a}}(\omega - i\omega). \tag{36}$$

After substitution in Eq. 20, the common factor $e^{i\bar{a}kt}$ cancels and we have the formula for k^2,

$$k^2 \equiv \frac{1}{\bar{a}^2}(\omega - i\alpha)^2 = \frac{\omega_n^2}{\bar{a}^2} + \frac{1}{\bar{p}E_n^2}\left\{\int \frac{\hat{h}}{\hat{\eta}_n}\psi_n\, dV + \oiint \frac{\hat{f}}{\hat{\eta}_n}\psi_n\, dS\right\}. \tag{37}$$

Take real and imaginary parts, and use the fact that all perturbations are small, so α/ω, $(\omega - \omega_n)/\omega_n \ll 1$, to give

$$\omega = \omega_n + \frac{\bar{a}^2}{2\omega_n \bar{p} E_n^2}\left\{\iint \frac{\hat{h}^{(r)}}{\hat{\eta}_n}\psi_n\, dV + \oiint \frac{\hat{f}^{(r)}}{\hat{\eta}_n}\psi_n\, dS\right\} \tag{38a}$$

$$\alpha_n = \frac{-\bar{a}^2}{2\omega_n \bar{p} E_n^2}\left\{\iint \frac{\hat{h}^{(i)}}{\hat{\eta}_n}\psi_n\, dV + \oiint \frac{\hat{f}^{(i)}}{\hat{\eta}_n}\psi_n\, dS\right\}. \tag{38b}$$

Equations 33a and b, 35a and b, and 38a and b are convenient recipes for computing the most important characteristics of linear behavior. They are valid, of course, only if linear coupling is absent.

A General Interpretation of Rayleigh's Criterion for Small Amplitude Motions

Rayleigh's criterion stated in the beginning of this chapter is concerned with the amount of energy transferred during one cycle of oscillation from heat addition to mechanical energy of the acoustical motions. The transfer takes place because of $p - v$ work. Heat added causes a local increase of temperature, decrease of density, and hence increase of specific volume. Hence, pdv is positive, and work is done on the gaseous medium. Thus the energy, and hence pressure, of the acoustic field increases. How this process actually proceeds depends on the details of the heat release and the details of its coupling to the medium, a matter we are not concerned with here. Rather, we shall treat the global aspects and construct a quantitative interpretation of the criterion.

A purpose here is to show that Rayleigh's criterion dealing with heat addition is really a special case. If all processes are accounted for, the principle is equivalent to the result of linear stability that a motion is unstable if the growth constant is positive. Therefore we begin with Eq. 20 for the time-dependent amplitude of the nth mode,

$$\frac{d^2\eta_n}{dt^2} + \omega_n^2 \eta_n = F_n. \tag{39}$$

This is the equation for a forced harmonic oscillator having "energy" $\mathscr{E}_n = (\dot{\eta}_n^2 + \omega_n^2\eta_n^2)/2$ which, within a constant multiplier is the acoustic energy of the nth mode. Energy flows to the mode at the rate $F_n \dot{\eta}_n$; thus at time t, the change of energy in one period $\tau_n = 2\pi/\omega_n$ of the oscillation is

$$\Delta\mathscr{E}_n(t) = \int_t^{t+\tau_n} F_n \dot{\eta}_n\, dt'. \tag{40}$$

For linear behavior, $\eta_n = \hat{\eta}_n e^{i\bar{a}kt}$ and $F_n = \hat{F}_n e^{i\bar{a}kt}$; substitution in Eq. 39 gives the formula for k^2,

$$k^2 = \frac{\omega_n^2}{\bar{a}^2} - \frac{\hat{F}_n}{\bar{a}^2} \tag{41}$$

from which the real and imaginary parts are

$$\omega^2 = \omega_n^2 - \hat{F}_n^{(r)}$$
$$\alpha_n = \frac{1}{2\omega_n}\hat{F}_n^{(i)}. \tag{42}$$

Now integrate the right-hand side of Eq. 40 by parts, giving

$$\Delta\mathscr{E}_n = [F_n\eta_n]_t^{t+\tau_n} - \int_t^{t+\tau_n} \eta_n \frac{dF_n}{dt'}dt'.$$

The first term vanishes because we consider steady oscillations (at most $F_n\eta_n$ changes by only a negligible amount in one period) and

$$\Delta\mathscr{E}_n = -\int_t^{t+\tau_n} \eta_n \frac{dF_n}{dt'}dt'. \tag{43a}$$

Let φ be the phase of F_n relative to the pressure oscillation,

$$F_n = |F_n|e^{i(\bar{a}kt+\varphi)}. \tag{43b}$$

Real quantities must be used in the right-hand side of Eq. 43a, so we set $\eta_n = \cos(\bar{a}kt) \approx \cos\omega_n t$ and $F_n = |F_n|\cos(\bar{a}kt + \varphi) \approx |F_n|\cos(\omega_n t + \varphi)$. Equation 43 is

$$\Delta\mathscr{E}_n = \omega_n \int_t^{t+\tau_n} \cos\omega_n t'|F_n|\sin(\omega_n t' + \varphi)\,dt'$$
$$= \omega_n \int_t^{t+\tau_n} |F_n|\cos\omega_n t'\{\cos\omega_n t'\sin\varphi + \sin\omega_n t'\cos\varphi\}\,dt'.$$

As in the discussion of time-averaging, we consistently assume that $|F_n|$ and φ are nearly constant during one cycle of the motion; only the first integral is nonzero and we find

$$\Delta\mathscr{E}_n = \omega_n|F_n|\frac{\tau_n}{2}\sin\varphi$$

But $\tau_n = 2\pi/\omega_n$ and $|F_n|\sin\varphi = \hat{F}_n^{(i)} = 2\omega_n\alpha_n$ by the equation 43b, so

$$\Delta\mathscr{E}_n = 2\pi\omega_n\alpha_n. \tag{44}$$

Thus if we take $\Delta\mathscr{E}_n$ to be the measure implied by Rayleigh's criterion, this equation establishes the equivalence between that principle and the general result of linear stability: if $\alpha_n > 0$ then $\Delta\mathscr{E}_n > 0$ and the mode in question is unstable. The result of course applies to any small-amplitude motion because we can synthesize an arbitrary motion with a superposition of modes (eqs. 19a and b).

Rayleigh's Criterion for Nonlinear Motions

It is commonly found that after a transient growth period, a combustion instability will reach a limiting form characterized by a waveform and frequencr spectrum nearly constant with time. As an idealization we may therefore assume that the system executes a true limit cycle in which the net energy $\Delta\mathscr{E}_n(t)$ added per cycle is zero. To confirm this claim, substitute Eqs. 25a and b into Eq. 40:

$$\Delta\mathscr{E}_n(t) = \omega_n \int_t^{t+\tau} F_n[A_n \cos \omega_n t'\, dt' - B_n \sin \omega_n t']\, dt'.$$

Because $A_n(t)$ and $B_n(t)$ are slowly varying, they can be taken outside the integral. This leaves integrals that can be identified with the right-hand sides of Eq. 25a, b if we choose $\tau = \tau_n$:

$$\Delta\mathscr{E}_n(t) = \frac{1}{\tau}\left[A_n \frac{dA_n}{dt} + B_n \frac{dB_n}{dt}\right] = \frac{1}{2\tau}\frac{d}{dt}(A_n^2 + B_n^2) = \frac{1}{2\tau}\frac{dr_n}{dt} \tag{45}$$

where $r_n + (A_n^2 + B_n^2)^{1/2}$ (Eq. 23) is the maximum amplitude of the nth mode. In the limit cycle, r_n is constant, so we find $\Delta\mathscr{E}_n = 0$ as asserted earlier.

This result provides no useful information, because all physical processes have been included in the calculation of $\Delta\mathscr{E}_n$. It is not possible to measure all the contributions contained on the right-hand side of Eq. 40, so it is useful to treat only a part that can in fact be observed, for example, heat addition as originally considered by Rayleigh. In recent years that has become a realizeable possibility in experimental work with measurements of the visible radiation emitted during combustion instabilities (Sterling 1987; Sterling and Zukoski 1987). The essential idea is that kinetic and combustion processes are sufficiently rapid that the radiation intensity is proportional to the instantaneous rate of heat release. Then simultaneous measurements of the pressure and radiation allow one to compute, to within a multiplicative constant, the portion of $\Delta\mathscr{E}_n$ due to heat release. Because this is most conveniently done when the oscillations are steady, the procedure must correctly take into account that the motions are nonlinear. The necessary calculations were first discussed by Culick (1987).

Only the homogeneous heat addition Q in the definition of $\mathscr{P}$ (Eq. 9) was treated previously. It is not presently known how the various contributions can be separated in a real situation, so we shall simply write Q_R for the bracketed terms,

$$\mathscr{P} = \frac{\bar{R}}{\bar{C}} Q_R$$

and

$$Q_R = Q + \delta Q_l + \delta\vec{u}_l \cdot \vec{\mathscr{F}}_l + \{(e_l - e) + \tfrac{1}{2}(\delta\vec{u}_l)^2\} w_l - \bar{C}_v \tau \Delta \cdot (\rho_l \delta\vec{u}_l). \tag{46}$$

Now write Eq. 16 as

$$h = -\frac{1}{\bar{a}^2}\frac{\bar{R}}{\bar{C}_v}\frac{\partial Q'_R}{\partial t} + g \tag{47}$$

so F_n (Eq. 21), consists of two parts:

$$F_n = F_n^Q + F_n^e \tag{48}$$

where

$$F_n^Q = \frac{\bar{R}/\bar{C}_v}{\bar{p}E_n^2}\int \psi_n \frac{\partial Q'_R}{\partial t}\,dV \tag{49}$$

$$F_n^e = -\frac{\bar{a}^2}{\bar{p}E_n^2}\left\{\int \psi_n g\,dV + \oiint \psi_n f\,dS\right\}. \tag{50}$$

The precise form of Q'_R is unimportant for the purposes here.

The definition Eq. 40 of $\Delta\mathscr{E}_n$ becomes

$$\Delta\mathscr{E}_n = \frac{\bar{R}/\bar{C}_v}{\bar{p}E_n^2}\int \psi_n\,dV \int_t^{t+\tau_n} \dot{\eta}_n \frac{\partial Q'_R}{\partial t}\,dt' + \int_t^{t+\tau_n} F_n^e \dot{\eta}_n\,dt'.$$

Integration of the first term by parts, assuming that $Q'_R\dot{\eta}_n$ has period τ_n, and setting $\ddot{\eta}_n \approx -\omega_n^2\eta_n$ leads to

$$\Delta\mathscr{E}_n = \frac{\bar{R}}{\bar{C}_v}\frac{\omega_n^2}{E_n^2}\int dV \int_t^{t+\tau_n} \frac{p_n}{\bar{p}}\frac{Q'_R}{\bar{p}}\,dt' + \int_t^{t+\tau_n} F_n^e \dot{\eta}_n\,dt' \tag{51}$$

where

$$p_n = \bar{p}\eta_n(t)\psi_n(\vec{r}) \tag{52}$$

is the pressure fluctuation in the nth mode. It is Q'_R that is assumed proportional to the radiation intensity according to earlier remarks. Hence the first term in Eq. 51 is the formal expression of the contribution covered by Rayleigh's criterion applied to the nth mode. Note that these calculations show why the fluctuation of Q'_R appears in the principle, and not its time derivative one might expect from its appearance as the source term in the wave equation 15 for the acoustic pressure.

Equation 51 provides a generalized statement of Rayleigh's principle: $\Delta\mathscr{E}_n$ is positive, and hence the nth mode is unstable, if the sum of the two terms is positive. It is neither necessary nor sufficient that the first term alone be positive for the mode to be unstable. If that part is positive, then one may conclude only that heat addition, as represented by Q'_R, encourages the instability. Thus even if this contribution to $\Delta\mathscr{E}_n$ should be found experimentally to be positive, one cannot incontrovertibly conclude that the mechanism of the observed instability is associated with the heat addition. It is necessary at least to estimate the contribution from the term containing F_n^e. In practice, that may in fact be negative—and likely often is—owing to the damping processes such as the exhaust nozzle.

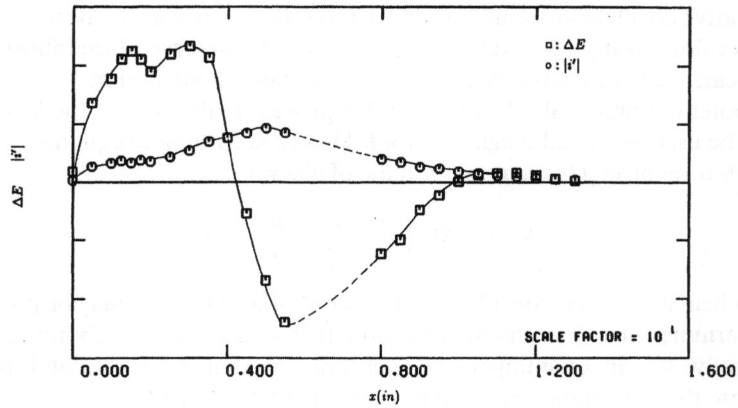

FIGURE 8.1

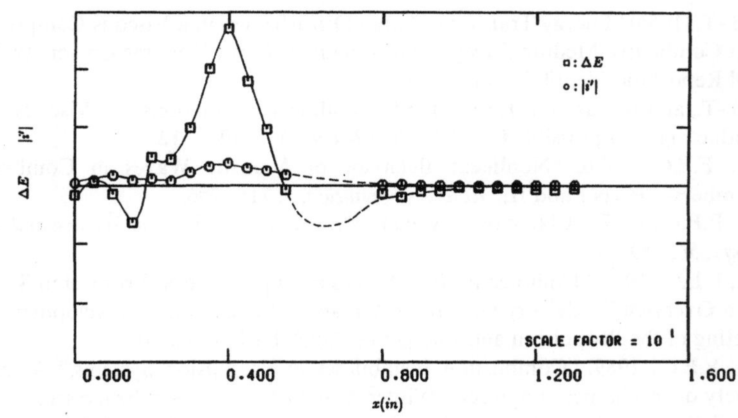

FIGURE 8.2

It appears that the best data available for the contribution from heat addition are those reported by Sterling (1987) and Sterling and Zukoski (1987). Both the fluctuating pressure and radiation intensity were measured simultaneously at many stations along the axis of a laboratory dump combustor having a cross section of 1 inch by 3 inches. All radiation in a one-quarter-inch length was collected at each location. Figures 8.1 and 8.2 show results obtained for two modes in the chamber (188 Hz and 535 Hz) obtained under different experimental conditions. In the first case, the maximum pressure amplitude is about $0.02\bar{p}$ and in the second nearly $0.09\bar{p}$. Correspondingly, $\Delta\mathscr{E}_n$ is nearly zero in the first case and substantially positive in the second.

Experimental uncertainty and incomplete knowledge of the relation between the radiation and heat addition prevent quantitative results but the qualitative conclusion seems correct: evaluation of Rayleigh's criterion in the manner described gives results consistent with the observed instabilities and it appears that volumetric heat addition is a major contribution.

A potential practical advantage of this procedure flows from the fact that it can be used as a local diagnostic tool. That is, according to equation 51 the contribution $\delta(\Delta\mathscr{E}_n)$ from a unit volume of chamber is

$$\delta(\Delta\mathscr{E}_n) = \delta V \frac{\bar{R}}{\bar{C}_v} \frac{\omega_n^2}{E_n^2} \int_t^{t+\tau_n} \frac{p_n}{\bar{p}} \frac{Q'_R}{\bar{p}} dt'$$

if only heat addition is considered. Thus with sufficient care it may be possible to determine where in a chamber the dominant cause of an instability resides. Naturally, for an unambiguous conclusion, one must be able at least to estimate the importance of those processes contained in F_n^e.

References

Chu, B.-T. (1956) "Energy Transfer to Small Disturbances in a Viscous Compressible Heat Conductive Medium," Dept. of Aeronautics, Johns Hopkins University Technical Report (no identifying number).

Chu, B.-T., and Kovasznay, L.S.G., 1957, "Nonlinear Interactions in a Viscous Heat-Conducting Compressible Gas," *J. Fluid Mech., 3*, 5, 494–512.

Culick, F.E.C., 1976, "Nonlinear Behavior of Acoustic Waves in Combustion Chambers—Parts I and II," *Acta Astronautica, 3*, 715–756.

Culick, F.E.C., 1987, "A Note on Rayleigh's Criterion," *Combustion Science and Technology, 56*, 159–166.

Culick, F.E.C., 1988, "Combustion Instabilities in Liquid-Fueled Propulsion Systems —An Overview," Advisory Group for Aerospace Research and Development, 72B Meeting of the Propulsion and Energetics Panel, Bath, England.

Culick, F.E.C., 1989, "Combustion Instabilities in Propulsion Systems," American Society of Mechanical Engineers, Winter Annual Meeting, San Francisco.

Culick, F.E.C., and Yang, V., 1990, "Prediction of Linear Stability in Solid Propellant Rockets," to appear in AIAA Progress Series Volume, *Nonsteady Burning and Combustion Stability of Solid Propellants.*

Doak, P.E., 1973, "Fundamentals of Aerodynamic Sound Theory and Flow Duct Acoustics," *J. Sound and Vibration, 28*, 3, 527–561.

Marble, F.E., 1963, "Dynamics of a Gas Containing Small Solid Particles," *Combustion and Propulsion*, 5th AGARD Colloquim, Pergamon Press, Braunschweig, Germany 175–213.

Putnam, A.A., 1971, *Combustion Driven Oscillations in Industry*, American Elsevier.

Lord Rayleigh, 1878a, "The Explanation of Certain Acoustic Phenomena," *Royal Institution Proceedings, VIII*, 536–542.

Lord Rayleigh, 1878b, *The Theory of Sound*, MacMillan and Co.; reprinted by Dover Publications, New York (1945).

Reynst, F.H., 1961, *Pulsating Combustion*, M.W. Thring, (Ed.) Pergamon Press, New York.

Sterling, J.D., 1987, "Longitudinal Mode Combustion Instabilites in Air Breathing Engines," *Ph.D. Thesis*, California Institute of Technology, Pasadena.

Sterling, J., and Zukoski, E.E., 1987, "Longitudinal Mode Instabilities in a Dump Combustor," *AIAA 25th Aerospace Sciences Meeting*, AIAA Paper No. 87-0220.

Tyndall, J., 1897, *Sound*, D. Appleton and Co., New York.

Yang, V., and Culick, F.E.C., 1990, "On the Existence and Stability of Limit Cycles for Transverse Acoustic Oscillations in a Cylindrical Combustion Chamber. I. Standing Modes," *Combustion Science and Technology*, Vol. 72, No. 1–3, p. 37.

Zinn, B.T., 1986, "Pulsating Combustion," In *Advanccd Combustion Methods*, F.J. Weinber (Ed.), Academic Press, London.

Zinn, B.T., and Powell, E.A., 1971, "Nonlinear Combustion Instability in Liquid Propellant Rocket Engines," *Proceedings of the 13th Symposium (International) on Combustion*, The Combustion Institute, 491–503.

II
Liquid Sprays

9
On the Anisotropy of Drop and Particle Velocity Fluctuations in Two-Phase Round Gas Jets

A. Tomboulides, M.J. Andrews, and F.V. Bracco

ABSTRACT: Drops and particles in steady fully developed gas jets exhibit anisotropic radial and axial velocity fluctuations and even models of two-phase flows that use isotropic $k - \varepsilon$ turbulence submodels generally reproduce the anisotropy. It is shown that the drop velocity fluctuations can be separated into radial fluctuations that depend mainly on gas turbulence and axial fluctuations that depend also on the drop radial motion across a mean drop velocity gradient. Drop equilibration time scales are defined and used to evaluate quantitatively the drop fluctuating velocities for different drop sizes, gas turbulence properties, and local mean drop velocity gradients.

Introduction

Anisotropic fluctuations of the radial and axial velocity exist in fully developed single-phase turbulent jets (Wygnanski and Fiedler 1969), in full-cone sprays (Andrews and Bracco 1989; Wu et al. 1969), and in round particle-laden jets (Shuen at al. 1985). Table 9.1 presents some measurements and computations of such fluctuations using an anisotropy parameter $A \equiv u'/v'$ where u' and v' are, respectively, the axial and radial RMS velocity fluctuations of either gas, particles, or drops. The table shows that the anisotropy of the Wygnanski and Fiedler gas jets ($A \approx 1.12$) is much less than that of the particle velocities of the particle-laden jets of Shuen et al. ($A \approx 3.5$), and the drops in the sprays of Wu et al. ($A \approx 2.5$). It is this large drop and particle anisotropy that is the subject of this chapter.

Particularly interesting are the computations of Shuen et al., that were performed with an isotropic $k - \varepsilon$ model and yet yielded anisotropic particle velocity fluctuations in good agreement with their measurements. The particle-laden jets of Shuen et al. comprised particles that were considerably larger and heavier than the drops in the Diesel-type sprays computed by Andrews and Bracco (1989), who used a similar $k - \varepsilon$ model and found isotropic drop-velocity fluctuations.

Hinze (1972) suggested the following mechanism for the anisotropy:

TABLE 9.1. Centerline anisotropy in jets and sprays.

Author	A_G	Comments
Wygnanski and Fiedler (1969)	1.12	Experiments: $A \equiv u'/v'$ $\rho_l/\rho_g = 1$
Shuen et al. (1985)	3.5	Computations: $A \equiv u'_p/v'_p$ Particle size = 79 μm $\rho_l/\rho_g = 2500.0$
Shuen et al. (1985)	5.0	Computations: $A \equiv u'_p/v'_p$ Particle size = 114 μm $\rho_l/\rho_g = 2500.0$
Wu et al. (1984) Case A	2.45	Measurement: $A \equiv u'_p/v'_p$ Drop size: not measured $\rho_l/\rho_g = 40.0$
Wu et al. (1984) Cases B & C	1.41	Measurement: $A \equiv u'_p/v'_p$ Drop size: not measured $\rho_l/\rho_g = 13.7$
Andrews and Bracco (1989)	1.0	Computations: $A \equiv u'_p/v'_p$ Drop size $\approx$ 20 μm $\rho_l/\rho_g = 40.0$

"When the mean velocity, and also the turbulence of the fluid, changes appreciably in adjacent regions, a particle coming from one region into the other may still remember the mean velocity and the turbulent motion it had in the first region. The result is that the turbulent spread of particles crossing the two regions will be different from that obtained with a response time of the particle small compared with the time scale of the change of its surrounding flow field. When we assume that the relative intensity of the turbulence does not differ much in the two regions, we may restrict ourselves with considering only the time scale for the rate of change of the mean flow. Let $\partial \bar{u}_g/\partial y$ be a mean-velocity gradient and v'_p the intensity of the particle turbulence in the y-direction, then the time scale for the rate of change of the mean velocity experienced by the particle when crossing the region considered may be estimated to be roughly

$$\left[\frac{v'_p}{\bar{u}_g} \cdot \frac{\partial \bar{u}_g}{\partial y}\right]^{-1} \text{"} \tag{1}$$

Hence a particle crossing the region will not assume the local mean velocity and exhibit additional fluctuations when

$$t_p \frac{v'_p}{\bar{u}_g} \cdot \frac{\partial \bar{u}_g}{\partial y} \gg 1 \tag{2}$$

Faeth (1986) referred to the same mechanism to explain the large anisotropy observed by Shuen et al. (1985). However the correctness, completeness, and accuracy of Hinze's suggestion do not seem to have been investigated. It is the purpose of this paper to show that the radial fluctuations of the drop (particle) velocity in full-cone sprays (round jets) are primarily due to the gas turbulence, and that the axial fluctuations are indeed influenced by Hinze's mechanism.

The objective will be achieved through schematic computations as well as through calculations of the far-field of steady, nonvaporizing, monodisperse spray jets. By the far-field of a full-cone spray, we mean the region downstream of the nozzle exit where most ($>$ 90 percent) of the injected momentum resides with the entrained gas, and the self-preserving profile of the mean axial drop and gas velocities have been achieved (Bracco 1985). Also in the far-field the volume occupied by the liquid is negligible in comparison with that occupied by the gas ($\theta \approx 1$). The computed mean velocities of the far-field have been shown to be insensitive to the physical and numerical accuracy of the near-field (Chatwani and Bracco 1985; Martinelli et al. 1985). Although our computations are for monodisperse sprays, the measurements of Wu et al. (1984) and the computations of Martinelli et al. (1985) suggest that the far-field starts around $x/d_n > 300$ to 400.

In the following sections we will outline the spray model used in the computations, consider separately radial and axial drop velocity fluctuations, and then summarize our conclusions.

The Spray Model

Introductory Comments

Our model of a spray jet originated with the stochastic parcel technique described by Dukowicz (1980) and extended by O'Rourke and Bracco (1980). The model uses a stochastic representation of drop motions in a spray, with a Eulerian description of the conservation laws governing the mean properties of the gas phase and a Lagrangian description for parcels of drops. Each parcel represents a group of drops that have the same physical properties, position, velocity, and current gas-eddy-velocity fiuctuation. Parcels are injected into the computational domain with the line source technique of Chatwani and Bracco (1985) to represent the finite core of the spray. The model then tracks parcels with Lagrangian equations that account for drag and gas phase turbulence. Gas turbulence is represented with the isotropic $k - \varepsilon$ model, and its effect on drops is included by associating with each parcel a gas velocity fluctuation for the eddy through which the parcel is moving. An isotropic gas velocity fluctuation is chosen at the moment of entry into an eddy, and subsequently drops experience the fluctuation via drag coupling. Sampling during the steady computation provides the computed velocity fluctuations.

We have concentrated on the coupling between droplet and gas velocities via drag and so for reference we give the Lagrangian equations for drop motion and the associated gas eddy submodel; the Eulerian gas equations of the model are available in Bracco (1985).

Lagrangian Equations of Drop Motion

The Lagrangian equations of motion for a drop in a turbulent gas field are for the axial, radial, and tangential directions, respectively:

$$\frac{du_p}{dt} = \frac{\Delta u}{t_p}$$
$$\frac{dv_p}{dt} = \frac{\Delta v}{t_p} \tag{3}$$
$$\frac{d(yw_p)}{dt} = \frac{y\Delta w}{t_p}$$

where the drag time scale $t_p = [C_D 3\rho_g|\Delta \underline{u}|/(8\rho_l r_p)]^{-1}$ and $\Delta \underline{u} = (\underline{\bar{u}}_g + \underline{u}'_g - \underline{u}_p)$.

For the present dilute sprays ($\theta \approx 1$), the drag coefficient C_D (O'Rourke 1980) is given by:

$$C_D(\theta, \mathrm{Re}_r) = \frac{24}{\mathrm{Re}_r}\left(1 + \frac{\mathrm{Re}_r^{2/3}}{6}\right) \tag{4}$$

with $\mathrm{Re}_r = 2\rho_g|\Delta \underline{u}|r_p/\mu_g$.

The Lagrangian equations (Eqs. 3) neglect pressure gradient, added mass, and Basset force terms because $\rho_l/\rho_g \gg 1$ in our sprays (Hinze 1972).

Drop-Eddy Interaction

The transit time of a drop through an eddy, t_r, is calculated from:

$$\int_t^{t+t_r} |\underline{\bar{u}}_g + \underline{\bar{u}}'_g - \underline{u}_p|\, dt = L_e \tag{5}$$

where L_e is the eddy size and is taken as the length scale $L_e = C_\mu^{3/4} k^{3/2}/\varepsilon$.

In addition, an eddy turn over time, t_e, is considered:

$$t_e = k/\varepsilon; \tag{6}$$

then the residence time of a drop in an eddy, t_d, is given by (Gosman and Ioannides 1981):

$$t_d = \min\{t_r, t_e\}. \tag{7}$$

Numerical Solution Procedure

The complexity of the spray model necessitates the use of a computational algorithm to obtain a solution. The model equations were solved by way of the spray code of O'Rourke (1981), who also gives details of the solution algorithm. The spray calculations described later used axisymmetry about the centerline of the spray and the 2-D computational grid shown in Figure 9.1. The grid is the same as that of Martinelli et al. (1985), where grid and time-step

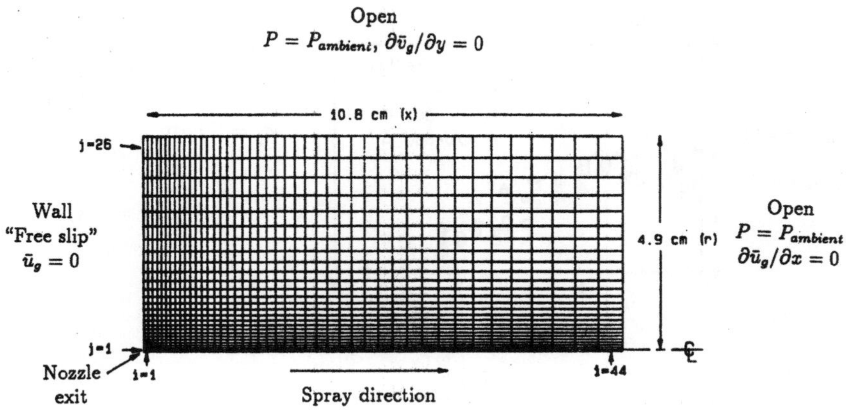

FIGURE 9.1. Computational grid and boundary conditions for Spray calculations.

independence were reported, as are the boundary conditions that are shown in Figure 9.1. The conditions were those of Case A of the experiments of Wu et al. (1984) ($P_g = 1.48$(MPa), $\rho_l/\rho_g = 40$, $\Delta P = 11$(MPa), $U_0 = 12.7$(cm/ms), liquid: n – hexane, $\rho_l = 665$ kg/m^3; gas: nitrogen at $T = 293$ K; $d_n = 127$ μm, $L_n/d_n = 4$), but as previously stated the computations used monodispersed drop sizes. Small initial values for k and ε of 1.7×10^{-8} were chosen to prevent underflow errors.

A spray computation began at the start of injection and continued until a steady state in the mean drop and gas velocity profiles was attained, typically 3/2 times the jet transit time, defined as the time taken for a drop to transit the axial length of the computational grid. The drop velocities were assumed to be statistically independent and at steady state, samples of drop velocities were periodically taken in each computational cell to obtain drop mean and RMS velocities in the axial and radial directions. A typical sample was taken over one jet transit time, with care taken to ensure the samples spanned the physical time scales of the problem, and contained some 200 data points.

Results

The results may be conveniently divided into radial and additional axial mechanisms. Radial drop velocity fluctuations are presented first and lead to consideration of supplementary effects in the axial direction.

Radial Drop Velocity Fluctuations

Figures 9.2 (a) and (b) show drop axial and radial RMS velocities from two calculations with our spray model that employed fixed drop diameters of 10 μm and 50 μm, respectively. Anisotropy in the 50 μm case is evident,

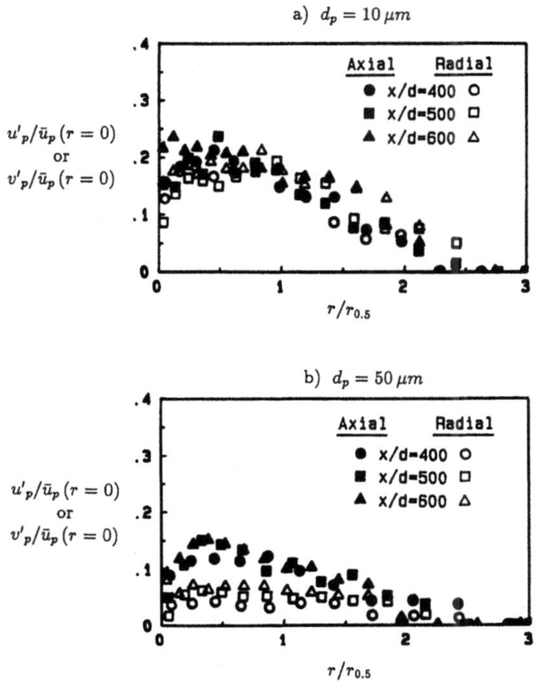

FIGURE 9.2. Drop axial and radial drop velocity fluctuations at $x/d_n = 300$; (a) $d_p = 10\ \mu$m; and (b) $d_p = 50\ \mu$m.

whereas the 10 μm drops exhibit almost isotropic velocity fluctuations. This suggests that the small drops ($\leq 10\ \mu$m) in our sprays follow the gas velocities, but a mechanism for anisotropy exists that is revealed by the larger drops.

Since the radial mean gas and drop velocities in the far field of a jet are much smaller than their respective fluctuations, the primary cause of radial drop velocity is the coupling of gas turbulence to drop motions via drag.

To better understand how drops respond to a gas fluctuation in our model we considered a simplified situation in which $\bar{u}_g = \bar{u}_p = 0$, negligible pressure gradient, and $t_e < t_r$; such conditions were verified in the far field of our sprays by drops whose diameter is less than 50 μm. Then we expect that:

$$\frac{v'_p}{u'_g} = \mathscr{F}(t_p/t_e). \tag{8}$$

In the far field ($\theta \approx 1$) the linearized drop relaxation time to the gas velocity is given by:

$$t_{p,0} = t_s \frac{1}{1 + \mathrm{Re}_{r_0}^{2/3}/6} \tag{9}$$

with $t_s = \dfrac{d_p^2}{18\nu_g}\dfrac{\rho_l}{\rho_g}$ the Stokes relaxation time, where Re_{r_0} is taken at the moment of entry into an eddy.

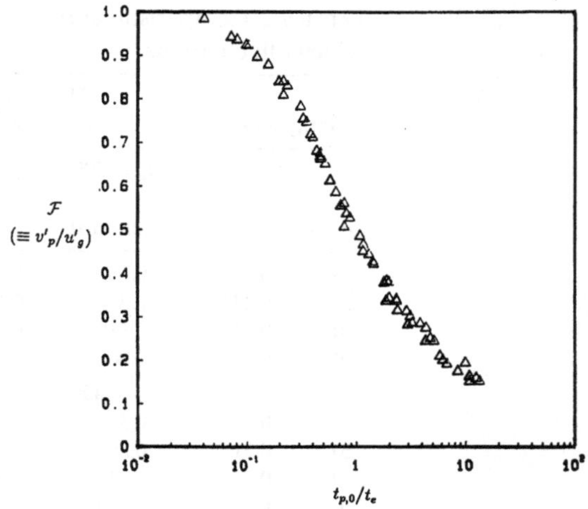

FIGURE 9.3. Drop equilibration in isotropic-homogeneous gas turbulence.

When $t_{p,0} \ll t_e$, drops respond quickly to a turbulent fluctuation and attain a high degree of equilibration with the gas velocity, whereas when $t_{p,0} \gg t_e$, drops are slow to respond and practically no equilibration results.

We evaluated the function $\mathcal{F}$ with a set of computations that used the following simplifying assumptions: a homogeneous isotropic gas field; mean gas and drop velocities set to zero; drops randomly distributed in a single, square, computational cell of side 1 cm.

With these assumptions, the velocity time history of 10^3 drops as they each successively entered and left, 50 model gas eddies were computed. Sampling of the drop velocities during a calculation furnished values for drop RMS velocities.

By varying independently the gas turbulence intensity, u'_g, and length scale, L_e, the function $\mathcal{F}$ was evaluated and is displayed in Fig. 9.3. The function shows that a good degree of equilibration with the gas velocity is attained by a drop for $t_{p,0}/t_e < 0.1$, and practically no equilibration for $t_{p,0}/t_e > 10$.

We now use our function $\mathcal{F}$ to check that drop radial velocity fluctuations are driven by gas turbulence in a monodisperse spray simulation. Four spray calculations were performed, each with a different drop size. From these calculations the maximum computed drop radial velocity fluctuations at $x/d_n = 300$ and 400 were taken with the gas fluctuation in the associated computational cell.

Data obtained from the spray calculations are summarized in Table 9.2 with the drop velocity fluctuation calculated with $\mathcal{F}$. The spray data and the corresponding values from our equilibration function, $\mathcal{F}$, compare well for all drop sizes, including those drops with $d_p > 50\ \mu$m where anisotropy occurs, and at both axial locations.

These results support our characterization of radial drop velocity fluctuations in monodisperse full-cone sprays as being caused by drag coupling with the underlying gas turbulence field.

TABLE 9.2. Comparison of radial drop-velocity fluctuations.

Spray model		Homogeneous isotropic model		
		$x/d_0 = 300$		
d_p (μm)	v'_p (cm/ms)	v'_p (cm/ms)	v'_p/u'_g	t_p/t_e
10	0.36	0.35	0.7	0.44
20	0.237	0.23	0.48	1.16
50	0.12	0.119	0.243	4.26
100	0.07	0.068	0.16	10.7
		$x/d_0 = 400$		
10	0.31	0.3	0.8	0.31
20	0.21	0.203	0.55	0.79
50	0.115	0.116	0.285	4.23
100	0.062	0.063	0.178	11.2

Axial Drop Velocity Fluctuations

Mechanism

Characterization of radial drop velocity fluctuations as being caused by gas turbulence, even when there are high levels of drop fluctuation anisotropy in an isotropic gas field, implies the existence of additional mechanisms in the axial direction that, for large drops, provide a significant supplementary axial velocity fluctuation. The supplementary mechanism we will explore is that of drop movement in the radial direction across the spray mean drop velocity profile as described by Hinze (1972) and Faeth (1986).

Hinze considered gradients of the mean gas velocity, however, the mean drop velocity gradient is used here because $\bar{u}_p = \bar{u}_g$ may be a poor assumption for larger drops (> 40 μm), and because drop velocity fluctuations are measured against the mean drop velocity. Thus we consider a mean drop velocity gradient $\partial \bar{u}_p/\partial y$, and the drop velocity fluctuation in the radial direction, v'_p. Then the time scale, t_m, for a change of u'_g in the mean velocity experienced by the drop when crossing the mean velocity gradient can be estimated as:

$$t_m \equiv \frac{u'_g}{v'_p} \bigg/ \frac{\partial \bar{u}_p}{\partial y}. \tag{10}$$

Qualitatively, when $t_p \ll t_m$ a drop crossing the region quickly adjusts to the local mean velocity; whereas when $t_p \gg t_m$, radial movement in conjunction with a poor response to the local mean conditions provides an additional axial drop velocity "fluctuation" for drops that come to the same location from different regions of the flow.

.A Calculation of Anisotropy with Stokes' Drag Law

To assess the proposed mechanism, a set of simplified "shear" computations was performed under the following conditions: a steady gas field of constant radial gradient in the mean velocity and no variation in the axial direction ($d\bar{u}_g/dy$ = constant); negligible pressure gradient; Stokes drag law (Eq. 9); a dilute drop field $\theta \approx 1$; a fixed drop size; no drop collisions; cyclic boundary conditions reintroduce drops that exit on the opposite boundary with a velocity shift that corresponds to the change in the mean velocity (this procedure preserves homogeneity of the problem); a steady homogeneous and isotropic gas turbulence properties; and initial drop velocities equal to the mean gas velocity.

The problem is displayed in Fig. 9.4. After releasing 10 drops randomly along the radial direction, their subsequent passage through gas eddies causes radial movement of drops across the shear and so generates anisotropic drop fluctuations. The calculation was run for several hundred t_s or t_e, depending on which is larger, with a time step Δt less than the smaller of $t_s/10$ or $t_e/10$. Samples were taken every 10 time steps, typically giving 10,000 data points from which statistics were formed. At steady state $\bar{u}_p = \bar{u}_g$ because homogeneity of the problem demands that there be invariance of mean velocities to coordinate translation and reflection. Hence in this problem the drops are equilibrated in the mean velocities but not in the fluctuating velocities.

Assuming the drop radial mean velocity is negligible, dimensional analysis indicates the following groups determine the problem:

Drop radial RMS velocities: v_p'/u_g', t_s/t_e.
Drop axial RMS velocities: u_p'/u_g', t_s/t_e, and $t_e d\bar{u}_p/dy$.

A series of calculations were performed for various u_g', t_e, and $t_e d\bar{u}_p/dy$. Figures 9.5 and 9.6 plot non-dimensionalized drop axial and radial velocity fluctuations against t_s/t_e as suggested by the dimensional analysis. The plots show that as $t_s/t_e \Rightarrow 0$ the drop RMS velocities equilibrate with the imposed gas RMS velocities. It is gratifying to observe that each set of points fall on a single curve in accord with the analysis, but unlike the radial plot in Fig. 9.5 the axial plot in Fig. 9.6 is a family of curves defined by $t_e d\bar{u}_p/dy = constant$, as shown in Fig. 9.7.

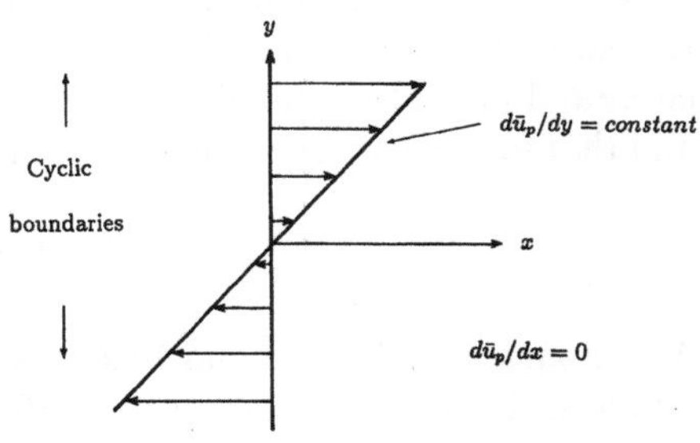

FIGURE 9.4. Schematic for the shear calculations.

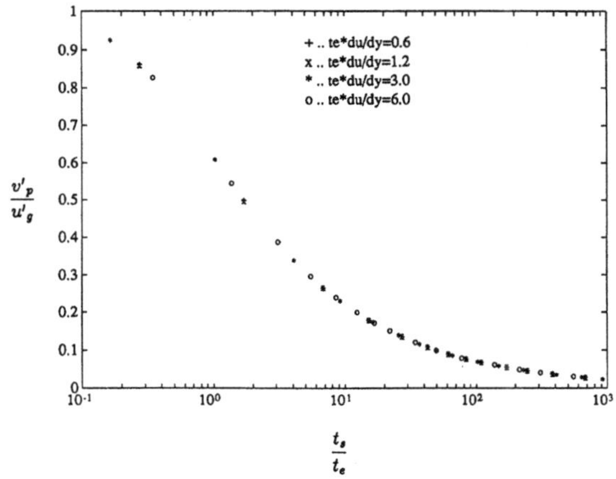

FIGURE 9.5. Radial drop RMS velocities for a Stokes drag law.

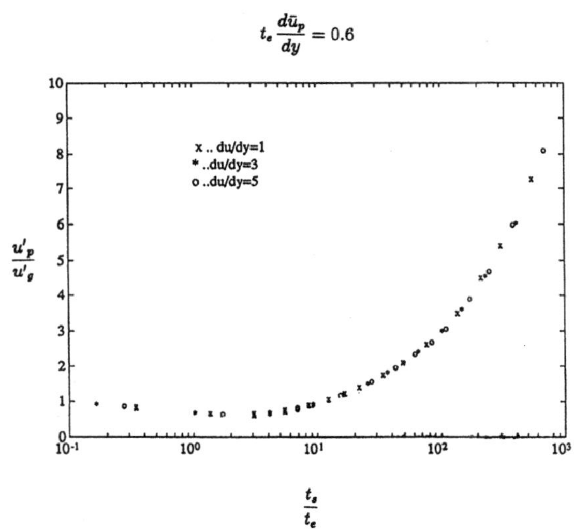

FIGURE 9.6. Axial drop RMS velocities for a Stokes drag law.

It would be satisfying to find a scaling that collapsed the family of curves for the drop axial RMS velocities to a single curve. A simple linear analysis suggested a scaling based on u'_p/v'_p and $t_s d\bar{u}_p/dy$ might collapse the family. Indeed as Fig. 9.8 shows, this scaling does collapse the family onto a single curve and shows that significant anisotropy ($u'_p/v'_p > 2$) occurs for $t_s d\bar{u}_p/dy > 2$, and that in this case the anisotropy parameter is given by:

$$\frac{u'_p}{v'_p} = 0.7\, t_s \frac{d\bar{u}_p}{dy} \tag{11}$$

A comparison of Eqs. 11 and 10 shows that the drag time scale also serves as a time scale for the radial transport effect.

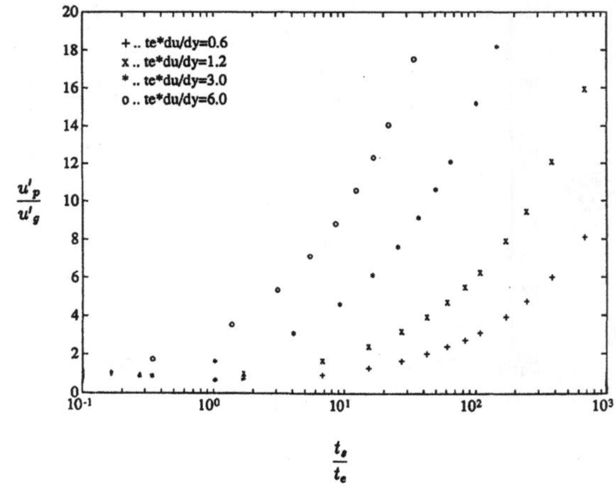

FIGURE 9.7. Family of axial drop RMS velocities curves for a Stokes drag law.

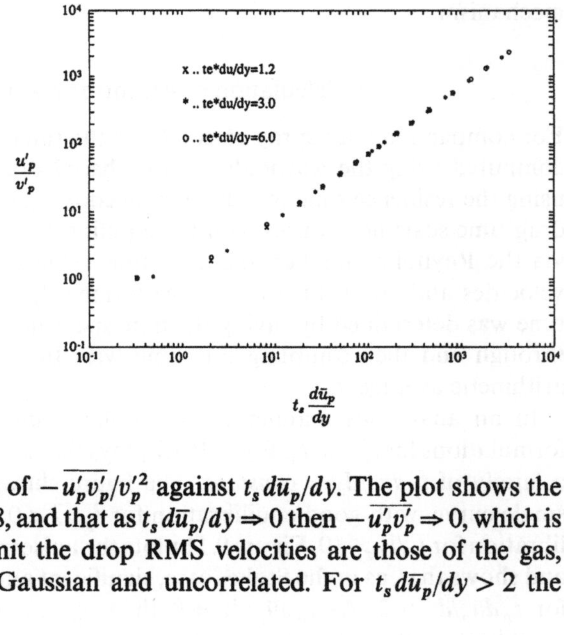

FIGURE 9.8. Anisotropy correlation of drop RMS velocities for a Stokes drag law.

Figure 9.9 shows a plot of $-\overline{u'_p v'_p}/v'^2_p$ against $t_s\, d\bar{u}_p/dy$. The plot shows the same collapse as in Fig. 9.8, and that as $t_s\, d\bar{u}_p/dy \Rightarrow 0$ then $-\overline{u'_p v'_p} \Rightarrow 0$, which is expected, since in this limit the drop RMS velocities are those of the gas, which in the model are Gaussian and uncorrelated. For $t_s\, d\bar{u}_p/dy > 2$ the correlation is given by:

$$-\overline{u'_p v'_p} = 0.5 D_p \frac{d\bar{u}_p}{dy} \tag{12}$$

where $D_p = v'^2_p t_s$ is the radial diffusivity of drop momentum in this shear calculation. For $t_s \frac{d\bar{u}_p}{dy} > 2$ division of Eq. 12 by Eq. 11 provides a constant value for the correlation function $-\overline{u'_p v'_p}/(u'_p v'_p)$ of 0.7.

Figures 9.5 and 9.8 completely describe the drop radial and axial RMS velocities for this Stokes drag law problem. The diffusivity D_p and Eq. 11 show

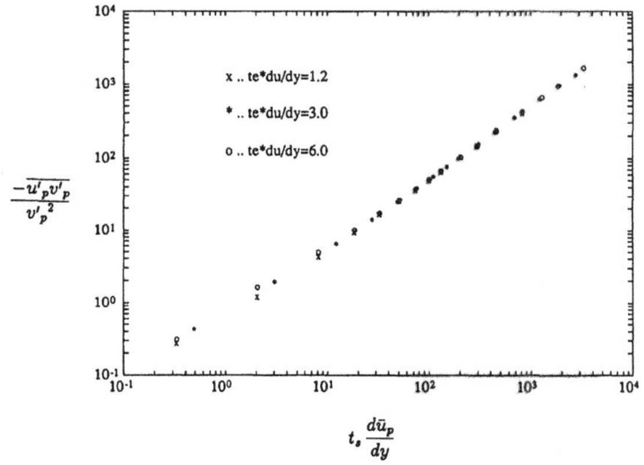

FIGURE 9.9. Stress correlation of drop RMS velocities for a Stokes drag law.

the important role that the drop radial RMS velocity plays in the anisotropic mechanism.

A Calculation of Anisotropy at Higher Re

For comparison with spray calculations the same "shear" problem has been computed using the Reynolds number based drag coefficient of Eq. 4, and using the residence time of a drop in an eddy, t_d, given by Eq. 7. Because the drag time scale now varies with the equilibration of drop and gas velocities via the Reynolds number, the drag time scale was sampled with the drop velocities and an arithmetic average formed, $\bar{t}_p$. An average drop residence time was determined by saving the time spent in the last eddy a drop passed through and then sampling this time with the drop velocities to form an arithmetic average, $\bar{t}_d$.

In an analogous manner to the Stokes case, and using the preceding formulations for $\bar{t}_p$ and $\bar{t}_d$, Fig. 9.10 displays the radial drop RMS velocities for a family of $\bar{t}_p d\bar{u}_p/dy = constant$. The figure shows a good collapse onto a single curve, with good equilibration for $\bar{t}_p/\bar{t}_d < 0.1$ and practically no equilibration for $\bar{t}_p/\bar{t}_d > 10$. Figure 9.11 plots the anisotropy u'_p/v'_p against $\bar{t}_p d\bar{u}_p/dy$ and shows that, as in the Stokes case, significant anisotropy ($u'_p/v'_p > 2$) occurs for $\bar{t}_p d\bar{u}_p/dy > 2$. As $\bar{t}_p d\bar{u}_p/dy \Rightarrow 0$ then $u'_p/v'_p \Rightarrow 1$ because drop velocities equilibrate well with gas velocities, and for $\bar{t}_p d\bar{u}_p/dy > 2$ the anisotropy is given by:

$$\frac{u'_p}{v'_p} = 0.5\, \bar{t}_p \frac{d\bar{u}_p}{dy}. \tag{13}$$

The correlation $-\overline{u'_p v'_p}/{v'_p}^2$ is plotted in Fig. 9.12 and is similar to the Stokes result with the correlation given by:

$$-\overline{u'_p v'_p} = 0.35\overline{D}_p \frac{d\bar{u}_p}{dy} \tag{14}$$

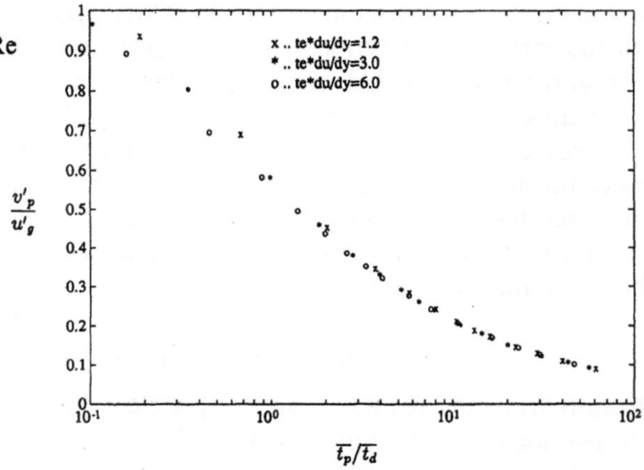

FIGURE 9.10. Radial drop RMS velocities for a high Re drag law.

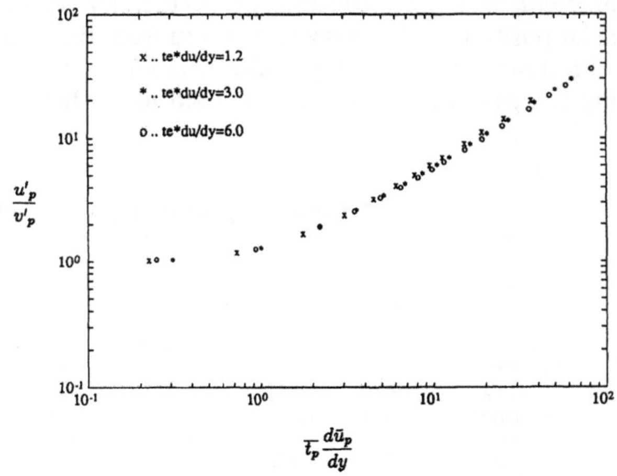

FIGURE 9.11. Anisotropy correlation of drop RMS velocities for a high Re drag law.

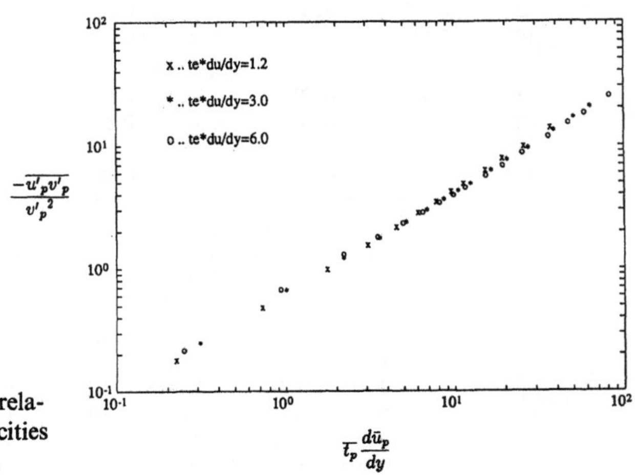

FIGURE 9.12. Stress correlation of drop RMS velocities for a high Re drag law.

where the drop radial momentum diffusivity, $\overline{D_p}$ given by $\overline{D_p} = v_p'^2 \overline{t_p}$. The value for the correlation function $-\overline{u_p' v_p'}/(u_p' v_p')$ when $\overline{t_p}\, d\overline{u}_p/dy > 2$ is again 0.7.

Figures 9.10 and 9.11 completely determine the anisotropy in this test problem and show the importance of the drop radial RMS velocities. The two test problems indicate that the scaling we have used reflects the physics of the mechanism embodied in the equations rather than the specific drag law and drop residence time formulations. So although the test problem results are specific to the model being used, the scaling of the drop RMS velocities may be more universal.

Comparison with Spray Calculations

Results from the regular drag law problem is compared in Table 9.3 with drop-velocity fluctuations computed from corresponding spray calculations at $x/d_n = 300$ and 400, which used different drop sizes, drop mean velocity gradient, and gas turbulence properties. The spray results at a particular axial position in the spray were taken from the computational cell with the largest mean drop velocity gradient, and it was this gradient and corresponding cell gas fluctuation that was used in the high Re drag law calculation.

TABLE 9.3. Comparison of axial and radial fluctuations.

	Axial			Radial		
	Spray	Shear		Spray	Shear	
			$x/d_n = 300$			
d_p (μm)	u_p' (cm/ms)	u_p' (cm/ms)	u_p'/u_g'	v_p' (cm/ms)	v_p' (cm/ms)	v_p'/u_g'
10	0.405	0.392	0.78	0.36	0.35	0.7
20	0.33	0.32	0.65	0.24	0.23	0.48
50	0.455	0.51	1.1	0.12	0.119	0.243
100	0.48	0.52	1.25	0.07	0.068	0.16
			$x/d_n = 400$			
10	0.33	0.328	0.87	0.31	0.3	0.8
20	0.255	0.248	0.67	0.21	0.203	0.55
50	0.362	0.38	0.93	0.115	0.116	0.285
100	0.36	0.372	1.05	0.062	0.063	0.178

Anisotropy: $A \equiv u_p'/v_p'$

	$x/d = 300$		$x/d = 400$	
$d_p(\mu\text{m})$	Spray	Shear	Spray	Shear
10	1.12	1.1	1.06	1.08
20	1.375	1.39	1.21	1.22
50	3.8	4.28	3.14	3.27
100	6.85	7.64	5.8	5.9

The radial drop-velocity fluctuations compare well at both axial locations, for all drop sizes and velocity gradients, confirming our previous radial results.

The table shows that the level of anisotropy A increases with drop size and velocity gradient. Comparison of axial drop velocity fluctuations computed from the spray model with the corresponding shear calculations shows them to be in good agreement, and most significantly, that the computed anisotropy from the spray and shear models shown in the bottom table compare well. The table shows that for large drops the relative axial drop velocity fluctuation (u_p'/u_g') increases with drop size because of radial drop motion coupled with slow equilibration to local conditions, whereas for radial drop velocity-fluctuations (v_p'/u_g') decreases because of poor equilibration.

Although the good agreement found here suggests that the primary mechanism for anisotropy in computed drop velocity fluctuations when using an isotropic model of gas turbulence is indeed radial motion across a mean velocity gradient, there may be additional secondary mechanisms that account for the small but increasing disparity shown in Table 9.3 between the spray and shear calculations of anisotropy for large drops. For example, the shear results that are compared with the spray results used a constant mean velocity gradient taken as the value in the computational cell of the spray, but drops in the spray experience a changing mean velocity gradient as they move radially and axially; consequently anisotropy in a computational cell of the spray is dependent on the anisotropy of surrounding cells. But our results indicate these effects may become significant at levels of large anisotropy.

Conclusions

Drop (particle) velocity fluctuations in steady fully developed turbulent round gas jets are caused by two primary mechanisms; first, drag on the drops couples them to the gas phase turbulence; and second, radial motion of drops across a mean drop velocity gradient with a slow equilibration rate to the local mean velocity provides an apparent additional axial fluctuation.

The magnitudes of the fluctuations are determined by the gas turbulence and by t_p/t_e, and $t_p\,d\bar{u}_p/dy$, which define the extent of the equilibration of a drop to the gas eddy and to a gradient in the mean drop velocity as the drop traverses the gradient.

Acknowledgments. The authors take the opportunity to thank Dr. J.M. MacInnes of the Engine Laboratory at Princeton, and Dr. J. Abraham of John Deere for their many useful discussions during this work. Funding for this work was provided by the Department of Energy, Office of Energy Utilization Research, Energy Conservation and Utilization Technologies Program (contract DE-AS-04-86AL33209, Mr. M.E. Gunn, Jr, contract monitor), General Motors Corp., Ford Motor Co., and Cummins Engine Co. Thanks are also due to the John von Neumann Supercomputer Center, and hence to the NSF, for providing a part of the computer time for the calculations presented.

Nomenclature

A = anisotropy parameter.
C_D = drag coefficient.
d_n = diameter of nozzle.
d_p = diameter of drop or particle.
$\mathscr{F}$ = equilibration function for a drop.
Re_r = drop Reynolds number ($= 2\rho_g r_p |\Delta \underline{u}|/\mu_g$).
r_p = radius of drop or particle.
$r_{0.5}$ = spray half-radius based on the average drop centerline velocity.
t_e = eddy turnover time.
t_m = characteristic time for drop radial motion.
t_p = drop characteristic relaxation time to the gas velocity.
U_0 = liquid velocity at injection.
u = axial velocity.
v = radial velocity.
w = tangential velocity.
x = axial distance from nozzle.
y = radial distance from centerline of spray.
μ = viscosity.
ν = kinematic viscosity.
ρ = density.
θ = volume-fraction of gas phase.

Subscripts

p = drop or parcel value.
g, l = gas and liquid/particle.

Superscripts

$(')$ = RMS value of quantity.
$(^{-})$ = time mean value of quantity.

References

Andrews, M.J., and Bracco, F.V., 1989, "On the Structure of Turbulent Dense Spray Jets," N.P. Cheremisinoff (Ed.), *Encyclopedia of Fluid Mechanics, 8.*

Bracco, F.V., 1985, "Modelling of Engine Sprays," *SAE paper 850394.*

Chatwani, A.U., and Bracco, F.V., 1985, "Computation of Dense Spray Jets," *ICLASS-85*, paper 1B/1/1, London.

Dukowicz., J.K., 1980, "A Particle-Fluid Numerical Model for Liquid Sprays," *J. Comp. Phys, 35.*

Faeth, G.M., 1986, "Turbulence/Drop Interactions in Sprays," *AIAA paper 86-0136.*

Gosman, A.D., and Ioannides, E., 1981, "Aspects of Computer Simulation of Liquid-Fueled Combustors," *AIAA paper 81-0323.*

Hinze, J.O., 1972, "Turbulent Fluid and Particle Interactions," *Prog. Heat and Mass Transfer, 6*, 433–452.

Martinelli, L., Bracco, F.V., and Reitz, R.D., 1985, "Comparisons of Computed and Measured Dense Spray Jets," *Progress in Astronautics and Aeronautics, 95.*

O'Rourke, P.J., 1981, "Collective Drop Effects on Vaporizing Liquid Sprays," *Ph.D. thesis*, MAE Dept., Princeton University.

O'Rourke, P.J., and Bracco, F.V., 1980, "Modelling of Drop Interactions in Thick Sprays and a Comparison with Experiments," *Publication 1980–9*, Institute of Mechanical Engineers, London, England.

Shuen, J.S., Solomon, A.S.P., Zang, Q.-F., and Faeth, G.M., 1985, "Structure of Particle-Laden Jets: Predictions and Measurements," *AIAA J., 23*, 396–404.

Wu, K.-J., Santavicca, D.A., Bracco, F.V., and Coghe, A., 1984, LDV Measurements of Drop Velocity in Diesel-Type Sprays," *AIAA J., 22*, 9.

Wygnanski, I., and Fiedler, H., 1969, "Some Measurements in the Self-Preserving Jet," *Journal of Fluid Mech., 38*, 3.

10
Unsteady, Spherically-Symmetric Flame Propagation Through Multicomponent Fuel Spray Clouds

G. CONTINILLO AND W.A. SIRIGNANO

ABSTRACT: The flame propagation through a fuel spray–air mixture in a spherical geometry is investigated by means of a one-dimensional unsteady analysis with a hybrid Eulerian-Lagrangian formulation. Finite-difference numerical schemes have been employed, with nonuniform grid spacing and an adaptive time step. Multicomponent sprays are considered. Emphasis is given to: the presence and role of diffusion and premixed flames; the movement of the droplets due to the expansion of hot gases and the resulting stratification; the effect of rapid vaporization of more volatile components; and the influence of the droplet size on droplet time history in a spray flame. More volatile fuels produce faster flame propagation. Nonuniform vapor fuel composition is generated due to the different volatilities of the components of the liquid fuel spray. Increasing the droplet size causes strong local deviation from the initially uniform equivalence ratio, due to the relative motion of the two phases. Flames generally have complex premixed and diffusion structures. Emphasis is given to flames propagating through unconfined domains.

Introduction

Structure and propagation mechanisms of laminar, one-dimensional flames in a premixed combustible mixture have been extensively studied (Williams 1985) and theoretical treatment using large activation energy asymptotics covers, at least for simplified chemistry, a wide range of conditions. The corresponding problem of two-phase, liquid-spray laminar flame propagation is more complicated, in that it involves a number of phenomena in addition to those already present in the gaseous case: droplet heat-up and vaporization; droplet drag for sufficiently large droplets; and liquid mass diffusion for multicomponent fuels.

As a consequence, only a limited set of situations can be covered by a closed-form mathematical approach. For example, when fuel prevaporization is negligible, the so-called heterogeneous flame propagation is found, treated in Williams (1985). More recentlv, off-stoichiometric situations have been analyzed by-means of matched asymptotic analysis (Lin and Law 1988), allowing for droplet pre- and post-vaporization. However, to perform such analysis (Lin and Law 1988), near-stoichiometric situations remain too complex and require separate study. Numerical analysis is therefore the only means currently available to treat most laminar spray flame propagation problems, even in simple geometrical situations.

An interesting situation is the unsteady propagation of a flame in an initially quiescent liquid spray. The flame propagates with spherical symmetry centered at the location of the ignition source. Some of the observed parameters e.g., pressure-time evolution in confined systems and flame propagation speed, are very sensitive to most of the physical and chemical parameters of the gas and the liquid phase, and this makes such systems suitable for the investigation of the relative importance of the factors of interest. The gas phase is spatially resolved on a scale smaller than the droplet spacing but larger than the droplet size; this allows a description of the flame structure that has not been attained by previous spray flame studies. On the other hand, the need for the development and validation of a droplet-scale model, to be used in the most complicated, multidimensional unsteady simulations, has been recognized (Sirignano 1988). For this purpose, these problems are particularly attractive for modeling, since, by choosing a simple geometry, they allow more details to be included in modeling droplet-gas interactions while requiring reasonable computational time for simulation of the whole phenomenon.

A numerical study of a one-dimensional closed combustor was conducted by Seth et al. (1980). This study employed a continuum (Eulerian) formulation for both the gas and the liquid phase; many physical variables were parametrically varied, such as fuel type, initial temperature, initial droplet size, stoichiometric ratio, along with some model parameters, such as the activation energy, the preexponential factor in the Arrhenius-type expression for the combustion rate, and the gas diffusivity. Subsequently, Aggarwal et al. (1983) presented a hybrid Eulerian-Langrangian formulation, which was used to investigate the behavior of different vaporization models (Aggarwal et al. 1984), and to predict and understand better some of the physics associated with the effect of droplet spacing, both for the ignition (Aggarwal and Sirignano 1985a) and the subsequent flame propagation (Aggarwal and Sirignano 1985b), for a physical system that was essentially the same as in Seth et al. (1980).

The cited studies, conducted for planar geometry, were limited to single-component fuels. Results of unsteady calculations conducted for a constant-pressure, nonreacting case (Aggarwal 1987) showed that significant differences are observed for multicomponent fuels. The current authors, in a previous study (Continillo and Sirignano 1988), computed flame propagation for a multicomponent fuel in a closed-volume configuration.

This study represents an extension of the previous work in that it refers to spherically symmetric geometry and multicomponent fuels in open environments. In addition, the variation of the specific heats in the gas phase is taken into account, and variable properties in the gas surrounding the droplet, as in Abramzon and Sirignano (1989), are also considered. Implicit evaluation of the chemical reaction contributions, nonuniform grid spacing, and automatic time-step control have been introduced.

The goal of this work is to gain deeper insight into some important aspects of the phenomenon, such as: the presence and respective roles of diffusion and premixed flames in spray flame propagation; the movement of droplets due to expansion of hot gases, which causes stratification; and the effect of rapid vaporization of more volatile components.

Mathematical Formulation

Gas-Phase Model

The mathematical model presented here has been formulated in order to describe open- as well as closed-volume configurations. This explains the presence of the time derivative of pressure in the model equations. The main assumptions made and the model equations for the gas phase, including balance of mass, species, momentum, and energy, are essentially the same as in Seth et al. (1980) but modified for a spherically symmetric system. The whole gas-phase problem is treated as unsteady. The flow is assumed to be one-dimensional, laminar, and spherically symmetric. It is assumed that the viscous dissipation rate is negligible and that the pressure is constant along the space coordinate. The model can account for pressure variations in time for confined deflagrations. In open environments, the lowest approximation gives a constant pressure. The gas mixture is assumed to .be thermally perfect. Binary diffusion coefficients for each pair of species are taken to be equal, and thermal mass diffusion is neglected. Fick's law for mass diffusion and Fourier's law for heat conduction are used. The diffusion coefficient is assumed to vary with temperature and pressure, in order to keep ρD constant. The thermal conductivity is directly related to the mass diffusivity through the assumption of constant unity Lewis number. Radiative heat transfer is neglected. The combustion chemistry is described by means of a single-step irreversible reaction of each fuel vapor species with oxygen.

Gas-phase equations include coupling terms accounting for mass, momentum, and energy exchanges between the two phases. The model equations are written for polydisperse sprays: droplets are subdivided into K groups, thus accounting for initial differences in diameter, temperature, and composition. The equations for the gas phase are:

Continuity:

$$\frac{\partial \rho}{\partial t} + \frac{1}{r^2}\frac{\partial}{\partial r}(r^2 \rho u) = \sum_{k=1}^{K} n_k \dot{m}_k \tag{1}$$

Species:

$$\frac{\partial}{\partial t}(\rho Y_j) + \frac{1}{r^2}\frac{\partial}{\partial r}(r^2 \rho u Y_j) - \frac{1}{r^2}\frac{\partial}{\partial r}\left(r^2 \rho D \frac{\partial}{\partial r} Y_j\right) = \sum_{k=1}^{K} \delta_{jF_m} n_k \dot{m}_k \varepsilon_{mk} + \dot{w}_j \tag{2}$$

$$j = F_1, \ldots, F_M, O, P$$

with

$$\dot{w}_{F_m} = G_m (Y_{F_m})^{\alpha_{F_m}} (Y_O)^{\alpha_{O_m}} \exp(-E_m/RT) \tag{3}$$

where

$$G_m = \frac{(m)_{F_m}}{(m)_{F_m}^{\alpha_{F_m}} (m)_{O}^{\alpha_{O_m}}} A_m \rho^{(\alpha_{F_m} + \alpha_{O_m})}$$

$$\dot{w}_O = (m)_O \sum_{m=1}^{M} \frac{\nu_{O_m} \dot{w}_{F_m}}{(m)_{F_m}} \tag{4}$$

$$-\dot{w}_P = (m)_P \sum_{m=1}^{M} \frac{\nu_{P_m} \dot{w}_{F_m}}{(m)_{F_m}} \tag{5}$$

Chemical reactions:

$$F_m + \nu_{O_m} O \rightarrow \nu_{P_m} P \qquad m = 1, \ldots, M$$

Energy:

$$\rho C_p \frac{\partial T}{\partial t} + \rho C_p u \frac{\partial T}{\partial t} - \frac{1}{r^2}\frac{\partial}{\partial r}\left(r^2 \rho D C_p \frac{\partial T}{\partial r}\right) - \frac{dp}{dt} = -\sum_{m=1}^{M} Q_m \dot{w}_{F_m} - \sum_{k=1}^{K} n_k \dot{q}_{Lk} \tag{6}$$

Momentum:

$$\frac{\partial p}{\partial r} = 0 \tag{7}$$

Ideal gas state equation:

$$p = \rho R T \sum_j \frac{Y_j}{(\bar{m})_j} = \frac{\rho R T}{(\bar{m})}. \tag{8}$$

Expressions for $\dot{m}_k$ and $\dot{q}_{Lk}$ will be considered in the droplet analysis.

Integrating Eq. 1 over $r^2\,dr$ yields

$$u(r) = \frac{1}{r^2 \rho(r)}\left[\int_0^r \sum_{k=1}^{K} n_k \dot{m}_k r^2\,dr - \frac{\partial}{\partial t}\int_0^r \rho r^2\,dr\right], \tag{9}$$

which, combined with Eq. 8 and using the boundary condition for the velocity at the spherical wall in the confined case, gives

$$\left[\int_0^R \frac{(\bar{m})}{RT} r^2\,dr\right]\frac{dp}{dt} + \left[\frac{\partial}{\partial t}\int_0^R \frac{(\bar{m})}{RT} r^2\,dr\right] p = \int_0^R \sum_{k=1}^{K} n_k \dot{m}_k r^2\,dr \tag{10}$$

In the case of the unconfined or open environment, Eq. 10 is not employed and the pressure remains uniform in space and constant in time as the first approximation. By using the transformation

$$\phi = Tp^{(1-\gamma)/\gamma} = Tp^{\Gamma} \tag{11}$$

and neglecting the derivatives of the specific heats, Eq. 6 can be rewritten as

$$\rho C_p\left[\frac{\partial\phi}{\partial t} + u\frac{\partial\phi}{\partial r} - \frac{1}{r^2}\frac{\partial}{\partial r}\left(r^2\rho D\frac{\partial\phi}{\partial r}\right)\right] = -p^{\Gamma}\left[\sum_{m=1}^{M} Q_m\dot{w}_{F_m} + \sum_{k=1}^{K} n_k\dot{q}_{Lk}\right] \tag{12}$$

so that the time derivative of the pressure disappears. After introducing Eq. 1, Eq. 2 can be rewritten as:

$$\rho\frac{\partial Y_i}{\partial t} + \rho u\frac{\partial Y_i}{\partial r} - \frac{1}{r^2}\frac{\partial}{\partial r}\left(r^2\rho D\frac{\partial Y_i}{\partial r}\right) = \sum_{k=1}^{K}(\delta_{iF_m}\varepsilon_{mk} - Y_i)n_k\dot{m}_k + \dot{w}_j. \tag{13}$$

The initial conditions for the gas-phase equations are:

$$\begin{gathered} Y_j(r,0) = Y_{j0};\quad \phi(r,0) = \phi_0 = T_0 p_0^{\Gamma_0} \\ u(r,0) = 0;\quad p(0) = p_0. \end{gathered} \tag{14}$$

The boundary conditions for the gas include a symmetry condition at the center of the sphere and zero-gradient conditions at the far boundary:

$$\begin{gathered} \left.\frac{\partial Y_i}{\partial r}\right|_{r=0} = 0;\quad \left.\frac{\partial Y_i}{\partial r}\right|_{r=R} = 0 \\ \left.\frac{\partial T}{\partial r}\right|_{r=0} = 0;\quad \left.\frac{\partial T}{\partial r}\right|_{r=R} = 0 \\ u(0,t) = 0. \end{gathered} \tag{15}$$

The boundary condition for the velocity at the far boundary is not necessary where the pressure is taken constant, that is, in the case of open environment.

Droplet Model

The transient behavior of a droplet is considered by means of a simplified model. The distribution of temperature and species concentrations inside the droplet are assumed to be spherically symmetric, and are calculated by means of a "conduction limit" model and a "diffusion limit" model, respectively. The diffusion coefficients are taken equal and constant, as for the thermal conductivity.

Species:

$$\frac{\partial Y_m}{\partial t} = \frac{D_l}{a^2}\frac{\partial}{\partial a}\left(a^2\frac{\partial Y_m}{\partial a}\right) \qquad m = 1, \ldots, M-1 \tag{16}$$

Energy:

$$\frac{\partial T}{\partial t} = \frac{Le_l D_l}{a^2}\frac{\partial}{\partial a}\left(a^2\frac{\partial T}{\partial a}\right) \tag{17}$$

Initial conditions:

$$\begin{aligned} Y_m(a,0) &= Y_{m0} \qquad m = 1,\dots, M-1 \\ T(a,0) &= T_{l0} \\ a_s(0) &= a_{s0} \end{aligned} \tag{18}$$

Boundary conditions:

$$\left.\frac{\partial T}{\partial a}\right|_{a=0} = \left.\frac{\partial Y_m}{\partial a}\right|_{a=0} = 0 \quad \text{(symmetry)} \tag{19}$$

$$\left.\frac{\partial Y_m}{\partial a}\right|_{a=a_s(t)} = \frac{\dot{m}[Y_{msl} - \varepsilon_m]}{4\pi a_s^2 \rho_l D_l}, \qquad m = 1,\dots, M-1 \tag{20}$$

$$\left.\frac{\partial T}{\partial a}\right|_{a=a_s(t)} = \frac{\dot{q}_L}{4\pi a_s^2 \lambda_l} \tag{21}$$

In the boundary conditions, $\dot{m}$ is the mass vaporization rate for a single droplet, $\dot{q}_L$ is the incoming heat flux to the droplet interior, and ε_m is the fractional vaporization rate of component m. This is a moving boundary problem. By means of the following variable transformation, it is recast into a fixed boundary problem:

$$\eta = \frac{a}{a_s(t)}, \qquad \eta_s = \frac{a_s(t)}{a_{s0}}. \tag{22}$$

For Eq. 16 we use the following temporal variable:

$$\tau_y = D_l \int_0^t \frac{d\theta}{a_s^2(\theta)} \tag{23}$$

so that

$$\frac{\partial Y_m}{\partial \tau_y} - \left(\frac{1}{\eta_s}\frac{d\eta_s}{d\tau_y}\eta\right)\frac{\partial Y_m}{\partial \eta} = \frac{1}{\eta^2}\frac{\partial}{\partial \eta}\left(\eta^2\frac{\partial Y_m}{\partial \eta}\right) \tag{24}$$

$$Y_m(\eta, 0) = Y_{m0} \tag{25}$$

$$\left.\frac{\partial Y_m}{\partial \eta}\right|_{\eta=0} = 0 \tag{26}$$

$$\left.\frac{\partial Y_m}{\partial \eta}\right|_{\eta=1} = \frac{\hat{\dot{m}}(Y_{msl} - \varepsilon_m)}{3\eta_s} \tag{27}$$

where

$$\hat{\dot{m}} = \frac{\dot{m}}{\frac{4}{3}\pi a_{s0}^3 \rho_l} \frac{a_{s0}^2}{D_l}. \tag{28}$$

For Eq. 17 we use the following temporal variable:

$$\tau_T = Le_l D_l \int_0^t \frac{d\theta}{a_s^2(\theta)}, \tag{29}$$

yielding

$$\frac{\partial \hat{T}}{\partial \tau_T} - \left(\frac{1}{\eta_s}\frac{d\eta_s}{d\tau_T}\eta\right)\frac{\partial \hat{T}}{\partial \eta} = \frac{1}{\eta^2}\frac{\partial}{\partial \eta}\left(\eta^2 \frac{\partial \hat{T}}{\partial \eta}\right) \tag{30}$$

where $\hat{T} = T/T_{l0}$

$$\hat{T}(\eta, 0) = 1 \tag{31}$$

$$\left.\frac{\partial \hat{T}}{\partial \eta}\right|_{\eta=0} = 0 \tag{32}$$

$$\left.\frac{\partial \hat{T}}{\partial \eta}\right|_{\eta=1} = \frac{1}{Le_l}\frac{\hat{\dot{q}}_L}{3\eta_s} \tag{33}$$

and

$$\hat{\dot{q}}_L = \frac{\dot{q}_L}{\frac{4}{3}\pi a_{s0}^3 \rho_l C_{pl} T_{l0}} \cdot \frac{a_{s0}^2}{D_l}. \tag{34}$$

The values of $\dot{m}$, $\dot{q}_L$, and ε_m are provided by the analysis of the thermal and diffusion processes in the gas phase surrounding each individual droplet.

This analysis and that of the droplet motion is taken from Abramzon and Sirignano (1989) and extended to a multicomponent vaporizing droplet. The so-called film theory is employed, and the effect of the Stefan flow on the thickness of the film is taken into account. The analysis of the liquid phase in this calculation does not follow that of Abramzon and Sirignano (1989), where an effective conductivity was employed. Only the models for the gas film and the droplet motion are used. According to this theory, the mass vaporization rate can be expressed as:

$$\dot{m} = 2\pi(\overline{\rho D})_g a_s Sh B_M \tag{35}$$

where

$$B_M = \frac{Y_{Fgs} - Y_{Ff\infty}}{1 - Y_{Ffs}} \tag{36}$$

is the Spalding mass transfer number, and

$$Sh = \frac{2a_s\left(-\dfrac{dY_{Fg}}{da}\right)^s}{Y_{Fgs} - Y_{Fg\infty}} \tag{37}$$

is the actual Sherwood number of an evaporating droplet. Here it is simply

$$Y_{Fg} = \sum_{m=1}^{M} Y_{F_m g}$$

It has been shown (Abramzon and Sirignano 1989) that

$$Sh = \left(2 + \frac{Sh_0 - 2}{F(B_M)}\right)\frac{\ln(1 + B_M)}{B_M} = Sh^* \frac{\ln(1 + B_M)}{B_M} \tag{38}$$

where

$$F(B) = (1 + B)^{0.7}\frac{\ln(1 + B)}{B} \tag{39}$$

and

$$Sh_0 = 1 + (1 + \mathrm{Re}\, Sc)^{1/3} F_1(\mathrm{Re}) \tag{40}$$

is the Sherwood number for a solid nonvaporizing sphere, with

$$F_1(\mathrm{Re}) = [\max(1, \mathrm{Re})]^{0.077}. \tag{41}$$

$\mathrm{Re} = 2\rho_\infty u a_s/\mu_f$ is the Reynolds number evaluated with the average (1/3 rule) film viscosity and the free-stream gas density, $Sc = \mu_f/(\overline{\rho D})_f$ is the Schmidt number evaluated at the average (1/3 rule) film conditions.

The total heat transferred into the droplet interior is (Abramzon and Sirignano 1989):

$$\dot{q}_L = \dot{m}\left[\frac{C_{pf}(T_\infty - T_s)}{B_T} - L(T_s)\right] \tag{42}$$

where

$$B_T = (1 - B_M)^\phi - 1 \tag{43}$$

with

$$\Phi = \left(\frac{C_{pFl}}{\bar{C}_{pf}}\right)\frac{Sh^*}{Nu^*}\cdot\frac{1}{Le_g}. \tag{44}$$

Sh^* comes from Eq. 38, Nu^* is the analogous

$$Nu^* = 2 + \frac{Nu_0 - 2}{F(B_T)}. \tag{45}$$

$Pr = \dfrac{C_{pf}\mu_f}{\lambda_f}$ is the Prandtl number evaluated at the average (1/3 rule) film conditions; $L(T_s)$ is the latent heat of vaporization for the total vaporizing fuel, which is expressed by

$$L = \sum_m \varepsilon_m L_m \tag{46}$$

where

$$L_m(T_s) = L_{\text{boil},m}\left(\frac{T_{\text{crit},m} - T_s}{T_{\text{crit},m} - T_{\text{boil},m}}\right)^{0.38}. \tag{47}$$

$T_{\text{crit},m}$ is the critical temperature for component m; $T_{\text{boil},m}$ is the boiling temperature for component m at normal conditions; and $L_{\text{boil},m}$ is the latent heat of vaporization of species m at normal conditions. ε_m, which also appears in the boundary conditions of Eq. 20, is the fractional vaporization rate for component m:

$$\varepsilon_m = Y_{mgs} + \frac{(Y_{mgs} - Y_{mg\infty})(1 - Y_{Fgs})}{(Y_{Fgs} - Y_{Fg\infty})} \tag{48}$$

The mass fractions at the surface in the gas phase, Y_{mgs}, are obtained by means of the phase equilibrium assumption and the Clausius-Clapeyron relationship for the saturated vapor pressure:

$$Y_{mgs} = \frac{X_{mgs}(m)_m}{\sum_j X_{jgs}(m)_j} \tag{49}$$

$$X_{mgs} = X_{mls}\frac{1}{p}\exp\left(-\frac{C_{1m}}{T_s} + C_{2m}\right). \tag{50}$$

The droplet motion is treated as one-dimensional, since the initial velocity of the droplet is set equal to zero; then gas and droplet velocities are parallel, and the scalar equation

$$\frac{dv}{dt} = \frac{3}{16}\frac{\mu_f}{\rho_l}\frac{(u_\infty - v)}{a_s^2}\text{Re}\, C_D \tag{51}$$

is sufficient, where

$$C_D = \frac{24}{\text{Re}}\left[1 + \frac{\text{Re}^{2/3}}{6}\right] \tag{52}$$

from Faeth (1977). The instantaneous position of a droplet, r_d, is calculated from

$$\frac{d(r_d)}{dt} = v. \tag{53}$$

Free stream values of temperature, pressure, species concentration, and velocity are those of the gas-phase model equations, evaluated at the current droplet location.

Nondimensionalization

The dimensional unknown quantities are nondimensionalized with respect to their initial values, whenever possible. The physical parameters are treated the same way. The following reference quantities are defined:

$$r_c = R; \quad t_c = \frac{r_c^2}{D_0} = \frac{R^2}{D_0}; \quad u_c = \frac{r_c}{t_c} = \frac{D_0}{R};$$

furthermore, the nondimensional number density $\hat{n}_k = n_k R^3$, the nondimensional heats of combustion $\hat{Q}_m = Q_m / C_{p0} T_0$, the nondimensional mass production rates $\hat{\dot{w}}_j = \dot{w}_j \cdot (R^2/\rho_0 D_0)$ are defined, and the following nondimensional equations are derived, remembering Eqs. 28 and 34. The superscript "^" is dropped for simplicity.

Continuity:

$$\frac{\partial \rho}{\partial t} + \frac{1}{r^2}\frac{\partial}{\partial r}(r^2 \rho u) = N_1 \sum_{k=1}^{K} n_k \dot{m}_k \tag{54}$$

where

$$N_1 = \frac{4}{3}\pi \frac{a_{s0}}{R} \frac{\rho_l}{\rho_0} \frac{D_l}{D_0} \tag{55}$$

Species (from Eq. 13):

$$\rho \frac{\partial Y_j}{\partial t} + \rho u \frac{\partial Y_j}{\partial r} - \frac{1}{r^2}\frac{\partial}{\partial r}\left(r^2 \frac{\partial Y_j}{\partial r}\right) = N_1 \sum_{k=1}^{K} (\delta_{jF_m} \varepsilon_{mk} - Y_j) n_k \dot{m}_k + \dot{w}_j \tag{56}$$

Energy (from Eq. 12):

$$\rho \frac{\partial \phi}{\partial t} + \rho u \frac{\partial \phi}{\partial r} - \frac{1}{r^2}\frac{\partial}{\partial r}\left(r^2 \frac{\partial \phi}{\partial r}\right) = -\frac{p^\Gamma}{C_p}\left(N_1 N_2 \sum_{k=1}^{K} n_k \dot{q}_{Lk} + \sum_{m=1}^{M} Q_m \dot{w}_{F_m}\right) \tag{57}$$

where:

$$N_2 = \frac{C_{pl} T_{l0}}{C_{p0} T_0} \tag{58}$$

Ideal gas state equation (from Eq. 8):

$$p = \rho \frac{(\bar{m})_0}{(\bar{m})} T \tag{59}$$

Velocity (from Eq. 9):

$$u = \frac{1}{\rho r^2} \int_0^r \left(N_1 \sum_{k=1}^{K} n_k \dot{m}_k - \frac{\partial \rho}{\partial t}\right) r^2 \, dr \tag{60}$$

Pressure (from Eq. 10):

$$\left[\int_0^1 \frac{r^2 \, dr}{T}\right] \frac{dp}{dt} + \left[\frac{\partial}{\partial t} \int_0^1 \frac{r^2 \, dr}{T}\right] p = N_1 \int_0^1 \sum_{k=1}^{K} n_k \dot{m}_k r^2 \, dr \tag{61}$$

The initial and boundary conditions for the gas phase become:

Initial conditions:

$$\begin{aligned} Y_j(r,0) &= Y_{j0} \\ \phi(r,0) &= 1 \\ u(r,0) &= 0 \\ p(0) &= 1 \end{aligned} \tag{62}$$

Boundary conditions:

$$\begin{aligned} &\left.\frac{\partial Y_j}{\partial r}\right|_{r=0} = 0; \quad \left.\frac{\partial Y_j}{\partial r}\right|_{r=R} = 0 \\ &\left.\frac{\partial T}{\partial r}\right|_{r=0} = 0; \quad \left.\frac{\partial T}{\partial r}\right|_{r=1} = 0 \\ &u(0,t) = 0 \end{aligned} \tag{63}$$

Equations 24 and 30 are already nondimensional, along with their initial and boundary conditions. Equation 35 becomes:

$$\dot{m} = \frac{3}{2}\frac{\rho_0 D_0}{\rho_l D_l}\eta_s Sh B_M. \tag{64}$$

Equation 43 becomes

$$\dot{q}_L = \dot{m}_L \frac{C_{p0}}{C_{pl}}\left[\frac{C_{pf}\left(\frac{T_0}{T_{l0}}T_\infty - T_s\right)}{B_T} - \frac{T_0}{T_{l0}}L\right] \tag{65}$$

having nondimensionalized L with respect to $C_{p0}T_0$.

Numerical Procedure

Gas-phase and droplet calculations are time-split. The source terms provided by the droplet calculations for the nth time level are used in the $(n+1)$th time level of the gas-phase calculations, after proper interpolation, and so forth. Explicit finite-difference schemes are employed to discretize the PDEs of the gas-phase model, except that for the chemical production terms for which a degree of implicitness is introduced, since they have been proved to be the most effective in such two-phase problems (Aggarwal et al. 1983). The pressure is evaluated from the integral Eq. 61, then the density from the ideal gas state Eq. 59, then finally the gas velocity from Eq. 60. Due to the coupling of the equations, the procedure is iterated until the desired level of convergence is achieved. Then the products and inert mass fractions can be calculated, since their equations are decoupled from the others. The time integration is performed until required.

The finite difference equations for the gas phase are obtained as follows. Since convective terms are present, a staggered computational grid is adopted (Fig. 10.1). A computational grid of 250 nodes has been employed for the gas-phase calculations.

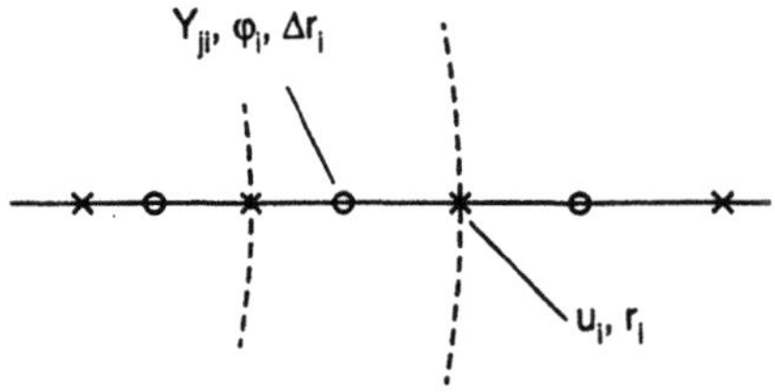

FIGURE 10.1. Computational grid.

Variable grid size can be used if necessary. This is especially useful for closed-volume calculations, due to the high gradients attained at the wall.

The schemes used for the balance equations are presented with reference to the following generic equation:

$$\frac{\partial \psi}{\partial t} + u\frac{\partial \psi}{\partial r} - \frac{1}{\rho r^2}\frac{\partial}{\partial r}\left(r^2\frac{\partial \psi}{\partial r}\right) = S_\psi + W_\psi \tag{66}$$

where S is the source term due to the evaporation and W is the production term. A single-step formula for the time derivative, an explicit upwind scheme for the convective term, a central three-point formula for the second-derivative term, and an explicit evaluation of the source term are employed, leading to the following explicit finite-difference equations:

$$\begin{aligned}\frac{\psi_i' - \psi_i}{\Delta t} &+ \left[\frac{u_{i-1} + |u_{i-1}|}{2\Delta r_{st,i-1}}(\psi_i - \psi_{i-1}) + \frac{u_i - |u_i|}{2\Delta r_{st,i}}(\psi_{i+1} - \psi_i)\right] \\ &- \frac{1}{\rho_i \overline{r_i^2}}\frac{1}{\Delta r_i}\left(r_i^2\frac{\psi_{i+1} - \psi_i}{\Delta r_{st,i}} - r_{i-1}^2\frac{\psi_i - \psi_{i-1}}{\Delta r_{st,i-1}}\right) = S_\psi + W_\psi\end{aligned} \tag{67}$$

where the prime denotes unknown time level, and where

$$\Delta r_{st,i} = \frac{\Delta r_i + \Delta r_{i+1}}{2} \tag{68}$$

$$\overline{r_i^2} = \frac{1}{4\pi\Delta r_i}\frac{4}{3}\pi(r_i^3 - r_{i-1}^3) = r_i r_{i-1} + \frac{\Delta r_i^2}{3} \tag{69}$$

$$r_i = \sum_{j=1}^{i} \Delta r_j. \tag{70}$$

In order to show how source terms S_ψ are calculated, it is necessary to specify how droplet groups are treated numerically. For each droplet group k, the average initial interparticle spacing, Δ_{0k}, is determined and used as a basis for discretization. In fact, each group is subdivided into Z_k discrete sets of N_{kz} droplets, which initially occupy the volume of Z_k concentric spherical shells of equal thickness, Δ_{0k}. Each discrete set of droplets is thus represented by a single droplet, identified by its group index k and initial location index z. Under the assumption of dilute spray (negligible interactions between droplets) the total number of droplets in each set is conserved.

The source term in Eq. 56 is computed as:

$$(S_{Y_j})_i = \frac{1}{\rho_i} N_1 \sum_{k=1}^{K} \sum_{z=1}^{Z} (\delta_{jF_m} \varepsilon_{mkz} - Y_{ji}) n_{kzi} \dot{m}_{kz} \tag{71}$$

and, for Eq. 57:

$$(S_\phi)_i = -\frac{p^\Gamma}{\rho_i C_{pi}} N_1 N_2 \sum_{k=1}^{K} \sum_{z=1}^{Z} n_{kzi} \dot{q}_L \tag{72}$$

where

$$\begin{aligned} n_{kzi} &= \frac{N_{kz}}{4\pi \overline{r_i^2} \Delta r_i} \quad && \text{if } r_{kz} \in]r_i, r_{i+1}] \\ n_{kzi} &= 0 && \text{if } r_{kz} \notin]r_i, r_{i+1}]. \end{aligned} \tag{73}$$

The production terms W_ψ are evaluated with some degree of implicitness. Equation 67 can be put in the following form:

$$\frac{\psi_i' - \psi_i}{\Delta t} = W + G + O(\Delta t), \tag{74}$$

that is,

$$\psi' = \psi \left(1 + \frac{W}{\psi} \Delta t \right) + G \Delta t + O(\Delta t^2).$$

Expanding the term in parenthesis according to

$$1 + x + O(x^2) = \frac{1}{1 - x}, \qquad x \in]-1, 1[$$

we have:

$$\psi' = \psi \frac{1}{1 - \dfrac{W}{\psi} \Delta t} + G \Delta t + O(\Delta t^2). \tag{75}$$

Developing and neglecting the higher-order terms:

$$\psi' - \frac{W}{\psi} \psi' \Delta t = \psi + G \Delta t + O(\Delta t)^2,$$

that is

$$\frac{\psi' - \psi}{\Delta t} = \frac{W}{\psi} \psi' + G + O(\Delta t). \tag{76}$$

The form

$$\frac{W}{\psi} \psi' \tag{77}$$

is adopted for the production terms in the balance equations for the fuels and the oxidizer: since $W < 0$, from Eq. 75 it is seen that a negative value for ψ' is never predicted, and no artificial adjustment is required when computing very low values for the species mass fractions. This feature is highly desirable where, as in the present case, chemical equilibrium calculations are not performed. For closed-volume calculations, the pressure is calculated from Eq. 61:

$$\frac{p' - p}{\Delta t} = \frac{\dfrac{-(I_1' - I_1)}{\Delta t} p' + I_2}{I_1} \tag{78}$$

where

$$I_1 = \int_0^1 \frac{r^2\, dr}{T} \tag{79}$$

and

$$I_2 = N_1 \int_0^1 \sum_{k=1}^{K} n_k \dot{m}_k r^2\, dr \tag{80}$$

are evaluated by means of a standard numerical scheme. The density is evaluated from Eq. 59:

$$\rho_i' = \frac{(\overline{m})_i}{(\overline{m})_0} \frac{p'}{T_i'} \tag{81}$$

where

$$T_i' = \phi_i'/(p')^\Gamma \tag{82}$$

and the velocity from Eq. 60:

$$u_i' = \frac{2}{(\rho_i + \rho_{i+1})} \frac{1}{r_i^2} \left[\frac{\rho_{i-1} + \rho_i}{2} r_{i-1}^2 u_{i-1}' + N_1 \overline{r_i^2} \Delta r \sum_{k=1}^{K} \sum_{z=1}^{Z} n_{kzi} \dot{m}_{kz} - \frac{\rho_i' - \rho_i}{\Delta t} \overline{r_i^2} \Delta r \right] \tag{83}$$

For closed-volume calculations, the right-hand sides of Eqs. 78, 81, 82, and 83 contain some unknown time-level terms, therefore iterations are needed. The iteration tests are performed on pressure and velocity.

The film-analysis eqs. 35 and 42, after substitutions from Eqs. 36, 38–41, and 43–50, constitute a system of coupled, nonlinear algebraic equations, in which T_∞, $Y_{m\infty}$, and p are the coupling terms with the external gas-phase equations, and Y_{mls} and T_s are the coupling terms with the liquid-phase equations. More precisely, if $r_k \in]r_i, r_{i+1}]$, we have:

$$T_\infty = T_i + \frac{r_k - r_i}{\Delta r_i} (T_{i+1} - T_i) \tag{84}$$

$$Y_{m\infty} = (Y_{F_m})_i + \frac{r_k - r_i}{\Delta r_i} [(Y_{F_m})_{i+1} - (Y_{F_m})_i]. \tag{85}$$

Thus, in principle, the film-analysis equations are coupled both to the liquid-phase problem and to the external gas-phase problem. An explicit formulation with the use of a fractional time step for the droplet calculations allows for solution until the droplet becomes very small; then the calculations are ceased and the residual liquid is instantaneously released to the gas phase.

The droplet motion Eq. (51) is solved as

$$\frac{v' - v}{\Delta t} = N_3 \mu_f \frac{(u_\infty - v')}{a_s^2} \operatorname{Re} C_D \tag{86}$$

where

$$N_3 = \frac{3}{16}\left(\frac{R}{a_0}\right)^2 \frac{\rho_0}{\rho_l} \tag{87}$$

and μ_f has been nondimensionalized with respect to $\rho_0 D_0$. The droplet position r_d is determined by means of

$$\frac{r_d' - r_d}{\Delta t} = \frac{v + v'}{2} \tag{88}$$

where v' is known from Eq. 86. If a droplet crosses the center of symmetry of the spherical reference frame, its position is kept positive while continuing the calculations. This corresponds to another droplet crossing the center of symmetry in the opposite direction.

The liquid-phase Eqs. 29 and 30 are solved by means of a finite-difference method with a nonuniform spatial grid and implicit schemes. Equation 30, put in conservative form and after the substitution

$$V_T = -\frac{1}{\eta_s}\frac{d\eta_s}{d\tau_T} \tag{89}$$

becomes

$$\frac{\partial T}{\partial \tau_T} + V_T\left[\frac{\partial(\eta T)}{\partial \eta} - T\right] = \frac{1}{\eta^2}\frac{\partial}{\partial \eta}\left(\eta^2 \frac{\partial T}{\partial \eta}\right). \tag{90}$$

Since V is always positive, the pseudo-convective term is approximated by means of a positive upwind scheme. The finite-difference equation is:

$$\begin{aligned} &\frac{T_i' - T_i}{\Delta \tau_T} + V_T\left[\frac{-\eta_{st,i} T_{i-1}' + \eta_{st,i+1} T_i'}{\Delta \eta_{st,i}} - T_i'\right] \\ &= \frac{1}{\overline{\eta_i^2}\Delta \eta_{st,i}}\left[\eta_{st,i+1}^2 \frac{T_{i+1}' - T_i'}{\Delta \eta_{i+1}} - \eta_{st,i}^2 \frac{T_i' - T_{i-1}'}{\Delta \eta_i}\right] \end{aligned} \tag{91}$$

where $\overline{\eta_i^2}$ is defined as $\overline{r_i^2}$ in Eq. 69. Equation 24 yields an equation that is analogous to Eq. 91.

The mathematical model is too complicated to allow for an exact evaluation of the numerical accuracy. However, some key problems can be identified, and proper steps can be taken to minimize inaccuracies.

A dynamically self-adjusted time step is chosen for gas-phase calculations.

The value is selected in order not to allow changes in the single node values greater than 5 percent over a time-integration step. The so-determined time step is always much smaller than would be required to ensure stability in an explicit formulation; that is why it was convenient to adopt explicit numerical schemes for gas-phase calculations. The so-determined time step is still too large to be used for droplet calculations, therefore a fractional time step is used in droplet calculations.

The droplet time step is also dynamically selected. By means of numerical tests we found that, for droplets of 100-μm diameter and an external temperature of 1000 K, a good value for the time step is $5.0 \cdot 10^{-5}$ sec. This value is then adjusted by making it increase with the radius squared and decrease with the square of the external temperature.

The solution is practically independent of the time step. However, space grid-size dependence exists in the results, mainly due to numerical diffusion arising from the combination of sharp gradients and variable gas velocity. It can be shown that the schemes employed for the convective terms (explicit upwind) are those that minimize artificial diffusion, if compared to other schemes of the same order accuracy.

Refining the gas-phase grid would improve accuracy with respect to the problems related to numerical diffusion; however, another difficulty would arise in terms of matching droplet to gas calculations. In fact, the gas-phase values of the variables at the droplet location are taken as values at infinity for the film analysis. Consistency requires that the grid size be larger than the expected size of the film surrounding the droplet. An estimate for the diameter of the film is about 10 droplet diameters, which means about 500 μm in the largest droplet case shown. This can be taken as a minimum limit for the space grid size in order not to invalidate the model formulation. Even when this is ensured, accuracy problems may arise. The main reason is that the external temperature used in droplet calculations is evaluated at a location obviously closer than infinity, hence it is lower with respect to the value at infinity that would result in that temperature value at that distance, in a spherically symmetric situation like that imagined in the film-theory analysis. This results in an overall underestimate of the droplet mass vaporization rate. A numerical correction suggested in Rangel and Sirignano (1989a) has been employed in these calculations. The computed mass vaporization rate is multiplied by a numerical correction factor:

$$\dot{m}_{\text{corr}} = \dot{m} \cdot \left(1 + \frac{\Delta r_i}{a_s}\right). \tag{92}$$

Results and Discussion

The influence of fuel composition, droplet size, equivalence ratio, and spray distribution has been investigated, and interesting phenomena have been predicted. The set of values employed for the properties and the initial conditions for the first case shown (base case) are reported in Table 10.1. For all of

TABLE 10.1. Values and expressions used for the most important physical parameters in the base case, F_1 = hexane, F_2 = decane.

Parameter	Value or Expression
a_{s0}	$2.5 \cdot 10^{-5}$ m
A_1	$3.8 \cdot 10^{11}$ s^{-1} (kg/m^3)$^{(1-\alpha_{F1}-\alpha_{O1})}$
A_2	$5.7 \cdot 10^{11}$ s^{-1} (kg/m^3)$^{(1-\alpha_{F2}-\alpha_{O2})}$
C_p	$C_{pa} + (C_{pF} - C_{pa})Y_{F_g}$
C_{pF}	$280 + 4.6T$ J/kg K
C_{pa}	1013 J/kg K for $300 < T < 400$
C_{pa}	$1013 + 0.195 \cdot (T - 400)$ J/kg K for $T > 400$
C_{pl}	$2.24 \cdot 10^3$ J/kg K
D_l	$8.25 \cdot 10^{-9}$ m^2/s
$E_1 = E_2$	$1.256 \cdot 10^8$ J/kg-mole K
$L_{\text{boil},1}$	$2.796 \cdot 10^5$ J/kg
$L_{\text{boil},2}$	$3.416 \cdot 10^5$ J/kg
Le_l	$1 \cdot 10^1$
p_0	$1 \cdot 10^5$ Pa
Q_1	$4.421 \cdot 10^7$ J/kg
Q_2	$4.476 \cdot 10^7$ J/kg
R	$5 \cdot 10^{-2}$ m
T_{b1}	$4.473 \cdot 10^2$ K
T_{b2}	$3.419 \cdot 10^2$ K
T_0	$3 \cdot 10^2$ K
$\alpha_{F_1} = \alpha_{F_2}$	0.25
$\alpha_{O_1} = \alpha_{O_2}$	1.5
γ	$C_p/(C_p - R)$
μ_0	$1.845 \cdot 10^{-5}$ kg/ms
ρ_l	$6.87 \cdot 10^2$ kg/m^3
χ	1
	(temperatures in K)

the cases reported and discussed here, ignition has been obtained by means of a source term in the energy equation. The source is active for 1 ms in a spherical region whose diameter is 1 cm, in the center of the spherical coordinate system, at the left in our representation. The flame propagates rightward in the figures to the outer regions. The domain of the calculation is a sphere of 5 cm radius. The calculations presented here are for an unconfined deflagration with the initial fuel spray/air mixture in a 3-cm spherical domain surrounded by a 2-cm layer of air. The ignition parameters have not been varied in this study.

Figure 10.2 shows the spatial profiles of gas temperature, vapor fuel, and oxygen mass fraction at successive times. The base case is a monodisperse stoichiometric spray of a liquid mixture of 50 percent hexane and 50 percent decane, with an initial droplet diameter of 50 μm. Due to the thermal expansion of the inner burning mixture, the flame reaches locations beyond the location of the outermost droplet in the cloud at the ignition time.

Strong local deviation from the original uniform stoichiometric equiva-

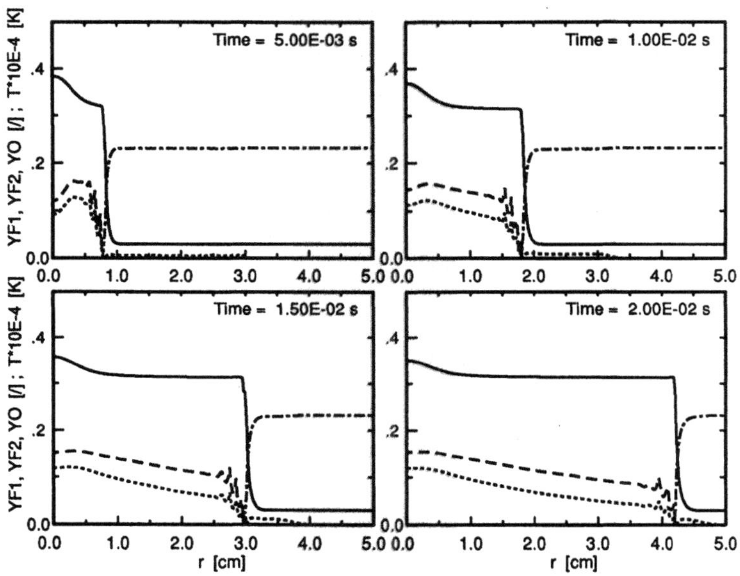

FIGURE 10.2. Temperature [T: solid line], Oxygen [YO: dot-and-dash], Hexane and Decane [YF1, dot and YF2, dash], as a function of the radial coordinate, at successive times. Base case. Initial droplet diameter: 50 microns; fuel composition: 50%–50%.

lence ratio is observed, due to the relative motion between the two phases. The droplets tend to remain behind when the gases accelerate outward due to the thermal expansion. The spherical geometry makes the gas velocity decay with the inverse radius squared. This means that the outer droplets have very little time to adjust to the gas velocity before they are reached by the flame. In fact, most of the deviation from the initial equivalence ratio is generated near the flame. This is because the droplets are passed by hotter, less dense gas, thus locally enriching the mixture. The extent of this deviation is related to various parameters: the droplet size, that is to say, the mechanical inertia; the fuel volatility; the combustion reaction rate.

The peaks in the fuel mass fraction profiles behind the flame, shown in Fig. 10.2, are due to the numerical representation of droplets in discrete sets. In fact, each droplet set is represented by a single droplet having average characteristics and located at the average distance from the center of the spherical gas-phase domain. However, those peaks indicate the ongoing vaporization of droplet passed by the flame. This is typical of the so-called heterogeneous regime, but the presence of the highly volatile component hexane, resulting in significant prevaporization ahead of the flame, gives the flame a premixed character too.

The results indicate the occurrence of a flame characterized by a rather complex structure. In the beginning, the fuel vapor formed ahead of the flame

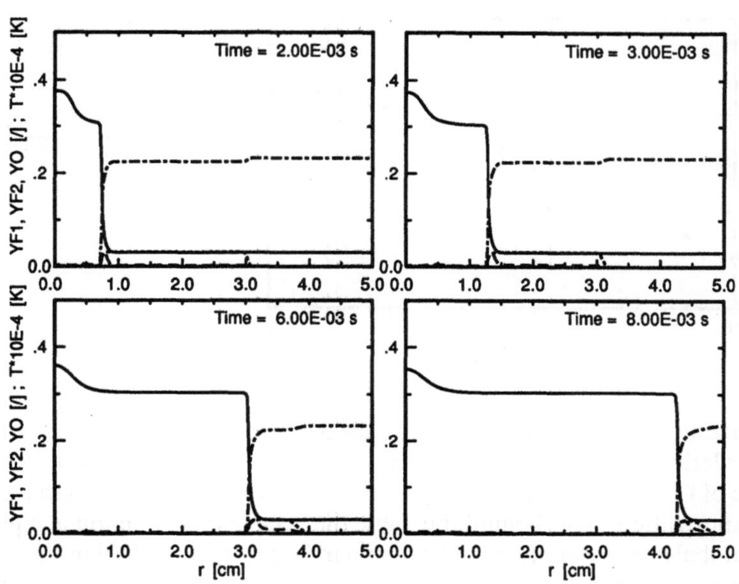

FIGURE 10.3. Temperature [T: solid line], Oxygen [YO: dot-and-dash], Hexane and Decane [YF1, dot and YF2, dash], as a function of the radial coordinate, at successive times. Initial droplet diameter: 6 microns; fuel composition: 50%–50%.

due to prevaporization is too little to account for the propagation of the flame, as observed, on the basis of a simple premixed-type mechanism. The fuel vapor formed behind the flame due to the vaporization of the droplets crossing the flame thus influences the flarne propagation. When the droplets are all passed by the flame, the flame splits into two flames, an inner, diffusion flame and an outer premixed-type flame.

Figure 10.3 illustrates the behavior of a spray of 6 μm initial diameter. Smaller droplets follow the gas motion much closer and, due to the higher surface/volume ratio, vaporize faster; therefore prevaporization in the fresh mixture is enhanced, giving the flame a premixed character. Calculations indicate that slower reaction rates would result in a slower flame propagation and in lower gas velocities, with less pronounced inertial effects, less deviation from the initially uniform equivalence ratio, and, again, a more premixed-type behavior.

Single-component spray calculations have been performed, giving rise to the same type of behavior already described. The multicomponent spray calculations are peculiar and it is interesting to observe how an uneven vapor fuel distribution is generated by the combustion of a 50/50 hexane/decane fuel spray. During the early vaporization occurring ahead of the flame in the fresh mixture the most volatile component (hexane) is mainly released; in the burned mixture, the residual vapor fuel results from the completion of the

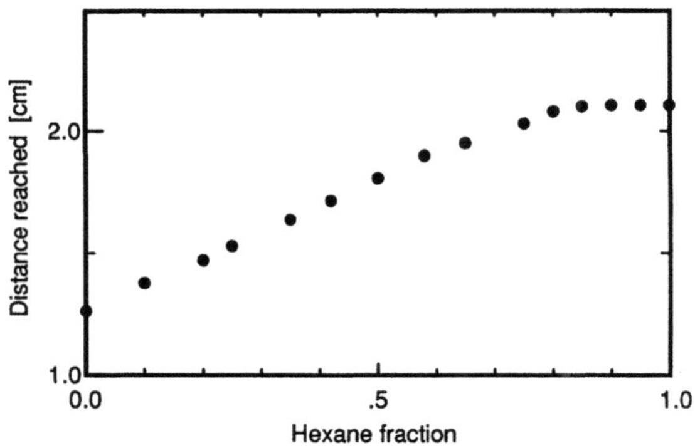

FIGURE 10.4. Distance reached by the flame after 10 ms as a function of fuel composition. Initial droplet diameter: 50 microns.

vaporization of the droplets crossing the flame front, which are richer in the less volatile component (decane). This is clearly seen in the second and third frame of Fig. 10.2, where hexane prevails ahead of the rightmost flame due to prevaporization; right behind the flame the decane mass fraction is higher.

A global parameter often used in laminar flame studies is the flame propagation speed, S_u. Since ours is an inherently unsteady configuration, S_u would be dependent on time, and accuracy problems would arise in numerically evaluating a time derivative. Therefore another parameter is used instead of S_u to characterize the speed of the flame. Since the ignition procedure is the same for all of the tests performed, the distance traveled by the flame front at a given time will give a good indication of the flame speed, provided that the flame is beyond the ignition region.

Figure 10.4 shows the distance reached by the flame in a spray cloud with droplets having 50 μm initial diameter, as a function of the liquid fuel composition. In these conditions, it is seen that the distance traveled by the flame increases linearly with the hexane fraction from 0.0 to 0.8, where it becomes fairly insensitive to the little decane content. This was expected since the less volatile component, decane, is mainly released towards the end of droplet lifetime, and with 50-μm droplets it is seen that not all the fuel participates in the combustion. The behavior changes, quite interestingly, when droplets are initially smaller. Figure 10.5 presents results for droplets having 17.5 μm initial diameter. Here it is seen that the flame propagation speed reaches a maximum for a certain intermediate blend of the two fuel components. It is also seen that, for this droplet diameter, pure hexane gives a slower flame than pure decane, as opposed to the previous case. This nonlinear behavior cannot be simply explained by means of a switch between a vaporization-controlled and a chemistry-controlled situation. Figure 10.6 reports the results of calculations conducted for pure hexane, pure decane, and a 50/50 mixture. The distance reached by the flame at a given time is evaluated as a function of the droplet initial diameter. Pure components are shown to have an "optimum" droplet size at which the flame speed is maximum, and this is in agreement with earlier

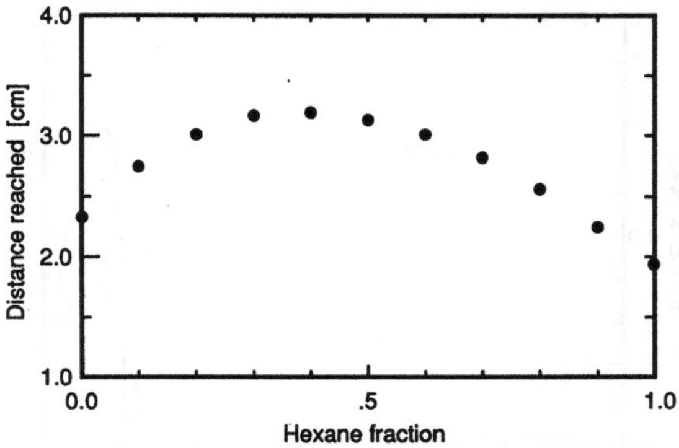

FIGURE 10.5. Distance reached by the flame after 5 ms as a function of fuel composition. Initial droplet diameter: 17.5 microns.

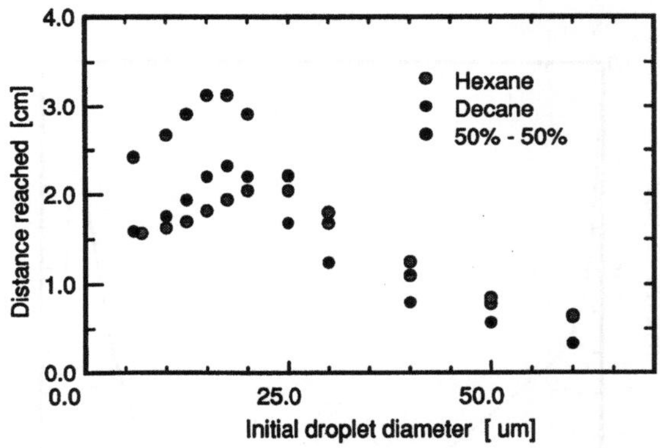

FIGURE 10.6. Distance reached by the flame after 5 ms as a function of droplet initial diameter.

computations concerning planar one-dimensional spray flames (Seth et al. 1980). This size value is higher for the most volatile fuel (hexane), as expected. In fact, at the "optimum" value droplets are predicted to cross the flame prior to complete vaporization, thus leading to fuel vapor build-up behind the flame. The multicomponent case also has an optimal droplet size for flame propagation but the surprise is that it exceeds the optimal values for either single-component case. The explanation can perhaps be found in the vaporization behavior of a multicomponent droplet, illustrated in Figs. 10.7 and 10.8 for single droplets from the calculations of Figs. 10.2 and 10.3, respectively. It is seen that, even for the smallest droplets, most of the decane vaporizes in the high-temperature zone, at the end of the dropet lifetime (when the peaks in the droplet mass vaporization rates are predicted). Thus, selective vaporization related to the different volatilities of the two fuel components produces the simultaneous presence of extensive prevaporization (of hexane) together with significant heterogeneity (decane-enriched droplets crossing the flame), thus enhancing flame propagation. This last argument also shows how impor-

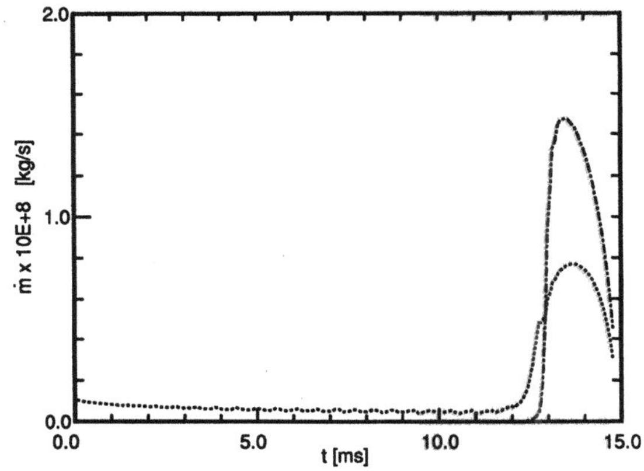

FIGURE 10.7. Droplet vaporization history. Mass vaporization rates: Hexane [dotted line] and Decane [dot-and-dash] as a function of time. Initial droplet location: 1.2 cm from the origin; initial liquid fuel composition: 50%–50%; initial droplet diameter: 50 microns.

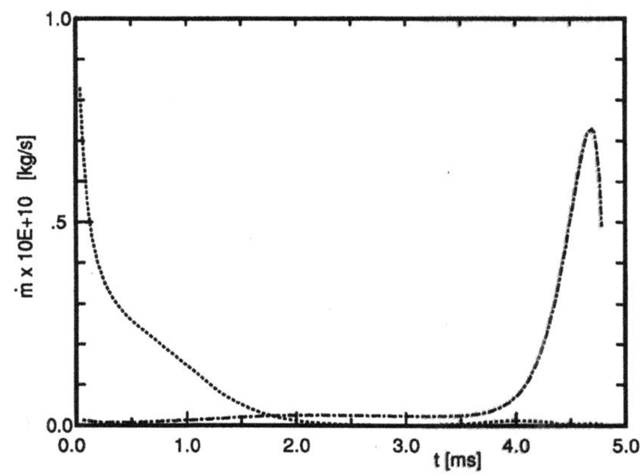

FIGURE 10.8. Droplet vaporization history. Mass vaporization rates: Hexane [dotted line] and Decane [dot-and-dash] as a function of time. Initial droplet location: 1.2 cm from the origin; initial liquid fuel composition: 50%–50%; initial droplet diameter: 6 microns.

tant it is to consider the structure of the flame and the droplet-gas slip effects, in order to fully understand spray flame phenomena.

Figures 10.7 and 10.8 also illustrate the different behavior of droplets having different sizes. The prevaporization of *n*-hexane for the 50-μm droplets (Fig. 10.7) occurs at a low, fairly constant rate (the slight decrease from the initial value is due to liquid surface cool-down to a quasi-steady value caused by the vaporization itself; the wiggles are a numerical effect due to a computed gas-phase velocity jitter resulting in small variations in the droplet Reynolds number). When the high-temperature zone associated with the flame encounters the droplet, a significant amount of liquid *n*-hexane is still present. The high surface-to-volume ratio of the 6-μm spray (Fig. 10.8) results in a much faster prevaporization. When the flame reaches the droplet, nearly all the hexane has evaporated, thus the separation of the two components is almost complete in this case.

Summary and Concluding Remarks

Unsteady spray combustion in spherically symmetric geometry has been studied. Unsteady droplet heat-up and vaporization for a multicomponent liquid fuel has been considered by means of a spherically symmetric "conduction-limit" and "diffusion-limit" model for the droplet interior; quasi-steady heat and mass transfer between liquid and gas-phase have been considered, with a spherically symmetric film model taking into account variable properties and effects of the Stefan flow. In the gas phase, combustion has been modeled by means of one single-step, irreversible chemical reaction for each one of the fuel components; heat conduction, mass diffusion, and gas motion with a low Mach number assumption have been considered; droplet motion has been taken into account.

The results show that, even for such a simple geometry, most of the characteristics of spray combustion phenomena are already present.

The droplet motion across gas regions having different densities results in local deviation from the initial equivalence ratio. As a consequence, the flame appears to have a complex diffusion-premixed character. The separation of diffusion and premixed flames has also been predicted with a model in which the gas density is taken to be constant, as shown in Rangel and Sirignano (1989b); gas expansion and relative motion are responsible for even larger-scale flame separation as those predicted herein. This effect becomes pronounced for initial droplet diameters in the order of 20 μm and larger, with the initial conditions used ($T = 300$ K, no relative velocity). It can be concluded that the effects of droplet motion are important in most practical spray situations and that more attention must be devoted to the modeling of the droplet drag.

The consideration of a multicomponent spray shows how a nonuniform fuel-vapor composition is caused by the different volatilities of the components. This result is in agreement with those of Aggarwal (1987), relative to a spray vaporization situation.

A two-component liquid fuel spray does not simply behave like a single-component liquid fuel of intermediate volatility in the nonheterogeneous spray-flame propagation regime. A mechanism has been indicated by which a flame propagates faster than for each of the single-component fuels in the same conditions. It would be interesting to seek an experimental confirmation of the predicted behavior. It is very likely related to the dual premixed and diffusion natures of the flame zone.

It should be also noted that in the present model, as in many spray combustion models, the preexponential factors of two parallel, independent reactions constitute the only discrimination between the two vapor fuel reactants; the actual chemistry is much more complex, especially for rich fuel mixtures, which have been shown to be locally present even in globally stoichiometric or lean sprays. For example, the consideration of pyrolysis reactions would give a much better description of the gas composition in the residual vapor fuels behind the flame, in which the presence of *n*-hexane and *n*-decane is clearly impossible at those temperatures. It is therefore concluded that, especially for multicomponent spray combustion, a more detailed description of the gas phase chemistry would be appropriate.

Nomenclature

a = radial coordinate in the droplet spherical reference frame.
A = preexponential factor in the gas-phase Arrhenius kinetic law.
C_D = drag coefficient.
C_p = specific heat at constant pressure.
C_1, C_2 = constants in the Clausius-Clapeyron relationship for the saturated vapor pressure.
D = diffusion coefficient.
E = activation energy in the gas-phase Arrhenius kinetic law.
K = total number of droplet groups.
L = latent heat of vaporization.
$\dot{m}$ = mass vaporization rate of a droplet.
(m) = molecular weight.
n = droplet number density.
N_{kz} = total number of droplets in group k, set z.
p = pressure.
Q = heat of combustion.
r = space coordinate in gas-phase calculations.
r_d = spatial position of a droplet.
R = radius of the vessel.
$\mathcal{R}$ = universal gas constant.
T = temperature.
T_s = surface temperature of a droplet.
t = time.
u = gas velocity.
v = droplet velocity.
$\dot{w}$ = mass production rate due to chemical reaction.
X = mole fraction.
Y = mass fraction.
Z_k = total number of droplet discrete sets in droplet group k.

Greek Symbols

α_{F_m}, α_{O_m} = nonunity exponents in the gas-phase reaction Arrhenius kinetic.
γ = ratio of the specific heats, C_p/C_v.
δ = Kroneker δ.
ε = fractional vaporization rate.
μ = gas viscosity.

ρ = density.
χ = equivalence ratio.

Subscripts

a = air.
boil = boiling point.
corr = corrected.
c = reference quantity.
crit = critical point.
F = fuel.
f = film.
g = gas phase.
i = space grid node.
j = species.
k = droplet group.
l = liquid phase.
m = fuel component.
p = product.
s = surface.
z = droplet discrete set.
0 = initial.

References

Abramzon, B., and Sirignano, W.A., 1989, "Droplet Vaporization Model for Spray Combustion Calculations," *Int. J. of Heat and Mass Transfer*, 32, 9.

Aggarwal, S.K., 1987, "Modelling of a Dilute Vaporizing Multicomponent Fuel Spray," *Int. J. of Heat and Mass Transfer*, 30, 1949–1961.

Aggarwal, S.K., Fix, G.J., Lee, D.N., and Sirignano, W.A., 1983, "Numerical Optimization Studies of Axisymmetric Unsteady Sprays," *J. Comput. Phy.*, 35, 229.

Aggarwal, S.K., and Sirignano, W.A., 1989a, "Ignition of Fuel Sprays: Deterministic Calculations for Idealized Droplet Arrays," *20th Symposium (International) on Combustion*, The Combustion Institute, 1773–1780.

Aggarwal, S.K., and Sirignano, W.A., 1985b, "Unsteady Spray Flame Propagation in a Closed Volume," *Combustion and Flame*, 62, 69.

Aggarwal, S.K., Tong, A.Y., and Sirignano, W.A., 1984, "A Comparison of Vaporization Models in Spray Calculations," *AIAA J.*, 22, 1448.

Continillo, G, and Sirignano, W.A., 1988, "Numerical Study of Multicomponent Fuel Spray Flame Propagation in a Spherical Closed Volume," *22d Symposium (International) on Combustion*, The Combustion Institute, 1941–1950.

Faeth, G.M., 1977, "Current Status of Droplet and Liquid Combustion," *Progress in Energy and Combustion Science*, 3, 191–224.

Lin, T.H., and Law, C.K., 1988, "Theory of Laminar Flame Propagation in Off-Stoichiometric Dilute Sprays," *Int. J. of Heat and Mass Transfer*, 31, 1023.

Seth, B., Aggarwal, S.K., and Sirignano, W.A., 1980, "Flame Propagation Through an

Air-Fuel Spray Mixture with Transient Droplet Vaporization," *Combustion and Flame*, 39, 149.

Sirignano, W.A., 1988, "An Integrated Approach to Spray Combustion Model Development," *ASME Winter Annual Meeting*, Dec. 7–12, 1986, Anaheim, Calif.; also, *Combustion Science and Tech.* 58, 1–3, 231–251.

Rangel, R.H., and Sirignano, W.A., 1989a, "An Evaluation of the Point Source Approximation in Spray Calculations," *Numerical Heat Transfer*, 16, 37–57.

Rangel, R.H., and Sirignano, W.A., 1989b, "Unsteady Flame Propagation in a Spray with Transient Droplet Vaporization," *22d Symposium (International) on Combustion*, The Combustion Institute, 1931–1940.

Williams, F.A., 1985, *Combustion Theory*, Benjamin-Cummins, Palo Alto, Calif.

III
Computational Fluid Dynamics

11
Efficient Solution of Compressible Internal Flows

M. Napolitano and P. De Palma

Abstract: This paper provides an efficient and accurate numerical method for computing compressible internal flows in two dimensions. The method has been designed to deal with subsonic-to-supersonic problems inside nozzles without shocks. However, it is found to be competitive also in the subsonic regime and can be easily adapted to compute transonic flows, by means of any suitable shock-fitting procedure. The perturbative lambda formulation Euler equations are considered and solved by a block-line-relaxation procedure. For the subsonic-to-supersonic flow case, of major interest here, an alternating direction block-line-Jacobi method is used to obtain the steady state solution in the subsonic region, up to the first computational column for which the longitudinal velocity component is supersonic at all gridpoints. Afterwards, a block-line-Gauss-Seidel method, which is essentially a downstream-marching implicit scheme using a quasi-Newton iteration for the nonlinear terms, is used to obtain a very fast convergence on each successive column of the computational domain. For the less interesting subsonic flow case, only the first part of the numerical procedure is necessary. The validity of the proposed technique is demonstrated for a well-documented nozzle-flow problem for both subsonic and subsonic-to-supersonic flow conditions.

Introduction

The continuous progress in the performance of aerodynamic and propulsion systems has relied mainly on costly and lengthy experiments, combined with the experience and skills of designers. In the last years, however, due to the exceptional progress in computer performance, as well as to the rapid growth of computational fluid dynamics (CFD), "numerical experiments" have been playing an increasing role in both the design and validation process of any new advanced piece or equipment, (see Jameson (1987) for an up-to-date review). In particular, airfoil design relies almost completely on very accurate and efficient methods for solving the Euler equations, and simulations of the flow field around an entire aircraft have already appeared in the literature (see, e.g., Baker (1987) and Volpe et al. (1987)).

In the last few years, the CFD group of the Istituto di Macchine ed Energetica of the University of Bari has been developing numerical methods for solving compressible inviscid flows (both external and internal) as well as incompressible viscous flows (see Napolitano (1986, 1988) for a comprehensive review of these two activities, respectively). For the case of compressible inviscid flows, of interest here, the work has been concerned with the so-called lambda formulation (Moretti 1979; Zanetti and Colsurdo 1981) and, more precisely, with developing efficient (implicit) integration schemes (Abbrescia et al. 1984; Dadone and Napolitano 1983, 1985b; Napolitano and Dadone 1985) and/or accurate formulations (Dadone and Napolitano 1985a, 1986). For the purpose of the present work, it is enough to briefly recall that, for the case of multidimensional flows, after developing alternating direction implicit (ADI) schemes (Dadone and Napolitano 1983, 1985a, 1985b, 1986). And various types of relaxation procedures of the line-Gauss-Seidel (LGS) type (Abbrescia et al. 1984; Napolitano and Dadone 1985) a semi-implicit method called Fast Solver has been singled out as the simplest and most efficient integration scheme (Dadone and Moretti 1988; Dadone et al. 1989). Actually, LGS methods were found to be superior to approximate factorization schemes of the ADI type essentially because they are more stable (due to the increased diagonal dominance of the linear systems to be solved) and characterized by a convergence rate much less sensitive to variations of the CFL number. However, with the advent of modern vector and parallel computers, the hard-to-vectorize LGS methods have been easily surpassed by the simpler and easy-to-vectorize Fast Solver.

However, there are still two main reasons which could provide a renewed interest in relaxation methods. Firstly, if one abandons the Gauss-Seidel relaxation process in favor of the simpler Jacobi one, ease of code vectorization is quickly recovered. Secondly, if one wants to combine a code based on the lambda formulation equations (used in all regions of smooth flow) with a code based on the Euler equations in conservation-law form (used to capture shocks and other discontinuities), as done, for example, by Pandolfi (1985), it may be more convenient to use the same type of integration scheme in both codes. Now, the Fast Solver exploits the unique features inherent to the lambda formulation equations (Dadone et al. 1989; Dadone and Moretti 1988) and cannot be applied to solve the Euler equations in conservation-law form. On the other hand, the latter have been solved quite successfully by means of relaxation schemes very similar to the one which is going to be presented in the following (see, e.g., Van Leer and Mulder 1984; Walters and Thomas 1987).

From the preceding considerations, it appears worthwhile to develop a new relaxation method of the line-Jacobi type for the lambda formulation Euler equations. Moreover, if one is interested in solving the flow inside two-dimensional nozzles at design conditions, namely with subsonic flow in the converging region and supersonic flow in the diverging one, such a procedure appears even more appealing, insofar as it becomes a fully implicit down-stream-marching method in the supersonic region, by simply choosing an infinite CFL number and taking into account the additional Gauss-Seidel terms.

This paper provides such a new technique and applies it to compute both subsonic and subsonic-to-supersonic flows inside a two-dimensional nozzle. In the following sections, after a brief review of the governing equations, of their spatial discretizations, and of the boundary condition treatment, the special features of the new method will be described in some detail, and finally, some interesting and novel results will be provided.

Governing Equations and Numerical Technique

In the present study only homentropic two-dimensional flows will be considered. The governing equations are expressed according to the perturbative version of the lambda formulation and written in a system of curvilinear orthogonal coordinates (Dadone and Napolitano 1985a, 1986):

$$\bar{C}_t + \bar{D}_t + \frac{v_1 + a}{h_1}\frac{\partial \bar{C}}{\partial q_1} + \frac{v_2}{h_2}\frac{\partial \bar{C}}{\partial q_2} + \frac{v_1 - a}{h_1}\frac{\partial \bar{D}}{\partial q_1} + \frac{v_2}{h_2}\frac{\partial \bar{D}}{\partial q_2}$$

$$= \frac{2\bar{v}_2}{h_1 h_2}\left(\frac{\partial h_2}{\partial q_1}\bar{v}_2 - \frac{\partial h_1}{\partial q_2}\bar{v}_1\right) - k1_{1,2}\bar{v}_1 - k2_{1,2}\bar{v}_2 - k3_1\bar{a} \tag{1}$$

$$\bar{E}_t + \bar{F}_t + \frac{v_1}{h_1}\frac{\partial \bar{E}}{\partial q_1} + \frac{v_2 + a}{h_2}\frac{\partial \bar{E}}{\partial q_2} + \frac{v_1}{h_1}\frac{\partial \bar{F}}{\partial q_1} + \frac{v_2 - a}{h_2}\frac{\partial \bar{F}}{\partial q_2}$$

$$= \frac{2\bar{v}_1}{h_1 h_2}\left(\frac{\partial h_1}{\partial q_2}\bar{v}_1 - \frac{\partial h_2}{\partial q_1}\bar{v}_2\right) - k2_{2,1}\bar{v}_1 - k1_{2,1}\bar{v}_2 - k3_2\bar{a} \tag{2}$$

$$\frac{1}{2}[\bar{C}_t - \bar{D}_t + \bar{E}_t - \bar{F}_t] + \frac{v_1 + a}{h_1}\frac{\partial \bar{C}}{\partial q_1} - \frac{v_1 - a}{h_1}\frac{\partial \bar{D}}{\partial q_1} + \frac{v_2 + a}{h_2}\frac{\partial \bar{E}}{\partial q_2} - \frac{v_2 - a}{h_2}\frac{\partial \bar{F}}{\partial q_2}$$

$$= -\frac{2\bar{a}}{h_1 h_2}\left(\frac{\partial h_2}{\partial q_1}\bar{v}_1 + \frac{\partial h_1}{\partial q_2}\bar{v}_2\right) - k4_{1,2}\bar{v}_1 - k4_{2,1}\bar{v}_2 - k5 \tag{3}$$

$$\bar{C} - \bar{D} - \bar{E} + \bar{F} = 0 \tag{4}$$

In Eqs. 1–4, the subscript t indicates partial derivatives with respect to time; q_1, q_2, h_1, and h_2 are the orthogonal curvilinear coordinates and the corresponding scale factors (Karamcheti 1966); v_1 and v_2 are the two velocity components; a is the speed of sound; and C, D, E, and F are the four bicharacteristic variables:

$$C = v_1 + \delta a \tag{5}$$

$$D = v_1 - \delta a \tag{6}$$

$$E = v_2 + \delta a \tag{7}$$

$$F = v_2 - \delta a \tag{8}$$

where $\delta = 2/(\gamma - 1)$, γ being the specific heats ratio of the perfect gas under consideration. Furthermore, the bars denote "perturbation-type" variables with respect to a suitable "incompressible-flow" solution (Dadone and Napolitano 1986) and the k coefficients, which are functions of such an "incompressible-flow" solution and of the coordinate system, are given in Dadone and Napolitano (1986).

All details about the derivation of Eqs. 1–3 are given in Dadone and Napolitano (1985b, 1986). Here, it is noteworthy to point out that Eqs. 1–3 are simply the two components of the Euler momentum equation and the continuity equation, written in terms or the bicharacteristic variables C, D, E, and F. Also, all of the derivatives appearing in these equations are easily seen to be associated with the advection of physical disturbances, so that appropriate upwind discretizations for the spatial derivatives can be chosen simply on the basis of the signs of the corresponding advection speeds, namely, the coefficients multiplying them (Dadone and Napolitano 1983, 1985b, 1986; Moretti 1979; Zannetti and Colasurdo 1981).

The numerical method employed in this study is based on the same time and space discretizations used in Dadone and Napolitano (1985b, 1986). Equations 1–3 are discretized and linearizerd in time by means of a two-level implicit Euler time stepping, using the delta form of Beam and Warming (1978), with only the derivatives of the bicharacteristic variables in Eqs. 1–3 and all the terms in Eq. 4 being evaluated at the new time level, to give:

$$\frac{\Delta C}{\Delta t} + \frac{\Delta D}{\Delta t} + \frac{v_1 + a}{h_1}\frac{\partial \Delta C}{\partial q_1} + \frac{v_2}{h_2}\frac{\partial \Delta C}{\partial q_2} + \frac{v_1 - a}{h_1}\frac{\partial \Delta D}{\partial q_1} + \frac{v_2}{h_2}\frac{\partial \Delta D}{\partial q_2} = \mathrm{RES}(1) \qquad (9)$$

$$\frac{\Delta E}{\Delta t} + \frac{\Delta F}{\Delta t} + \frac{v_1}{h_1}\frac{\partial \Delta E}{\partial q_1} + \frac{v_2 + a}{h_2}\frac{\partial \Delta E}{\partial q_2} + \frac{v_1}{h_1}\frac{\partial \Delta F}{\partial q_1} + \frac{v_2 - a}{h_2}\frac{\partial \Delta F}{\partial q_2} = \mathrm{RES}(2) \qquad (10)$$

$$\frac{1}{2}\left[\frac{\Delta C}{\Delta t} - \frac{\Delta D}{\Delta t} + \frac{\Delta E}{\Delta t} - \frac{\Delta F}{\Delta t}\right] + \frac{v_1 + a}{h_1}\frac{\partial \Delta C}{\partial q_1} - \frac{v_1 - a}{h_1}\frac{\partial \Delta D}{\partial q_1} + \frac{v_2 + a}{h_2}\frac{\partial \Delta E}{\partial q_2} - \frac{v_2 - a}{h_2}\frac{\partial \Delta F}{\partial q_2} = \mathrm{RES}(3) \qquad (11)$$

$$\Delta C - \Delta D - \Delta E + \Delta F = \mathrm{RES}(4) \qquad (12)$$

In Eqs. 9–12 RES(N) is a shorthand notation for the steady-state part (i.e., the residual) of Eq. (N), evaluated at the old iteration (time level); Δt is the time step; and ΔC, ΔD, ΔE, and ΔF are the variations of $\bar{C}$, $\bar{D}$, $\bar{E}$ and $\bar{F}$, respectively, between the new and old iteration. It is noteworthy that Eq. 12 is always used to eliminate ΔF in favor of ΔC, ΔD, and ΔE and that all of the spatial derivatives in Eqs. 9–11 are discretized by means of upwind differences (according to the signs of the advection speeds). In this study, first-order-accurate two-point differences are used, due to the extreme intrinsic accuracy of the perturbative lambda formulation. A 3×3 block-pentadiagonal system of linear algebraic equations is thus to be solved at every time step.

Of course, more nonzero diagonals appear if second-order accuracy is sought, unless a simple deferred-correction strategy is employed, as in Dadone and Napolitano (1985b, 1986). There, the derivatives in the LHS implicit operator are evaluated using two-point first-order-accurate upwind differences, whereas three-point second-order-accurate upwind differences are employed in the RHS steady-state residuals.

The treatment of boundary conditions in the present scheme is identical to that of Dadone and Napolitano (1985b, 1986). In short, the boundary conditions are all and only those required by the physics of the problem. At the inlet gridpoints, the total enthalpy and the direction of the velocity vector are prescribed; at the solid wall gridpoints, the direction of the velocity vector is prescribed so as to satisfy the flow tangency condition; at the outlet gridpoints, finally, the pressure is prescribed for the case of subsonic outflow conditions, whereas no boundary condition is needed for supersonic outflow conditions. These physical boundary conditions are then complemented by appropriate linear combinations of Eqs. 9–11, containing only spatial derivatives associated with physical disturbances arriving at the boundary gridpoint under consideration from inside the computational domain. At every boundary gridpoint, 3 equations for the 3 variables (ΔC, ΔD, and ΔE) are thus available, without any need for numerical boundary conditions (Dadone and Napolitano 1985b, 1986).

Subsonic Flow Solver

In the subsonic flow region, the upwind differences in the LHS of Eqs. 9–11 are such that the linear system to be solved at every time step has a block-pentadiagonal structure. Therefore, for computational convenience, the system is solved approximately by an alternating-direction block-line-Jacobi (BLJ) relaxation procedure, which requires solving only block-tridiagonal systems along each row and column of the computational domain. With respect to the well-known block-ADI factorization procedure of Dadone and Napolitano (1985b, 1986), the present approach is just as simple and can be implemented on vector and parallel computers just as easily, but is considerably more efficient and robust with a convergence rate much less sensitive to the value of the time step.

Supersonic Flow Solver

The subsonic flow solver just described can be used, as it is, also for supersonic flow regions, where the streamwise component of the velocity, v_1, is greater than the speed of sound a. However, for such a region, due to the upwind differences used in the LHS of Eqs. 9–11, the linear system to be solved at every time step has a block-tetradiagonal structure and can be reduced, without any approximation, to a series of smaller easy-to-solve block-tridiagonal systems (one for each column of the computational grid), by bringing the ΔC, ΔD, and ΔE variables corresponding to the previous column to the RHS. This amounts to employing a downstream-marching block-LGS method, which, if the time step is set to infinity, becomes a fully implicit quasi-Newton method and converges on the nonlinear terms within four to six iterations. In more detail, for each column of the computational domain, Eqs. 9–11 are solved coupled together to provide the unknowns ΔC, ΔD, and ΔE. The solution is updated to provide new values of all the coefficients in the equations and the process is repeated until the RHS residual is reduced to a suitably small value.

Combination of the Subsonic and Supersonic Solvers

For transonic flows, with a small supersonic region embedded inside a large subsonic one, the subsonic flow solver, combined with an appropriate shock-fitting procedure, should be used throughout. However, if one is concerned with solving subsonic-to-supersonic internal flows, such as nozzle flows, where the supersonic region is larger than the subsonic one, a significant reduction of computer time can be achieved by combining the two aforementioned solvers, as follows. The subsonic flow solver is applied to the subsonic region up to and including the first supersonic computational column, until convergence is achieved. The supersonic flow solver is then applied on each successive computational column; a few iterations being sufficient to converge on the nonlinear terms.

In more detail, the procedure is employed as follows. The initial condition is taken to be the appropriate one-dimensional flow solution. Namely, the v_1 velocity component and a are taken to be constant along each column of the computational domain and v_2 is zero throughout. The value of the longitudinal gridpoint index, i, corresponding to the first computational column for which $v_1 > a$ at all j locations ($i = i\,\text{sup}$) is then determined by a trivial search algorithm and a two-sweep BLJ iteration is performed over the entire computational region characterized by $i \leq i\,\text{sup}$. These steps are repeated until a satisfactory convergence level is achieved. Notice that, in this procedure, no downstream boundary condition is needed, since the last column gridpoints (for which $i = i\,\text{sup}$) are always characterized by a supersonic longitudinal velocity component v_1. Finally, the time step is set to infinity and the downstream-marching block-LGS (supersonic flow) solver is used to compute each successive column, with just a few iterations needed to achieve a satisfactory convergence.

One word of caution is in order. The proposed methodology is expected to perform very well when the initial condition, based on a quasi-one-dimensional flow solution, predicts the position of the sonic line reasonably well, so that the value of $i\,\text{sup}$ does not change at all, or changes very little during the first, more costly part of the computation process. In this respect, the use of a grid aligned with the potential flow streamlines, as in this study, could be an important ingredient for the success of the proposed methodology.

Results

The present method has been applied to compute two different flow fields inside a nozzle proposed by Moretti (private communication). The geometry is obtained by means of a conformal mapping, which provides a very suitable orthogonal computational grid, as well as its scale factors and the "incompressible-flow" solution required by the perturbative lambda formulation.

Figure 11.1 provides the nozzle geometry as well as one of the three computational grids used in this study.

Subsonic flow conditions, characterized by an isentropic Mach number at the lower right corner equal to 0.15 were considered at first. The present results are given in Fig. 11.2 as the Mach number distributions along the lower and upper walls of the channel, indicated by LW and UW, respectively. The solid lines refer to the finest grid (65×17 gridpoints) solution whereas the symbols refer to the coarsest grid (17×9 gridpoints) solution. The results obtained on the intermediate grid of Fig. 11.1, containing 33×13 gridpoints, are omitted insofar as they coincide with the most accurate ones, within plotting accuracy. The results in Fig. 11.2, which are identical to those obtained in Dadone et al. (1989), where the same equations and spatial discretization are used, confirm once more the extreme accuracy of the perturbative lambda formulation, the very coarse grid results practically coinciding with the "exact" solution. It is noteworthy that such a remarkable accuracy is lost if one solves the same problem using the same grids and first-order-accurate upwind differences, but the standard (nonperturbative) form of the equations. Figure 11.3 shows, for example, the upper wall Mach number distributions obtained using the standard equations and the three aforementioned grids (symbols), together with the "exact" finest-grid perturbative results of Fig. 11.2 (solid line). It appears that all three sets of results are very inaccurate, the only encouraging feature being that the truncation error decreases when the grid is refined. The standard lambda formulation is an accurate method in most applications; these poor results must be due to the particular geometry, for which, to the authors' knowledge, only the perturbative lambda formulation has been able to produce satisfactory results using reasonable grids and first-order-accurate distances.

The convergence history of the method is given in Fig. 11.4, where the loga-

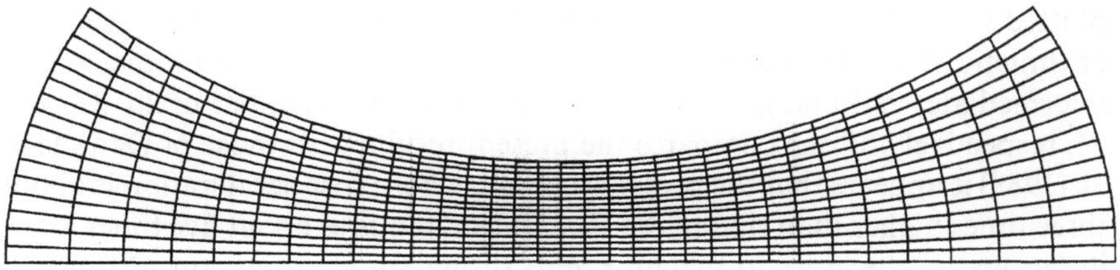

FIGURE 11.1. Nozzle geometry and intermediate computational grid.

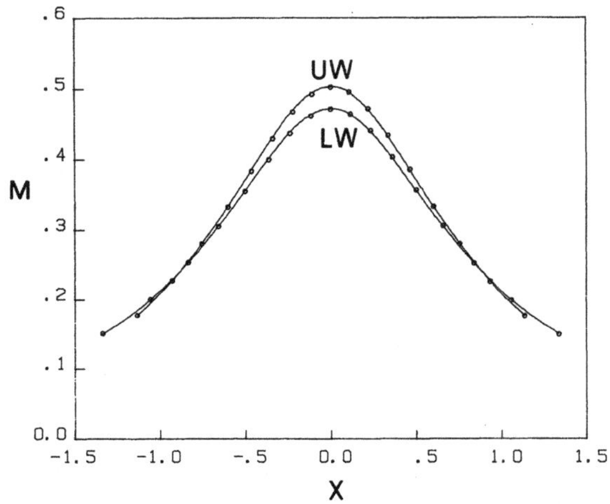

FIGURE 11.2. Subsonic-flow Mach number distributions along the upper and lower walls of the nozzle.

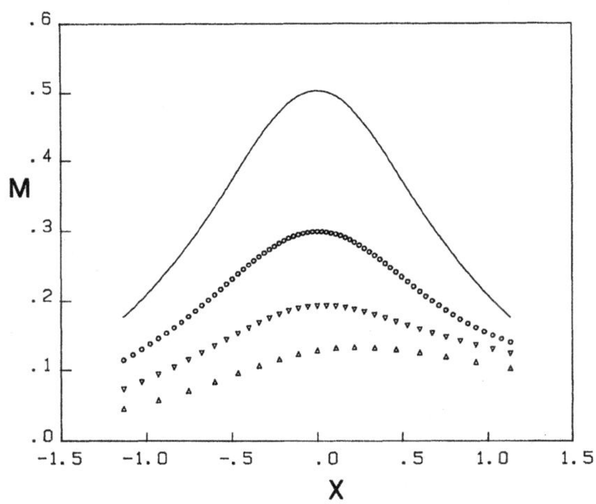

FIGURE 11.3. Comparison of the nonperturbative results versus the "exact" perturbative solution.

rithm of the $L1$ norm of the residuals of the three governing equations is plotted versus the number of iterations. Machine zero, using single-precision arithmetic, is achieved within about 50, 80, and 180 iterations on the coarse. intermediate, and fine grid, with a longitudinal CFL number of 20, 27, and 35, respectively. If one compares the present results with those of Dadone et al. (1989) it appears that the present method, whose cost per iteration is about three times higher than that of the Fast Solver, requires about five times fewer iterations to converge to machine zero (using simple precision arithmetic)

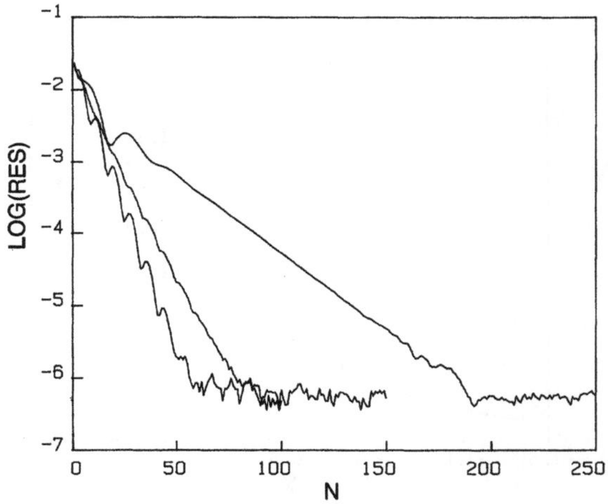

FIGURE 11.4. Subsonic-flow convergence histories.

Therefore the proposed approach, although certainly less simple, is competitive with the Fast Solver of Dadone et al. (1989). Incidentally, the finest-grid calculation requires about 2 CPU seconds per iteretion on an HP 9000/840S minicomputer.

Subsonic-to-supersonic flow conditions were then considered, using the same geometry and a downstream pressure of 0.063 (normalized with respect to the inlet total pressure). The numerical results are given in Fig. 11.5, as the Mach number distributions along the lower and upper walls of the nozzle. Once more, only the solutions obtained on the finest and coarsest grids are given, the agreement being not quite perfect, due to the lower effectiveness of the perturbative approach in the supersonic region.

In order to demonstrate the efficiency of the method in this case, the convergence histories of the three calculations are given in Fig. 11.6, again as the logarithm of the $L1$ norm of the residuals of the three equations, plotted versus the work. One work unit is the CPU time required by one iteration in the subsonic zone. One has to bear in mind that the entire subsonic flow region (up to and including the first entirely supersonic column) is computed at first, by means of a global relaxation procedure. The first part of the convergence history corresponds to such a calculation, with the $L1$ norm being computed over this entire region and the number of work units coinciding with the number of (two-sweep global) iterations. Then, a block-LGS method is used to converge on each successive supersonic column. Figure 11.6, accordingly, shows the reduction of the $L1$ norm of the residuals computed over each supersonic column, one after another, and thus takes a saw-tooth shape. Notice that each supersonic column calculation requires four to six iterations, each employing significantly less than a work unit. Therefore, the apparently strange shape of the convergence-history curves is indeed correct and the

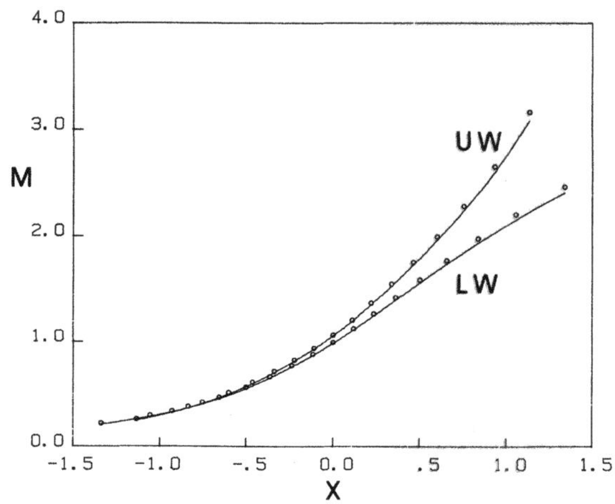

FIGURE 11.5. Supersonic-flow Mach number distributions along the upper and lower walls of the nozzle.

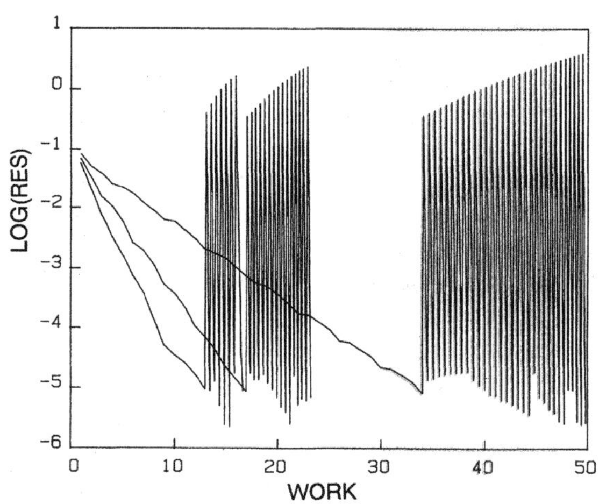

FIGURE 11.6. Supersonic-flow convergence histories.

chosen representation is believed to be very informative. From Fig. 11.6, in fact, one can easily count the number of supersonic columns and compare the relative computational cost of the subsonic and supersonic regions.

In conclusion, a new method has been developed for computing two-dimensional compressible internal flows accurately and efficiently. The method, designed to compute subsonic-to-supersonic flows, has been shown to be competitive with respect to a recently developed state-of-the-art scheme, when applied to solve subsonic and possibly transonic flows.

Acknowledgement. This work has been supported by CNR-PFE2 Grant no. 88.1301.59. The authors are grateful to A. Dadone and A. Lippolis, for fruitful discussions during the entire duration of this research, and to one of the referees, for his useful comments.

Nomenclature

a	= speed of sound.
C, D, E, F	= bicharacteristic variables defined in Eqs. 5–8.
$\bar{C}, \bar{D}, \bar{E}, \bar{F}$	= perturbative bicharacteristic variables, namely, differences between the "compressible-flow" variables and the corresponding incompressible-flow variables.
h_1, h_2	= scale factors of the orthogonal curvilinear coordinate system.
$k1_{1,2}, \ldots, k5$	= coefficients depending on the geometry and the incompressible-flow solution.
q_1, q_2	= orthogonal curvilinear coordinates.
RES(N)	= residual of Eq. (N).
t	= time.

Greek Letters

γ	= specific heat ratio.
δ	= gas constant $\delta = 2/(\gamma - 1)$.
∂	= partial derivative sign.
Δt	= time step.
$\Delta C, \Delta D, \Delta E, \Delta F$	= Variations of $\bar{C}, \bar{D}, \bar{E}, \bar{F}$ between the new and old iteration.

Subscripts

t	= time derivative.

References

Abbrescia, B., Dadone, A., and Napolitano, M., 1984, "Implicit Lambda Schemes for Cascades Flows," *Inst. of Mech. Eng.*, paper no. C62/84.

Baker, T.J., 1987, "Three Dimensional Mesh Generation by Triangulation of Arbitrary Point Sets," *Proceedings of the AIAA 8th CFD Conference*, Honolulu, Hawaii, June 9–11, 255–271.

Beam, R.M., and Warming, R.F., 1978, "An Implicit Factored Scheme for the Compressible Navier-Stokes Equations," *AIAA J.*, 16, April, 393–402.

Dadone, A., Fortunato, B., and Lippolis, A., 1989, "A Fast Euler Solver for Two- and Three-Dimensional Internal Flows," *Computers and Fluids*, 17, 25–37.

Dadone, A., and Moretti, G., "Fast Euler Solver for Transonic Airfoils, Part I: Theory," *AIAA J.*, 26, April, 409–416.

Dadone, A., and Napolitano, M., 1983, "An Implicit Lambda Scheme," *AIAA J.* 21, Oct., 1391–1399.

Dadone, A., and Napolitano, M., 1985a, "Accurate and Efficient Solutions of Compressible Internal Flows," *J. of Propulsion and Power*, 1, 456–463.

Dadone, A., and Napolitano, M., 1985b, "An Efficient ADI Lambda Formulation." *Computers and Fluids*, 13, 383–395.

Dadone, A., and Napolitano, M., 1986, "A Perturbative Lambda Formulation," *AIAA J.* 24, March, 411–417.

Jameson, A., 1987, "Successes and Challenges in Computational Aerodynamics," *Proceedings of the AIAA 8th CFD Conference*, Honolulu, Hawaii, June 9–11, AIAA Paper 87-1184, AIAA CP 874, 1–35.

Karamcheti, K., 1966, *Principles of Ideal Fluid Aerodynamics*, Wiley.

Moretti, G., 1979, "The λ-Scheme," *Computers and Fluids*, 7, 191–205.

Napolitano, M., 1986, "Simulation of Compressible Inviscid Flows: The Italian Contribution," 10th International Conference on Numerical Methods in Fluid Dynamics, Beijing, June. In *Lecture Notes in Physics*, 264, Springer-Verlag, 47–56.

Napolitano, M., 1988, "Efficient Solution of Two-Dimensional Steady Separated Flows," International Symposium on Computational Fluid Dynamics, Sydney, August, 1987; In G. deVahl Davis and C. Fletcher (Eds.), *Computational Fluid Dynamics*, North-Holland, 89–102.

Napolitano, M., and Dadone, A., 1985, "Implicic Lambda Methods for Three-Dimensional Compressible Flows," *AIAA J.*, 23, Sept., 1343–1347.

Pandolfi, M., 1985, "The Merging of Two Different Ideas: A Shock Fitting Performed by a Shock Capturing," *International Symposium on Computational Fluid Dynamics*, Tokyo, Sept.

Van Leer, B., and Mulder W.A., 1984, "Relaxation Methods for Hyperbolic Equations," *Delft University of Technology Report no.* 84–20.

Volpe, G., Siclari, M.J., Jameson, A., 1987, "A New Multigrid Euler Method for Fighter-Type Configurations," *Proceedings of the AIAA 8th CFD Conference*, Honolulu, Hawaii, June 9–11, 627–646.

Walters, R.W., and Thomas, J.L., 1987, "Advances in Upwind Relaxation Methods," In *State-of-the-Art-Surveys on Computational Mechanics*, ASME Publication.

Zannetti, L., and Colasurdo G., 1981, "Unsteady Compressible Flows: A Computational Method Consistent with the Physical Phenomena," *AIAA J.*, 19, July, 851–856.

12
An Upwind Formulation for Hypersonic Nonequilibrium Flows

M. PANDOLFI AND S. BORRELLI

ABSTRACT: We investigate the interaction between fluid dynamics and nonequilibrium chemistry for air in the hypersonic regime and propose a methodology for achieving the numerical prediction of flows of this kind. Transport phenomena, leading to the viscosity, thermal conductivity, and diffusion of chemical species are neglected so the fluid dynamics is described by the Euler equations. The chemical nonequilibrium is based upon a classical 5 species and 17 reactions model. The flux-difference splitting formulation is assumed to be the basis of the algorithm for the Euler equations and is extended to include the nonequilibrium chemical phenomena. Some numerical experiments are presented about the chemical relaxation occurring behind a shock in a nozzle and the attention is focused on the effects of the Damköhler number.

Introduction

The physics of hypersonic flows is rather complicated. Shock waves and contact surfaces characterize the basic features of the fluid dynamics. The viscosity is responsible for wall effects such as boundary layers and separations that often extend over a wide region of the physical domain. The thermal conductivity contributes largely to the energy balance because of the strong temperature gradients. Finally vibrational and chemical processes take place with relaxation times either similar (nonequilibrium flow) or quite different from fluid-dynamic times (frozen or equilibrium approximations).

The Euler equations describe the basic fluid dynamics. The additional terms included in the Navier-Stokes equations and related to the viscosity and thermal conductivity, complete the description of the fluid motion. Thermodynamic and chemical models lead to the description of the relaxation phenomena and contribute to the closure of the system of governing equations. Matching and interaction of the fluid dynamics with the relaxations characterize the hypersonic flow with respect to the supersonic regime, where only fluid dynamics comes into the physical picture.

In order to understand the interactions between the fluid dynamics and the chemical relaxations, we can carry out investigations based upon relatively simple modeling of the actual physics. Therefore we assume the Euler equations for the description of the fluid dynamics, we approximate the thermodynamical relaxation for the vibration with a fixed level of excitation (in particular a half-excited level), and we interpret the chemistry with a model of 5 species (O, N, NO, O_2, N_2) and 17 reactions. We point out that, even within such a simple description of the actual phenomenology, useful and practical information can be drawn from the appropriate numerical analysis.

In the following, we present the basic governing equations and the *flux-difference splitting* (FDS) formulation for the nonequilibrium flow. Then we discuss the approximate solution of the Riemann problem, on which the FDS formulation is founded and report on the integration scheme. Finally we present some numerical experiments, pointing out the role and effects of the Damköhler number.

The Equations

We consider the Euler equations for the quasi-one-dimensional flow in a duct and adopt for the description of the real gas effects (finite rate equations) the model given in Park (1985) and largely used in literature.

Since we are interested in developing a procedure that allows a correct numerical capturing of discontinuities, we write the conservative form of the governing equations. By introducing the usual notation and indicating the cross-section area along the duct with $A(x)$, we have:

$$W_t + F_x + G = 0 \tag{1}$$

where $W = (wA)$, $F = (fA)$, and $G = (gA)$.

The vectors w, f, and g are defined as:

$$w = [\rho_1, \rho_2, \rho_3, \rho, \rho u, e]^T$$

$$f = [\rho_1 u, \rho_2 u, \rho_3 u, \rho u, (p + \rho u^2), u(p + e)]^T$$

$$g = \left[\Omega_1, \Omega_2, \Omega_3, 0, -p\left(\frac{A_x}{A}\right), 0\right]^T.$$

The first three equations of the system (Eq. 1) refer to the species O, N, NO, denoted, respectively, by $i = 1, 2, 3$. The diffusion of the species is neglected and only convections and productions are considered. The partial density ρ_i is related to the mixture density ρ and to the mass concentration Y_i or to the molar one q_i by:

$$\rho_i = \rho Y_i = \rho \mu_i q_i \tag{2}$$

where μ_i denotes the molecular mass. The concentrations Y_4 of the molecular oxygen O_2 and Y_5 of the molecular nitrogen N_2 follow from the conservation of the atomic species:

$$Y_4 = Y_{4\infty} - \frac{\mu_4}{2}\left(\frac{Y_1}{\mu_1} + \frac{Y_3}{\mu_3}\right) \tag{3}$$

$$Y_5 = Y_{5\infty} - \frac{\mu_5}{2}\left(\frac{Y_2}{\mu_2} + \frac{Y_3}{\mu_3}\right). \tag{4}$$

The ∞ values refer to the upstream undisturbed air concentrations ($Y_{1\infty} = Y_{2\infty} = Y_{3\infty} = 0$). The rates of production of the species Ω_i appear in the vector g. They are evaluated on the basis of formulas and constants suggested in Park (1985). With $R_i = R/\mu_i$ being the constant elasticity of the i species, the specific heat at constant pressure is given by:

$$c_{pi} = \tfrac{5}{2}R_i \quad \text{(for O, N)} \tag{5}$$

$$c_{pi} = (\tfrac{7}{2} + \tfrac{1}{2})R_i \quad \text{(for NO, } O_2, N_2\text{)}. \tag{6}$$

The additional 1/2 for the molecules refers to the approximation of the half-excited vibration.

The equation of state of the gas mixture is given by:

$$\frac{p}{\rho} = \sum_{i=1}^{5} R_i Y_i T \tag{7}$$

and the enthalpy is defined as:

$$h = h_{for} + \sum_{i=1}^{5} c_{pi} Y_i T \tag{8}$$

Here h_{for} represents the total heat of formation:

$$h_{for} = \sum_{i=1}^{3} Y_i h_i^o \tag{9}$$

where h_i^o is the heat of formation of the i species.

From the previous equations we obtain:

$$h = \gamma\left(\frac{e}{\rho} - \frac{u^2}{2}\right) + (\gamma - 1)h_{for} \tag{10}$$

Here γ represents the ratio of specific heats and is defined as:

$$\gamma = \frac{\sum_{i=1}^{5} c_{pi} Y_i}{\sum_{i=1}^{5} (c_{pi} - R_i) Y_i}. \tag{11}$$

It is also convenient to write:

$$p = \frac{\gamma - 1}{\gamma}\rho(h - h_{for}) \tag{12}$$

$$T = \frac{p}{\rho}\frac{1}{\sum_{i=1}^{5} R_i Y_i} \tag{13}$$

Finally we remind the reader that the definition of the frozen speed of sound a_f is given by:

$$a_f^2 = \frac{h_\rho}{1/\rho - h_p} = \gamma \frac{p}{\rho}. \tag{14}$$

All terms in these equations are defined. From the updated value of the vector W (Eqs. 1) and the cross-section area A, we obtain ρ_1, ρ_2, ρ_3, ρ, u, and e. Then we evaluate the concentrations Y_i from Eqs. 2, 3, and 4. The total heat of formation comes from Eq. 9 and the enthalpy from Eq. 10. Finally the pressure and temperature follow from Eqs. 12 and 13. The rates of reaction Ω_i are evaluated from Y_i, ρ, and T, according to the suggestions given in Park (1985).

The Flux-Difference Splitting Formulation

The system of Eq. 1 is hyperbolic. It is then convenient to put into evidence the propagation of signals in the x-t domain, along appropriate rays (characteristic lines). So we recognize that the evolution in time, at a given point, is provided by the merging of information that is convected along the rays reaching the point. The formulation represents the step where we work out the proper arrangement of the governing equations and give a suitable interpretation of the initial data. Often we specify *upwind* formulation to emphasize the role of the convection of the signals.

The *flux-vector splitting* or *flux-difference splitting* (FDS) formulations are used within the context of the conservative approaches. Despite the similarity in the denomination, the two formulations are rather different. In the present investigation, we base our analysis on the FDS formulation. We refer the reader not familiar with this formulation to Pandolfi (1989) for a comparative review of different FDS approaches.

First, we will consider only the Euler equation for a perfect nonreacting gas. The FDS formulation is founded upon the interpretation of the initial data as constant values distribution of the flow properties over the cell that extends about any computational point. A discontinuity appears at the interface between two neighboring cells. The evolution in time, of such a discontinuity, provides information useful to proceed to the splitting of the difference of the flux between the neighboring points and to evaluate a proper *upwind* approximation of F_x in Eq. 1. The prediction of the collapse of the initial discontinuity is obtained through the solution of a Riemann problem. Up to this point, we have only reported the original suggestions proposed in Godunov (1959).

It is well known how expensive the development of the exact solution of the Riemann problem turns out to be and how little we can profit from these exact results when, in the following step, we introduce them into a numerical scheme. Therefore approximate solutions have been proposed in the literature, in order to simplify the solution of the Riemann problem and reduce the computational efforts without penalizing the quality of the final numerical results.

Two approximate solvers are quite popular. They have been developed and presented some years ago and have since been tested in a large variety of applications. In one of them (Roe 1981), the solution of the Riemann problem is carried out on the basis of a linearized version of the Euler equations. In the other (Osher and Solomon 1982) the nonlinearity of the original equations is retained, but shock waves are approximated with isentropic compression fans; moreover the waves generated by the collapse of the discontinuity are approximated with those that could generate the initial discontinuity by merging together at the interface. A third approximate solution has been proposed by one of the authors of this chapter. It is somehow located between the previous two and is perhaps closer to the second one. Such an approximate solver is presented in a detailed form in Pandolfi (1984) for the perfect nonreacting gas. Here it is extended to the flow of a reacting gas.

With reference to Fig. 12.1, we approximate the concentrations of the species after the collapse of the discontinuity to be frozen. Therefore, the initial concentrations in regions a and b, given by the initial data and generally different from each other remain unchanged through the *acoustic* waves (I, III), respectively in regions c and d. This approximation is added to the original assumption of considering isentropic the evolution through the *acoustic* waves and agrees with it. In conclusion, we describe the solution of the Riemann problem on the basis of the following equations:

$$q_{it} + uq_{ix} = 0 \qquad (i = 1, 2, 3) \tag{15}$$

$$p_t + up_x + \rho a_f^2 u_x = 0 \tag{16}$$

$$u_t + uu_x + \frac{p_x}{\rho} = 0 \tag{17}$$

$$h_t - \frac{p_t}{\rho} + u\left(h_x - \frac{p_x}{\rho}\right) = 0. \tag{18}$$

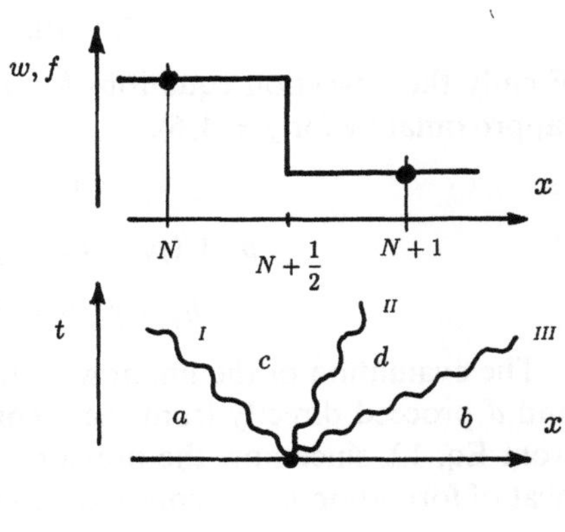

FIGURE 12.1. Interpretation of the initial data and collapse of the discontinuity.

The frozen speed of sound that appears in the continuity equation is defined in Eq. 14. The set of Eqs. 15, 16, 17, and 18 represents the quasi-linear form of the governing conservative form of the system in Eq. 1 that describes the nonequilibrium flow. However, the chemical process is assumed to be frozen since the rate of the reactions Ω_i is set equal to zero. A proper arrangement of these quasi-linear equations leads to the equations that express the advection of signals:

$$R_{jt} + \lambda_j R_{jx} = 0. \tag{19}$$

Here λ_j represents the slope of the characteristic rays:

$$\lambda_j = u \qquad (j = 1, 2, 3, 6)$$
$$\lambda_4 = u - a_f$$
$$\lambda_5 = u + a_f$$

and dR_j is the corresponding signal:

$$dR_j = dY_j \qquad (j = 1, 2, 3)$$
$$dR_4 = dp - \rho a_f\, du$$
$$dR_5 = dp + \rho a_f\, du$$
$$dR_6 = dh - \frac{dp}{\rho}.$$

On the basis of the advection equations, we can evaluate the flow properties in regions c and d. With reference to Fig. 12.1, we consider Eq. 19 for $j = 1, 2, 3, 5, 6$, and we note that through the wave I, we have (approximately for $j = 5, 6$):

$$Y_{ic} = Y_{ia} \qquad (i = 1, 2, 3)$$
$$p_c + (\rho_a a_{fa})u_c = p_a + (\rho_a a_{fa})u_a$$
$$h_c - p_c/\rho_a = h_a - p_a/\rho_a.$$

On the contact surface (wave II), we impose the usual continuity of the pressure and velocity:

$$p_c = p_d; \qquad u_c = u_d$$

Finally the advection equations for $j = 1, 2, 3, 4, 6$, through wave III, give (approximately for $j = 4, 6$):

$$Y_{id} = Y_{ib} \qquad (i = 1, 2, 3)$$
$$p_d + (\rho_b a_{fb})u_d = p_b + (\rho_b a_{fb})u_b$$
$$h_d - p_d/\rho_b = h_b - p_b/\rho_b.$$

The evaluation of the unknown values Y_1, Y_2, Y_3, p, h, and u in regions c and d proceed directly from these conditions. The density is then obtained from Eq. 12, since now the pressure and enthalpy are known and the total heat of formation h_{for} is computed from Eq. 9.

We now look at the direction of propagation of each wave and identify the region (one among a, b, c or d) that extends in time, at the location of the interface $x_{N+1/2}$, after the collapse of the discontinuity. For example, in the case of Fig. 12.1, such a region is the region c. Then we define the flux-vector at the interface, on the basis of the flow properties that pertain to this region. Always in the particular case of Fig. 12.1, we have:

$$f_{N+1/2} = [\rho_{1c}u_c, \rho_{2c}u_c, \rho_{3c}u_c, \rho_c u_c, (p_c + \rho_c u_c^2), u_c(p_c + e_c)]^T.$$

The most interesting cases occur when a sonic transition shows up within one of the fans that describe acoustic waves I and III. In these cases one characteristic is vertical, being either $u = a$ or $u = -a$, and we can have a sonic expansion or a compression fan approximating a shock that is at rest or moving slowly. Such configurations are shown respectively in Fig. 12.2 and Fig. 12.3. As reported in Pandolfi (1984), the splitting is now operated not only among waves, but also inside that wave in which the sonic characteristic is embedded. The procedure is carried out as follows.

First of all, we predict the value (*) on the sonic transition. Let us presume that this occurs inside wave I. The advection equations (Eq. 19) for $j = 1, 2, 3, 5, 6$ relate flow properties at this sonic point with values in region a:

$$Y_i^* = Y_{ia} \qquad (j = 1, 2, 3)$$

$$p^* + (\rho_a a_{fa})u^* = p_a + (\rho_a a_{fa})u_a$$

$$h^* - p^*/\rho_a = h_a - p_a/\rho_a.$$

Then we impose the condition of the sonic point on the vertical characteristic:

$$u^* = a_f^*.$$

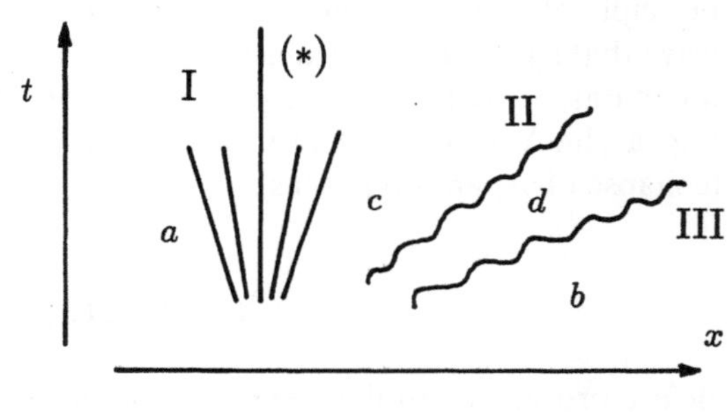

FIGURE 12.2. Expansion wave with the sonic transition (*).

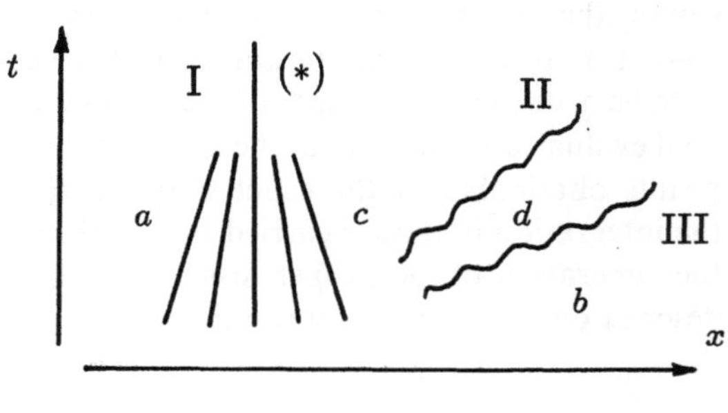

FIGURE 12.3. Compression wave with the sonic transition (*).

The frozen speed of sound comes from Eqs. 12 and 14:

$$a_f^* = \sqrt{(\gamma - 1)(h^* - h_{\mathrm{for}})}.$$

Here γ and h_{for} have the same values as in region a because the concentrations do not change throughout wave I. So the flow properties at ($*$) are obtained. Finally, we compute the flux f^*:

$$f^* = [\rho_1^* u^*, \rho_2^* u^*, \rho_3^* u^*, \rho^* u^*, (p^* + \rho^* u^{*2}), u^*(p^* + e^*)]^T.$$

For the wave pattern shown in Fig. 12.2 (expansion fan), the flux at the interface is defined as:

$$f_{N+1/2} = f^*$$

For the case of Fig. 12.3 (a compression fan that approximates a shock), we evaluate the fictitious intermediate value that respects the domains of dependence suggested by the splitting:

$$f_{N+1/2} = f_N + (f_c - f^*)$$

or, with the equivalent form:

$$f_{N+1/2} = f_{N+1} - (f_b - f_d) - (f_d - f_c) - (f^* - f_a).$$

The explicit introduction of the sonic characteristic ($*$) and the related splitting inside the wave leads to two positive features in the numerical results (Pandolfi 1984). The first is represented by the neat and sharp description of numerically captured shock waves even on the basis of a plain first-order scheme, and without the need of introducing any artificial viscosity and related parameters. The second feature is given by the natural way of preventing the formation of expansion shocks and violations of the second principle of thermodynamics. In particular the splitting of the content of a wave that shows a sonic transition, in two contributions that act in opposite directions, helps the piling up of compression waves (in the attempt of simulating a shock wave) and promotes the spreading of the expansion waves (collapse of expansion shocks).

The Integration

Before proceeding to the integration, let us emphasize a particular point. In solving the Riemann problem we have introduced some approximations. They consist mainly in assuming isentropic the acoustic waves and to freeze the chemistry during the collapse of discontinuity. Therefore, the splitting and the final evaluation of the flux at the interfaces can be somehow different from the results obtainable by the exact solution. However, let it be reminded that the integration in time is carried out on the complete equations (Eq. 1), with the generation of the proper dissipation through shock waves and the full description of the chemical reactions.

In the present investigation we describe our numerical procedure based upon a first-order-accuracy scheme. The updating is obtained as follows:

$$(wA)_N^{K+1} = (wA)_N^K - \frac{Dt}{Dx}(f_{N+1/2}^k A_{N+1/2} - f_{N-1/2}^k A_{N-1/2}) - (gA)_N^K Dt. \quad (20)$$

At moderate values of the Damköholer number, that is, for a nonequilibrium flow far from equilibrium conditions and nearly close to the frozen one, stable numerical solutions are provided by the fully explicit integration scheme shown earlier. However, by increasing the Damköhler number and approaching the equilibrium conditions, the source terms in the chemical equations (the first three of Eq. 1) induce severe numerical instabilities. The problem is known and expected. The solution is easily found by carrying out an implicit evaluation of the source terms that express the rate of reactions, with reference to their dependence on the concentrations of the species. Therefore we proceed with a half-implicit procedure (implicit evaluation for the source term and explicit for the convection one), adopted for the chemical equations, while we follow a fully explicit algorithm for the last three of Eq. 1, that refer to the fluid dynamics.

Numerical Experiments

We have performed some numerical examples on the basis of the methodology presented earlier. The studied problem is simple, but quite significant to check the accuracy of the numerical results and the robustness of the algorithm and to provide physical insights on some aspects of flows in nonequilibrium.

We have considered a divergent nozzle with the exit cross-section area double to the inlet one. The conditions at the inlet are prescribed: $M_\infty = 20$, $p_\infty = 10$ (N^2/m) and $T_\infty = 250$ (°K). The air is assumed to be a mixture of molecular oxygen and nitrogen ($Y_4 = 0.233$ and $Y_5 = 0.767$). The static pressure at the exit is prescribed at the level of about 4200 (N^2/m). This value brings to flow configurations with a shock wave inside the nozzle that separates the front supersonic region from the rear subsonic one. The length l_∞ of the nozzle has not yet been defined. It will assume different values, to simulate the effects of the Damköhler number.

Computations have been performed for $l_\infty = 0.001$, 0.010, 0.100, and 1.000 (m). We expect the shock be located at different stations; further downstream the value of l_∞ becomes larger. Nevertheless, we anticipate that the flow conditions just ahead of it do not change much. So the relaxation process behind the shock will be governed by the same values of thermodynamical properties (density and temperature, in particular) and take place over the same physical distance in the different cases. We call this distance l_{ch} and assume it as reference length of the chemical relaxation. The ratio of the reference lengths is proportional to the Damköhler number.

In the case of a long nozzle, say $l_\infty = 1.000$ (m), the Damköhler number is rather high. Since the nonequilibrium chemical process evolves over distances (l_{ch}) that are finite but very small with respect to the fluid dynamic distance (l_∞), the global picture of the flow looks very similar to the equilibrium configuration. On the other hand, if we consider a very short nozzle, say $l_\infty = 0.001$ (m), the Damköhler number tends to vanish. Now the resldence time of the gas inside the nozzle is so small that no time is allowed for the chemical reactions to appreciably change the concentrations of the species. Therefore, the flow picture is very close to the frozen configuration.

We emphasize that in both cases the gas behind the shock goes into the same relaxation, with the same chemical reference length (l_{ch}). The resemblance with the equilibrium and frozen configurations is only related to the fluid dynamic distance (l_∞), the length of the nozzle, over which we observe the phenomenon. For large l_∞ the nonequilibrium process (that extends for l_{ch}) is confined to a very small fraction of the nozzle and an equilibrium-like configuration appears all over the nozzle. At small l_∞, the reactions do not develop appreciably along the nozzle and we do not observe changes in the concentrations of the species, just as in the frozen flow.

We predict steady flow configurations on the basis of the partial differential equations (PDE) of the system (Eq. 1), according to the *time-dependent technique*. As an initial configuration, we have assumed the Eulerian flow for the nonreacting perfect gas. It will be convenient to compare these numerical results with the exact solutions. These benchmark solutions are obtainable by setting the time derivative (W_t) equal to zero in Eq. 1. The resulting system of ordinary differential equations (ODE) is easily integrated along the only independent variable left, that is, x. Starting from the inlet station, where all the properties are prescribed, the flow looks very smooth in the supersonic region, and the integration here can be carried out with only a few points that is a relatively large step along x. The concentrations do not change because of the low temperature. We terminate the supersonic region at the shock, where we apply the Rankine-Hugoniot conditions in order to have the properties on the high-pressure side of it. The concentrations still remain unchanged through the shock. Finally we integrate the ODE from the shock, down to the exit and here we use a number of points (or steps along x) large enough to preserve accuracy, especially in the relaxation region just behind the shock. The location of the shock that we have previously assumed will be detemined by an iterative procedure, in order to match the pressure computed at the end of the nozzle with the prescribed exit pressure. Even if these solutions are obtained by the numerical integration of the ODE along x, they can be considered *exact* because a very large number of points (rather fine step in x) can be used. The computational times required to achieve these exact solutions are negligible.

We now consider the results obtained by integrating in time the PDE of the system (Eq. 1). For $l_\infty = 0.001$ (m), we anticipate that the Damköhler number is so small that no difference is detectable with the frozen flow configuration. Therefore, only the cases of $l_\infty = 0.010$, 0.100, and 1.000 (m) will be discussed.

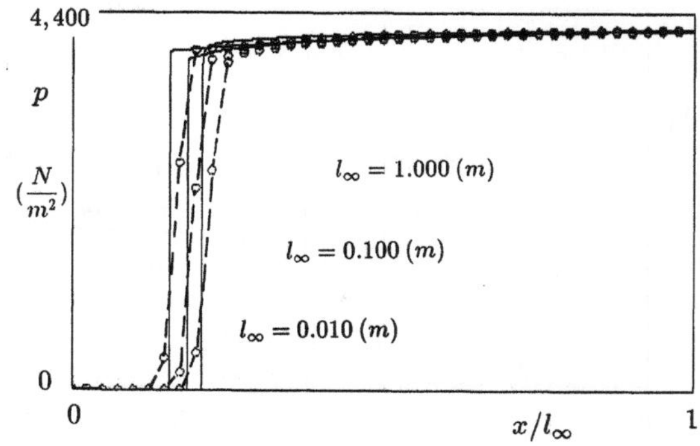

FIGURE 12.4. Distribution of the pressure at different lengths of the nozzle.

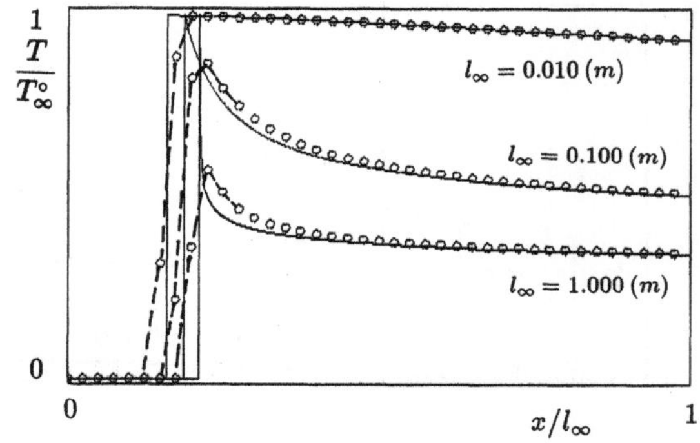

FIGURE 12.5. Distribution of the temperature at different lengths of the nozzle.

The pressure distribution is shown in Fig. 12.4. The symbols refer to the numerical results from the integration of the PDE (41 points are used along the nozzle) and the solid lines to the exact solution from the ODE. We point out the sharp description of the numerically captured shock as well as the agreement with the exact solutions, features that seem to be not penalized by the effects of the nonequilibrium. The shock location moves downstream with larger l_∞, an expected trend related to the development of the dissociation within the nozzle and related involved energy.

The distribution of the temperature is shown in Fig. 12.5. For $l_\infty = 0.010$ (m) the temperature decreases fairly behind the shock because of the weak development of the chemical reactions. At $l_\infty = 0.100$ (m), we have a typical nonequilibrium flow; the chemical reactions develop along the nozzle and the temperature behind the shock decreases strongly, but continuously. Finally, for $l_\infty = 1.000$ (m), the relaxation is concentrated in a small thickness of the normalized abscissa x/l_∞, just behind the shock. Here, the numerical results follow badly the exact solution, which presents a very steep decreasing of the temperature followed by an almost flat distribution. The theoretical peak of the temperature at the shock is also lost in the numerical results and the chemical relaxation tends to be swallowed in the structure of the numerically captured shock.

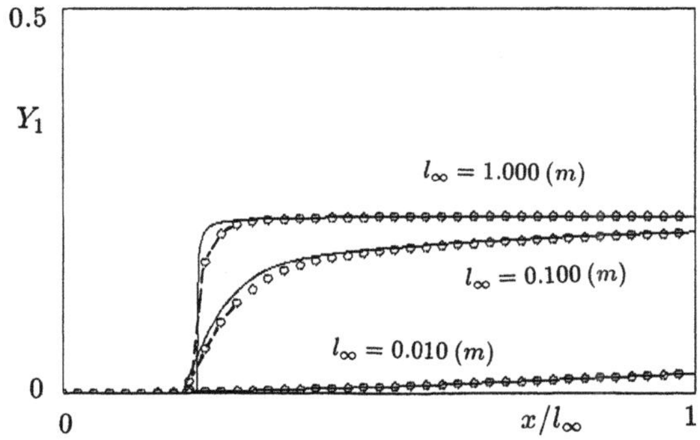

FIGURE 12.6. Concentration of the atomic oxygen (Y_1) at different lengths of the nozzle.

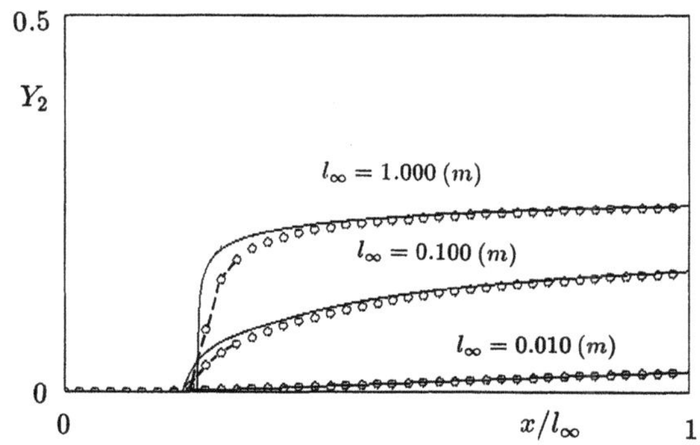

FIGURE 12.7. Concentration of the atomic nitrogen (Y_2) at different lengths of the nozzle.

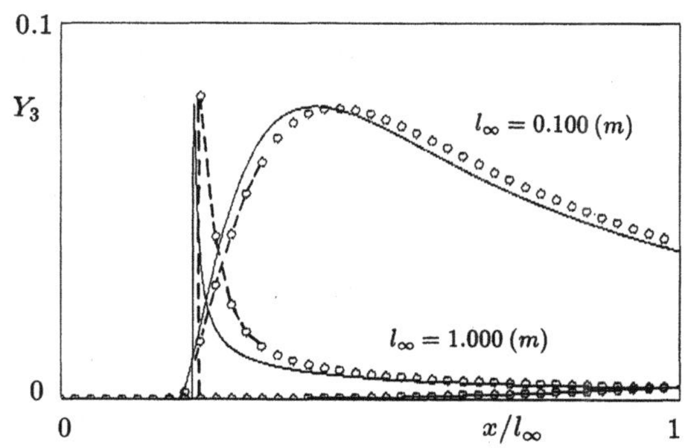

FIGURE 12.8. Concentration of the nitric oxide (Y_3) at different lengths of the nozzle.

These distributions of the temperature are related to the production of the new species (O, N, NO) behind the shock. The concentrations Y_1 and Y_2 of the atomic oxygen and nitrogen and Y_3 of the nitric oxide are, respectively, shown in the Figs. 12.6, 12.7, and 12.8. As is well known, the distribution of the concentration of the atomic nitrogen is similar to that of the atomic oxygen, but develops more slowly and later along the nozzle. The production of the atomic species promotes the increasing of the total heat of formation and the temperature falls. Finally, we note the typical overshooting of the concentration Y_3 of the nitric oxide (Fig. 12.8). For $l_\infty = 0.010$ (m), the maximum does not appear inside the nozzle as it does for the larger lengths. At $l_\infty = 0.100$ (m), the overshooting is smoothly distributed all over the nozzle, while for $l_\infty = 1.000$ (m) is reduced to the usual sharp peak behind the shock.

The agreement of the numerical results from the PDE and the exact solutions obtained by the ODE deteriorates with the increasing of the Damköhler number. This is due to the lack of computational points in describing the region of the chemical relaxation when this zone is embedded inside a much longer nozzle. Improvements can be obtained by using more computational points or by adopting more accurate schemes, since the region of nonequilibrium is described by differential equations and cannot be captured numerically, as can be done for a discontinuity such as a shock wave.

Conclusions

We have presented a numerical method for the prediction of nonequilibrium flows in the hypersonic regime. The procedure is based upon an upwind formulation for the conservative form of the fluid dynamic equations. Here we have focused the attention on the central core of the methodology and its application to the one-dimensional problem.

The results we have shown indicate a promising extension to multidimensional problems. Actually, such an extension has almost been fullfilled. A large number of satisfactory predictions have been obtained for the flow about a double-ellipse body. The results will be presented at the Workshop on *Hypersonic Flows for Reentry Problems (INRIA, GAMNI-SMAI)*, where a special session is planned on this problem. The comparison among the contributions proposed by many researchers will be fruitful in assessing the reliability of the procedures.

Nomenclature

a = speed of sound.
c_p = specific heat at constant pressure.
e = total internal energy per unit volume.
h = enthalpy.
p = pressure.
q = molar concentration of species.
R = universal gas constant.
T = temperature.
t = time.
u = velocity.
x = x coordinate.
Y = mass fraction of species.

γ = specific heat ratio.
μ = molecular weight.
ρ = density.
ω = chemical production rate species.

Subscripts

f = frozen.
i = of the ith species.
j = of the jth characteristic ray.

References

Godunov, S.K., 1959, "A Finite Difference Method for the Numerical Computation of Discontinuous Solutions of the Equations of Fluid Dynamica," *Math. Sb., 47.*

Osher, S., and Solomon, F., 1982, "Upwind Difference Schemes for Hyperbolic Systems of Conservation Laws," *Mathematics of Computations, 38.*

Pandolfi, M., 1984, "A Contribution to the Numerical Prediction of Unsteady Flows," *AIAA J., 22,* 5.

Pandolfi, M., 1989, "On the Flux Difference Splitting Formulation," *Notes on Numerical Fluid Mechanics, 24,* Vieweg.

Park, C., 1985, "On Convergence of Computation of Chemically Reacting Flows," *AIAA Paper-85-0247,* Jan.

Roe, P.L., 1981, "Approximate Riemann Solvers, Parameters Vectors and Difference Schemes," *J. of Computational Physics, 43.* 1981.

13
Numerical Methodologies for the Compressible Navier-Stokes Equations for Two-Phase Flows

F. GRASSO AND V. MAGI

ABSTRACT: In the present work, the attention is focused on two-phase flows containing liquid droplets. The physics of sprays and numerical methodologies for the solution of the compressible Navier-Stokes equations for two-phase flows are reviewed, and a novel numerical approach is presented. Applications of the method to sprays in constant pressure ambient and in confined volumes are reported, and results are compared with available experimental data.

Introduction

In the present paper we focus our attention on two-phase flows containing liquid droplets. In particular we examine the physics of full and hollow cone sprays, their structure and modeling. The sprays of interest are assumed to be made out of a liquid phase dispersed into the gas (Bracco 1985; Dukowicz 1980; O'Rourke 1981).

According to the classification of O'Rourke (1981), four different regimes can be identified during the lifetime of such a spray: (1) churning; (2) thick; (3) thin; and (4) very dilute. The churning flow regime occurs very near the injector, where liquid ligaments are still present and the spray cannot be assumed as a disperse phase. However, churning flow is still a subject of research for understanding the physics. Hence attention here is focused on sprays that evolve starting from the thick-spray regime, whereby the effects of the churning flow are introduced via simplified models (Bracco 1985; Chatwani and Bracco 1985; Magi 1987; Magi and Grasso 1985; O'Rourke 1981; O'Rourke and Amsden 1987; Reitz and Diwakar 1986) that will be described in later sections. Thick-spray regimes are characterized by liquid particles dispersed in a continuous gas phase, and droplet-gas as well as drop-drop interactions are very important on account of relatively high mass and volume occupied by the liquid phase. In thin-spray regimes, the liquid-to-gas volume ratio is very small (typically ≤ 0.1) so that drop-to-drop interactions are negligible and the droplets can be treated separately. The very thin-spray regimes consist of very small droplets of negligible mass and volume, and therefore their presence practically does not influence the gas phase.

Different approaches have been followed for the modeling and computation of such sprays. However, they all amount to solve Williams' spray equation, developed in the late 1950s (Williams 1965). A deterministic solution of such an equation has been followed by various authors. For example, Gupta and Bracco (1978) and Haselman and Westbrook (1978) obtained a finite difference solution of the spray equation in the phase space identified by physical coordinates, velocity components, size and temperature of droplets. However such an approach has its major drawback in the enormous computer storage requirement.

Dukowicz (1980) was the first to follow a stochastic approach by solving Williams' spray equation with a Monte Carlo technique: the spray is assumed to be composed of parcels, each representing a class of particles having the same properties (size, position, velocity, temperature, etc.), obtained by sampling the distribution of drops near the injector.

Following the idea of Dukowicz, several papers have appeared. O'Rourke (1981) has contributed to develop a model to account for drop-drop interaction. The atomization and injection processes have been dealt with in different ways by different authors. Reitz and Diwakar (1986) have suggested injecting liquid blobs the same size of the nozzle exit diameter. The breakup of these large drops occurs whenever stability criteria based on Weber number inequalities are not satisfied, as suggested by Nichols (1972). Chatwani and Bracco (1985) have assumed the existence of an intact core length that acts as a line source of drops. O'Rourke and Amsden (1987) suggested the so-called Taylor analogy breakup (TAB) model, implying an analogy between an oscillating distorting droplet and a spring mass system.

One of the most important processes controlling the evolution of sprays in gases is turbulence. On account of large velocity gradients typical of flows of gases containing liquid particles, and due to the presence of drops occupying a large fraction of the gas volume, in principle one must account for gas- and liquid-phase turbulence effects, and a turbulent interaction mechanism between the dispersed phase and the carrier fluid. Regarding the gas phase turbulence in general two approaches are followed: (1) constant diffusivity models (Hotchkiss and Hirt 1972; Margolin 1978), and (2) two-equation turbulence models (Grasso 1981; Launder and Spalding 1974), usually based on the use of $k - \varepsilon$ models, where k is the turbulence kinetic energy and ε its dissipation rate. The use of constant-diffusivity models has its advantages amounting to simplicity, and drawbacks mainly because of the gross simplification in describing the physics of turbulence, particularly in the vicinity of regions characterized by strong gradients (El-Tahry 1985). A $k - \varepsilon$ model allows to partly overcome the drawbacks typical of the models previously described and it is shown to have good prediction capabilities (El-Tahry 1985; Grasso 1981). However, drawbacks stem from the gradient-diffusion assumption and the isotropic definition of diffusion coefficients. Regarding the turbulence effects on the particles, the main effect is particle diffusion (Dukowicz

1980; El-Tahry 1985; Elghobashi and Abou-Arab 1983; Grasso 1981; Launder and Spalding 1974; Margolin 1978). The latter is generally accounted for by introducing a term equivalent to an external force that modifies the drag acting on the particles: a fluctuating component, selected randomly from an isotropic Gaussian distribution with root mean square deviation (r.m.s.) equal to $\sqrt{2/3k}$, is added to the mean gas velocity. Recently Andrews and Bracco (1989) have addressed the problem of anisotropy effects on smaller droplets by redistributing the turbulence kinetic energy according to experimentally measured drop fluctuations. There have been several attempts to predict the turbulent interaction mechanism between the two phases. Elghobashi and Abou-Arab (1983) developed a turbulence model that accounts for the direct effects of the particles on the gas turbulence. However Martinelli et al. (1984) argue that these effects are limited to the region near the injector and are negligible in the far field where the mass of the drops is small compared to the entrained gas.

The solution of the Reynolds averaged compressible Navier-Stokes equations, reformulated to account for droplet-gas interactions, generally employs a Lagrangian description of the liquid phase to avoid particle numerical diffusion (Aggarwal et al. 1984; Dukowicz 1980) and an Eulerian formulation for the gas phase.

Existing methods to solve these equations fall in two categories: explicit and implicit methods. The former suffer of severe Courant number stability restrictions on the allowed time step. Implicit methods eliminate in principle time-step stability restrictions; however, for time-dependent problems, time-step restriction may be dictated by the physics.

A class of schemes that falls in between these two large categories is the one of semi-implicit schemes that treat implicitly the pressure contribution. In order to evaluate the pressure terms, two different approaches are generally followed: one that uses a relaxation technique for the pressure field, and one that solves an algebraic Poisson pressure equation rigorously derived by imposing mass conservation (Grasso and Magi 1985; Magi 1987; Magi and Grasso 1985). For example, *CONCHAS* (Butler et al. 1979), *KIVA* (Amsden et al. 1985), *SIMPLE*, and *SIMPLER* (Patankar and Spalding 1972) techniques belong to the first type of approaches; *PISO* (Issa 1985), *EPISO* (Watkins et al. 1986), and the method of the present authors belong to the second type.

In the present paper a discussion of the mathematical model is given and the numerical methodology is outlined. Applications of the method to several complex flow configurations such as sprays in ambient and in confined volume resembling diesel-type environments are discussed, and conclusions given.

Mathematical Model

The gas phase is treated in a Eulerian fashion by use of the compressible Reynolds averaged Navier-Stokes equations reformulated to account for mass, momentum, and energy exchanges due to the interaction between gas and liquid phases. These exchanges are modeled according to O'Rourke (1981). The liquid is assumed to be a phase dispersed into the gas and is treated in a Lagrangian fashion.

Gas Phase

The governing equations for the gas phase, assumed to be nonreacting, are written in strong conservation form. The Reynolds averaged equations are

• kth species mass equation

$$\frac{d}{dt}\int_V \rho^k \, dV + \oint_S \underline{n}\cdot \rho^k \underline{u} \, dS = \oint_S \underline{n}\cdot \rho_g D \nabla Y^k \, dS - \gamma \frac{d}{dt}\left\{\sum_p \frac{4}{3}\pi r_p^3 \rho_l \left[\iiint f \, dr_p \, d\underline{u}_p \, dT_p\right]\right\} \tag{1}$$

where $\gamma = 1$ for volatile component and $\gamma = 0$ for nonvolatile component.

• Momentum equation ($\underline{m} = \rho_g \underline{u}$)

$$\frac{d}{dt}\int_V \underline{m} \, dV + \oint_S \underline{n}\cdot \underline{\underline{mu}} \, dS = \oint_S (\underline{n}\cdot \underline{\underline{\sigma}} - \underline{n}p) \, dS - \frac{d}{dt}\left\{\sum_p \frac{4}{3}\pi r_p^3 \rho_l \underline{u}_p \left(\iiint f \, dr_p \, d\underline{u}_p \, dT_p\right)\right\} \tag{2}$$

• Total energy equation ($E_g = \rho_g \tilde{E}_g$)

$$\begin{aligned}
&\frac{d}{dt}\int_V E_g \, dV + \oint_S \underline{n}\cdot E_g \underline{u} \, dS + \oint_S \underline{n}\cdot \theta p \underline{u} \, dS \\
&\quad = -p\frac{d}{dt}\int_V \theta \, dV + \oint_S \underline{n}\cdot\left(\lambda_g \nabla T + \sum_k h_k \rho_g D \nabla Y^k + \underline{\underline{\sigma}}\cdot \underline{u}\right) ds \\
&\qquad - \sum_p \left\{\frac{d}{dt}\left(\frac{4}{3}\pi r^3 \rho_l h_l\right) + \left[\frac{1}{2}u_p^2 \frac{d}{dt}\left(\frac{4}{3}\pi \rho_l r_p^3\right)\right.\right. \\
&\qquad \left.\left. + \frac{4}{3}\pi r_p^3 \rho_l K_p \underline{u}_p \cdot (\underline{u} + \underline{u}' - \underline{u}_p)\right] \iiint f \, dr_p \, d\underline{u}_p \, dT_p\right\}
\end{aligned} \tag{3}$$

• Turbulence kinetic energy ($K = \rho_g k$)

$$\frac{d}{dt}\int_V K \, dV + \oint_S \underline{n}\cdot K \underline{u} \, dS = \oint_S \underline{n}\cdot \frac{\mu}{\sigma_k} \nabla k \, dS + \int_V (G - E) \, dV \tag{4}$$

• Dissipation of turbulence kinetic energy ($E = \rho_g \varepsilon$)

$$\frac{d}{dt}\int_V E \, dV + \oint_S \underline{n}\cdot E \underline{u} \, dS = \oint_S \underline{n}\cdot \frac{\mu}{\sigma_\varepsilon} \nabla \varepsilon \, dS + \int_V (C_1 G - C_2 E + C_3 K \nabla\cdot \underline{u}) E/K \, dV \tag{5}$$

- Equation of state

$$p = \rho_g RT \sum Y^k/W^k \tag{6}$$

where

$$\begin{aligned}
\underline{\underline{\sigma}} &= \mu(\nabla \underline{u} + \nabla \underline{u}^T) - \tfrac{2}{3}(\mu \nabla \cdot \underline{u} + K)\underline{\underline{I}} \\
\underline{\underline{R}} &= \mu_t(\nabla \underline{u} + \nabla \underline{u}^T) - \tfrac{2}{3}K\underline{\underline{I}} \\
G &= \underline{\underline{R}} : \nabla \underline{u} \\
\mu &= \mu_{lam} + \mu_t \\
\mu_t &= 0.09\rho_g k^2/\varepsilon \\
\rho_g D &= \mu \\
\lambda_g &= c_{p_g}\mu
\end{aligned}$$

Liquid Phase

The liquid phase is assumed to be made out of parcels that represent classes of particles having the same properties (Bracco 1985; Dukowiez 1980; O'Rourke 1981). The droplet equations of motion are written in Lagrangian form, thus the droplets can be tracked in time as they pass through the gas. The effects of the thrust imparted to the drop by asymmetry in the vaporization are neglected on account of the high Reynolds number. Furthermore, it is assumed that the internal circulation velocity is small compared to the relative velocity. Likewise the effects of the gas shear stress acting on the drop surfaces are neglected for high Reynolds numbers. Hence the equations are.

- Momentum equation

$$\frac{d}{dt}\underline{u}_p = K_p(\underline{u} + \underline{u}' - \underline{u}_p) - \frac{1}{\rho_l}\nabla p \tag{7}$$

where the coefficient K_p and the correlation used for the drag coefficient are (O'Rourke 1981):

$$\begin{aligned}
K_p &= \frac{3}{8}\frac{\rho_g}{\rho_l}\frac{|\underline{u} + \underline{u}' - \underline{u}_p|}{r_p} C_D(\theta, \mathrm{Re}) \\
C_D &= \frac{24}{\mathrm{Re}}\left(\theta^{-2.65} + \frac{\mathrm{Re}^{2/3}}{6}\theta^{-1.78}\right)
\end{aligned} \tag{8}$$

with

$$\mathrm{Re} = \frac{\rho_g |\underline{u} + \underline{u}' - \underline{u}_p| r_p}{\mu_{lam}}$$

The effects of gas-phase turbulence on the droplets are expressed via a turbulent fluctuation $\underline{u}'$ that is selected randomly from an isotropic Gaussian probability distribution for the instantaneous gas velocity with r.m.s. equal to $\sqrt{2/3k}$.

- Exchange rate equations: Heating and evaporation of droplets are calculated by solving energy and mass conservation in Lagrangian form

$$-\frac{1}{3r_p^2}\frac{d}{dt}(\rho_l r_p^3) = \frac{\lambda_g}{c_{p_g} r_p} B Nu_g(\theta, \mathrm{Re}, B)/2 \tag{9}$$

$$\frac{d(c_l T_p)}{dt} = \frac{3}{2}\frac{\lambda_l Nu_l}{\rho_l r_p^2}(T_s - T_p) + \frac{c_l T_s - c_l T_p}{\rho_l r^3}\frac{d}{dt}(\rho_l r_p^3) \tag{10}$$

where B is the drop transfer number and is defined as

$$B = \frac{Y_{vs} - Y_v}{1 - Y_{vs}} = \frac{c_{p_g}}{L(T_s)}\left[(T_g - T_s) - \frac{Nu_l \lambda_l}{Nu_g \lambda_g}(T_s - T_p)\right].$$

These equations differ from the ones reported in O'Rourke (1981); liquid thermodynamic properties here are allowed to vary with temperature. The correlations used for the gas-phase Nusselt number Nu_g and the latent heat of vaporization $L(T)$ are, respectively,

$$Nu_g = \left[2\theta^{-1.75} + 0.6\left(\frac{\mathrm{Re}}{\theta}\right)^{1/2} Pr^{1/3}\right]\frac{\ln(1 + B)}{B}$$

$$L(T) = -\frac{R}{W}\left[\frac{d(\ln p)}{d(1/T)}\right]\beta$$

where W and β are, respectively, the molecular weight of the liquid and the compressibility factor.

Liquid Injection Model

A line source drop injection technique is used (Andrews and Bracco 1989; Chatwani and Bracco 1985; Magi and Grasso 1985) to account for the intact liquid core. Starting from the beginning of the injection, the tip of the line source penetrates into the gas at a velocity that is a fraction of the injection velocity until it reaches the steady-state length. To evaluate the spray angle, the size of injected drops, and the length of the intact core the following equations are used:

$$\tan\frac{\theta}{2} = \frac{1}{A_\theta} 4\pi \left(\frac{\rho_g}{\rho_l}\right)^{1/2} f_m^*$$

$$\bar{d}_{p,0} = B_d \frac{2\pi\sigma_l}{\rho_g V_T^2} \lambda_m^*$$

$$\frac{x_c}{d_j} = C_c \left(\frac{\rho_l}{\rho_g}\right)^{1/2} f_{m}^{*}{}_{-1}$$

where f_m^* and λ_m^* are known functions of $\dfrac{\rho_l \sigma_l^2}{\rho_g \mu_l^2 V_T^2}$ (O'Rourke 1981).

The constants A_θ, B_d, and C_c are determined experimentally. It is also assumed that the intact core is a straight cone with length x_c and base diameter d_j. The instantaneous length of the line source is determined assuming that the tip moves at 70 percent of the mass-mean injection velocity (V_{inj}) and the droplets are injected at a position randomly selected on the drop-generating surface. The initial drop axial velocity is taken as V_{inj}, and the radial one is selected so that the computed spray angle satisfies the equations.

Collision Model

A probability that a drop (a) undergoes a collision with a neighboring drop and (b) is obtained from a Poisson distribution function that uses the following collision frequency (O'Rourke 1981):

$$\nu_{ab} = f_a f_b \pi (r_a + r_b)^2 |\underline{u}_a - \underline{u}_b|.$$

The outcome of a collision (leading to either coalescence or separation) is measured by the probability of coalescence once a collision has occurred; the following expression is used (O'Rourke 1981):

$$\eta_c = \min(2.4 g(\xi)/\mathrm{We}, 1)$$

where

$$\mathrm{We} = \rho_l |\underline{u}_a - \underline{u}_b|^2 \frac{r_a}{\sigma_l} \quad \text{if} \quad r_a \leq r_b$$

and

$$\xi = \frac{r_a}{r_b}; \quad g(\xi) = \xi^3 - 2.4\xi^2 + 2.7\xi.$$

Numerical Model

Existing methods for solving the Reynolds averaged compressible Navier-Stokes equations for two-phase flows fall in two categories: explicit and implicit methods. The former suffer of severe Courant number stability restrictions on the allowed time step, however they are computationally very efficient and can exploit parallel and/or vector algorithms. Implicit methods eliminate in principle time step stability restrictions, however, the computational efficiency may degrade depending upon the algorithm used for matrix inversions. Furthermore, for time-dependent problems, the time-step restriction may be dictated by the physics.

A class of schemes that falls in between these two large categories is the one of semi-implicit schemes (Issa 1985); the main feature is the implicit treatment of the pressure contribution in the momentum and energy equations. A close examination of the governing equations written in characteristic form shows that the pressure forces and pressure work are indeed responsible for Courant stability restriction (Casulli and Greenspan 1984). In order to evaluate the pressure contribution, different approaches are generally followed. A relaxation technique, originally devised by Chorin (1966) and modified by Hirt and Cook (1972), is used for obtaining the pressure, contribution in ICE-type methods (Rivard et al. 1975). This approach is revisited in *CONCHAS* and then *KIVA* (Amsden et al. 1985; Butler et al. 1979), in which an acoustic subcycling method is implemented, consisting of an explicit treatment of the pressure gradient and compression terms in the momentum and energy equations, with a time step that satisfies sound speed stability restrictions and of which the main time step is a multiple (Haselman 1980).

The *SIMPLE* and *SIMPLER* methods fall in the category of iterative methods; they were originally developed by Patankar and Spalding (1972) and used by Gosman (Watkins et al. 1986) for the solution of flows in internal combustion engines. Other approaches have been developed that use an algebraic pressure equation rigorously derived by imposing mass conservation. The pressure implicit splitting operators (*PISO*) method (Issa 1985) uses the splitting of operators in the solution of the momentum and pressure equations, yielding an order of accuracy that depends on the number of operator splittings used. The solution is advanced in time by use of a predictor-corrector time-stepping algorithm.

In the present work a fully implicit finite volume numerical method is developed. In this method the pressure is used as a dependent variable. A discretized pressure equation is derived from the discretized form of the continuity and momentum equations, to ensure that no spurious terms are introduced and with the constraint of the equation of state (Grasso and Magi 1985; Issa 1985; Magi 1985; Magi and Grasso 1985).

The equations are discretized in space by one-sided differences, backward first-order temporal discretization is used and a Jacobi-type iterative procedure is employed to avoid the inversion of large matrices. The method of solution for the pressure equation is based on Stone's strongly implicit iterative method (1968).

The flow region is subdivided into computational control volumes (cells). In Fig. 13.1 a typical cell in the computational space is shown; the velocity components are staggered with respect to the cell center; all other variables ($p, \rho, T, E,$ etc.) are located at the center of the cell. Such a choice of grid staggering is particularly convenient when sonic propagation is treated implicitly (Grasso and Magi 1985; Stewart and Wendroff 1984).

The solution is advanced in time by the following sequence of steps:

1. The liquid spray equations are solved explicitly by means of a Newton-type iterative procedure.
2. A Jacobi-type iterative procedure is used to solve the gas-phase equations and drop velocities are updated to account for the pressure effects.

FIGURE 13.1. Computational cell.

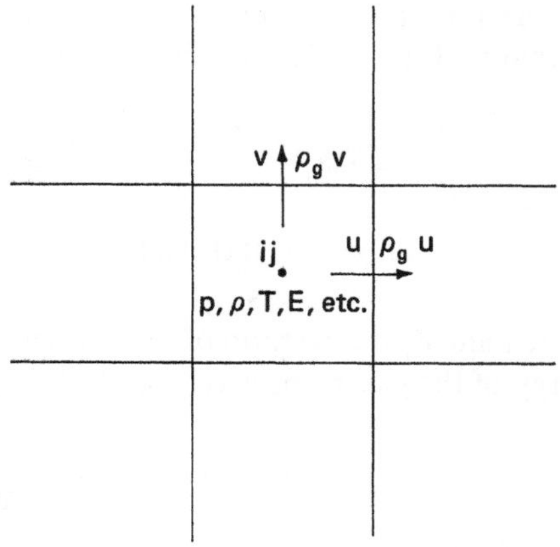

3. The $k - \varepsilon$ equations are solved by using the most updated values of density and velocity of the gas.

In the following, the discretized gas-phase equations are given in a symbolic operator form (Grasso and Magi 1985):

- Continuity equation

$$\rho_g + \tau B \underline{m} = \rho_g^n + \tau \dot{\rho}_s \tag{11}$$

- Momentum equation

$$\left[1 + \tau \sum_p \left(\frac{d_p m_p K_p}{\rho_l (1 + \tau K_p)}\right)\right] \underline{m} + \tau B\left(\frac{\underline{m}\underline{m}}{\rho_g}\right) - \tau\left[1 - \sum_p \left(\frac{d_p m_p K_p}{\rho_l (1 + \tau K_p)}\right)\right] B^T p$$
$$= \underline{m}^n + \tau D_{\underline{m}} + \tau \underline{m}_s^n \tag{12}$$

- Energy equation

$$E_g + \tau B\left[\underline{m}\left(\frac{E_g}{\rho_g} + \frac{p}{\rho_g}\right)\right] + p\Delta\theta = E_g^n + \tau D_E + \tau \dot{E}_s \tag{13}$$

- kth species equation

$$\rho^k + \tau B\left(\frac{\underline{m}\rho^k}{\rho_g}\right) = \rho^{k,n} + \tau D_k + \tau\gamma\dot{\rho}_s \tag{14}$$

- Momentum equation of droplets

$$(1 + \tau K_p)\underline{u}_p - \tau K_p \frac{\underline{m}}{\rho_l} - \frac{\tau}{\rho_l} B^T p = \underline{u}_p^n \tag{15}$$

where τ is the time step, and B, $-B^T$, and D represent, respectively, the discretized forms of the divergence, gradient, and diffusion operators; $\Delta\theta$ represents the change of the void fraction in a single time step; and $\dot{\underline{m}}_s$, $\dot{E}_s$, and $\dot{\rho}_s$ are spray contributions.

Equation 11 is replaced by the following equation for pressure obtained by combining Eqs. 11, 12, and the equation of state:

$$\left\{\frac{I}{(\bar{\gamma}-1)\tilde{e}_g}+\tau^2 BM_1^{-1}\left(I-\sum_p \frac{d_p m_p K_p}{\rho_l(1+\tau K_p)}\right)B^T\right\}p$$

$$=\rho_g^n+\tau^2 BM_1^{-1}B\left(\frac{\underline{m}\underline{m}}{\rho_g}\right)-\tau BM_1^{-1}(\underline{m}^n+\tau D_{\underline{m}}+\tau \underline{m}_s^n) \tag{16}$$

where $\bar{\gamma}$ and $\tilde{e}_g$ are the ratio of specific heat coefficients and the specific internal energy of the gas, respectively, and M_1 is the following matrix

$$M_1 = I+\tau\sum_p \frac{d_p m_p K_p}{\rho_l(1+\tau K_p)}. \tag{17}$$

On account of the high nonlinearity of Eqs. 12–16, the solution can be obtained either by using a (quasi-)linearization method or by an iterative procedure. Here a method of solution based on successive iterations is proposed. The algorithm here is outlined by casting the system of Eqs. 12–16 in matrix form. Let

$$M=\left\{\frac{I}{(\tilde{\gamma}-1)\tilde{e}_g}+\tau^2 BM_1^{-1}\left(I-\sum_p \frac{d_p m_p K_p}{\rho_l(1+\tau K_p)}\right)B^T\right\} \tag{18}$$

then

$$\begin{bmatrix} M & 0 & 0 & 0 \\ -\tau M_1^{-1}\left[I-\sum_p \frac{d_p m_p K_p}{\rho_l(1+\tau K_p)}\right]B^T & (I-D_{\underline{m},\underline{m}}) & 0 & 0 \\ \Delta\theta+\tau B\left(\underline{m}\frac{o}{\rho_g}\right) & \tau B\left(o\frac{E_g}{\rho_g}\right) & (I-D_{E,E}) & 0 \\ 0 & \tau B\left(o\frac{\rho^k}{\rho_g}\right) & 0 & (I-D_{k,k}) \end{bmatrix}^{\nu} \begin{bmatrix} p \\ \underline{m} \\ E_g \\ \rho^k \end{bmatrix}^{\nu+1}$$

$$= \begin{bmatrix} \tau^2 BM_1^{-1}\left[B\left(\underline{m}\frac{\underline{m}}{\rho_g}\right)-D_{\underline{m}}\right] \\ -\tau M_1^{-1}B\left(\underline{m}\frac{\underline{m}}{\rho_g}\right)+\tau D_{\underline{m}}-\tau D_{\underline{m},\underline{m}}\underline{m} \\ \tau D_E-\tau D_{E,E}E_g \\ \tau D_k-\tau D_{k,k}\rho^k \end{bmatrix}^{\nu} + \begin{bmatrix} \rho_g-\tau BM_1^{-1}(\underline{m}+\tau\underline{m}_s)+\tau\dot{\rho}_s \\ \underline{m}+\tau D_{\underline{m}}+\tau\dot{\underline{m}}_s \\ E_g+\tau\dot{E}_s \\ \rho^k+\tau\gamma\dot{\rho}_s \end{bmatrix}^{n}.$$

It is important to point out that the diagonal elements of the Jacobian of the diffusion operator are treated implicitly (the symbolic representation is $D_{\underline{m},\underline{m}}$, $D_{E,E}$, $D_{k,k}$) as suggested by Issa (1985).

At each time level the iteration is started by an initial guess for temperature. Inspection of the above systems of equations shows that the implicit operator matrix is lower triangular. This property is exploited in the computational strategy: once the pressure is obtained, momentum, species densities, and total energy are updated. This process is repeated until the r.m.s. error in temperâture is less than a given number; for a convergence analysis of such a procedure refer to (Grasso and Magi 1985). Once the convergence criterium is satisfied, Eq. 15 is solved for $\underline{u}_p$.

Finally the equations for the turbulence kinetic energy and its dissipation rate are solved by using the most recent flow variables. The equations are discretized as done for the transport equations described earlier, however, for the sake of conciseness the finite volume form is not reported.

Results

The computational methodology described here has been validated by extensive comparison with experimental measurements of liquid sprays injected in ambient. Moreover, the method has been applied to simulate the fluid dynamics of vaporizing liquid sprays in confined volumes, resembling diesel-type conditions.

In Table 13.1 the experimental conditions for two cases of full cone sprays in ambient corresponding to a low gas pressure case (A) and a high gas pressure case (B) are given.

The grid (Fig. 13.2) uses rectangular cells whose size is in geometrical progression in both axial and radial direction starting from the injector nozzle location for a total of 44 × 26 cells.

For both test cases the initial conditions are:

1. The ambient gas is assumed to be quiescent with pressure and temperature corresponding to the values of Table 13.1.
2. For the spray the liquid injection model is used enforcing the experimental injection velocity and mass flow rate.

At the boundaries the following conditions are imposed:

1. Symmetry conditions along the centerline.
2. At the left, solid wall conditions (no-slip, adiabatic).
3. At the top, the pressure is set equal to the free-stream value.

TABLE 13.1. Spray in ambient.

Case	p(MPa)	ρ_l/ρ_g	V_{inj}(m/s)	x/d*
A	1.48	39	127	300, 400, 600, 800
B	4.24	13.7	127	400, 500, 600

liquid: hexane; gas: nitrogen; gas and droplet temperature: 300 K

* Measurement position from Wu et al. (1984).

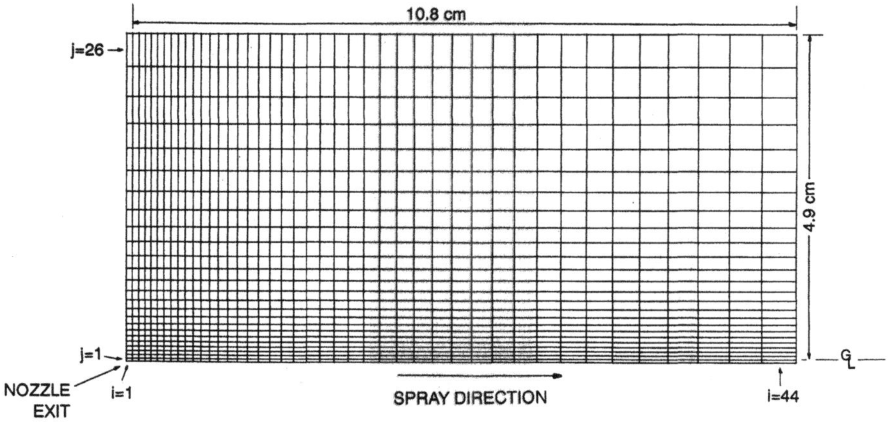

FIGURE 13.2. Computational domain.

4. At the right, uniform axial flow conditions.
5. For the spray the symmetry condition requires that for each particle crossing the axis of symmetry a new particle (having the same thermodynamic properties and velocity components specularly reflected) is introduced.

In Figs. 13.3–5 the evolution of the spray for case A is shown by means of mean gas velocity vector plots, particle distribution, turbulent kinetic energy, and dissipation contour lines at three different times ($t = 0.5$ms, 1ms, 2ms). Note the entrainment of the gas due to the axial spray injection, the high level of turbulence kinetic energy, and dissipation within the region of the highest velocity gradients located near the tip of the spray. The computed Sauter mean radius compares well with the one computed by using the *LDEF* (Lagrangian drop Eulerian fluid) code of O'Rourke (1981). The penetration of the spray can be also deduced by the distribution of the centerline gas velocity versus x at different times. The logarithmic distribution of gas and droplet velocities normalized by the injection velocity versus the logarithm of the axial position (scaled by the factor $d_j(\rho_l/\rho_g)^{1/2}$), are shown in Figs. 13.6 and 13.7.

Defining the steady-state configuration by use of an ensemble-averaging, we recover a universal far-field behavior as in incompressible gas jets as found by Wygnanski and Fiedler (1969). At steady state the computed mean axial droplet velocity at different axial locations shows good agreement with the measured one as presented in Fig. 13.8. The latter also shows the computed mean axial gas velocity.

For case B we only show the steady-state results in terms of mean axial gas and drop velocities in comparison with experiments (Fig. 13.9).

Comparing the computational efficiency of the proposed method with that implemented in the *LDEF* code, we find that we can use a time step of an order of magnitude greater, consequently the number of iterations per time step does increase; however not in the same factor, thus mantaining the cost-effectiveness of the proposed methodology.

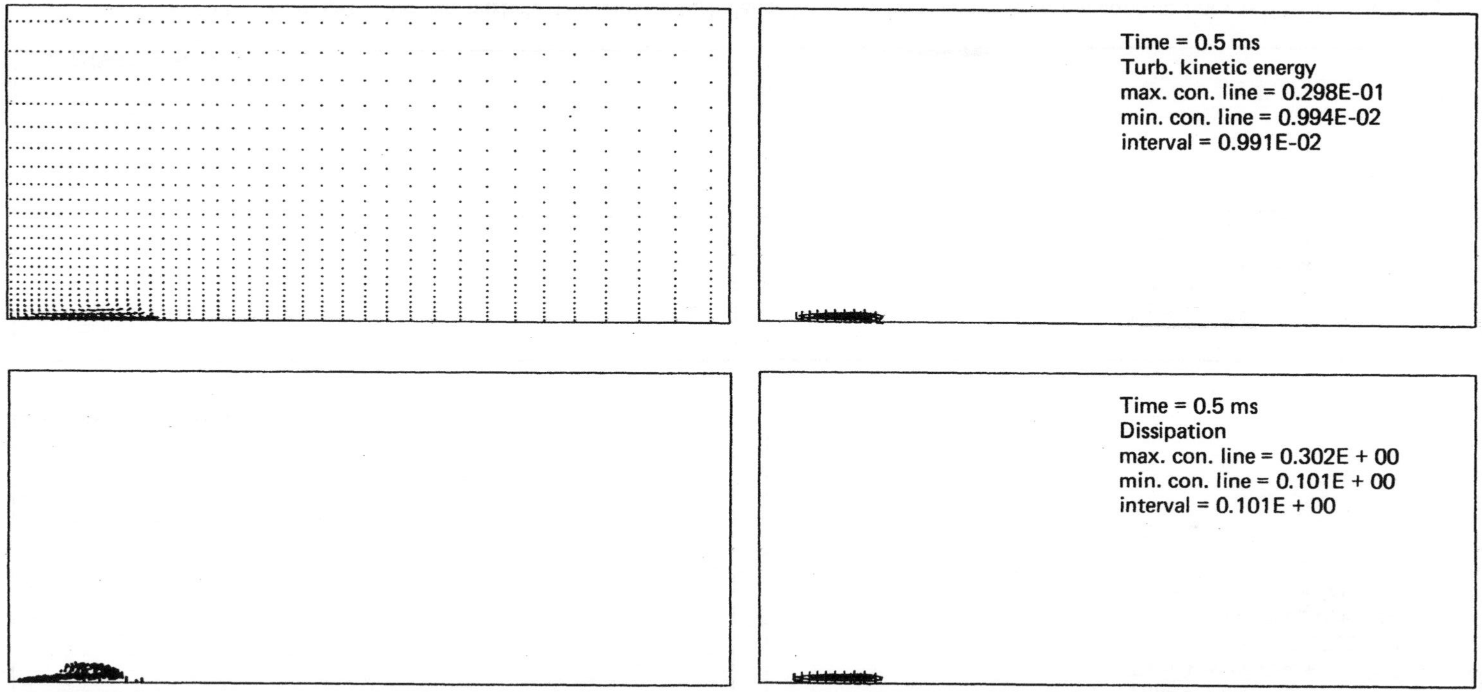

FIGURE 13.3. Vector mean gas velocities, particle distribution, turbulence kinetic energy, and dissipation contour lines at time 0.5 ms (case A).

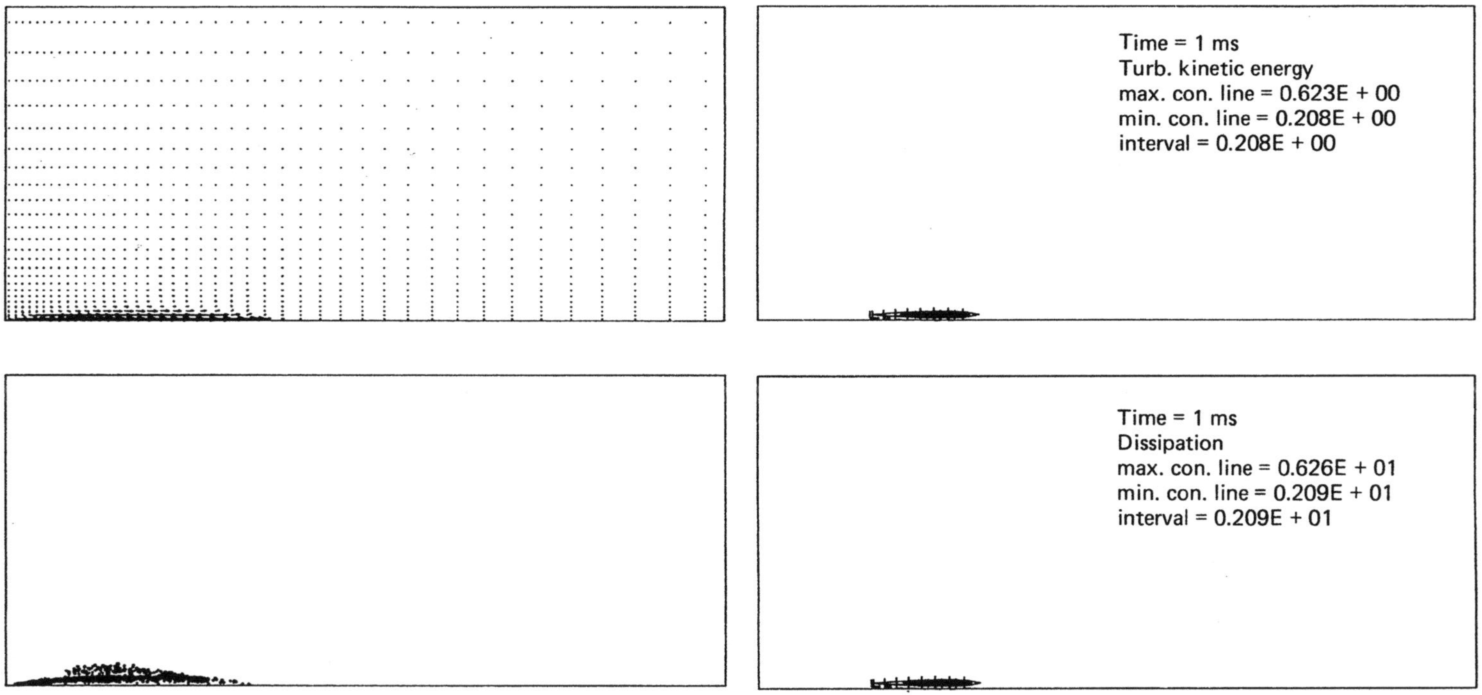

FIGURE 13.4. Vector mean gas velocities, particle distribution, turbulence kinetic energy, and dissipation contour lines at time 1 ms (case A).

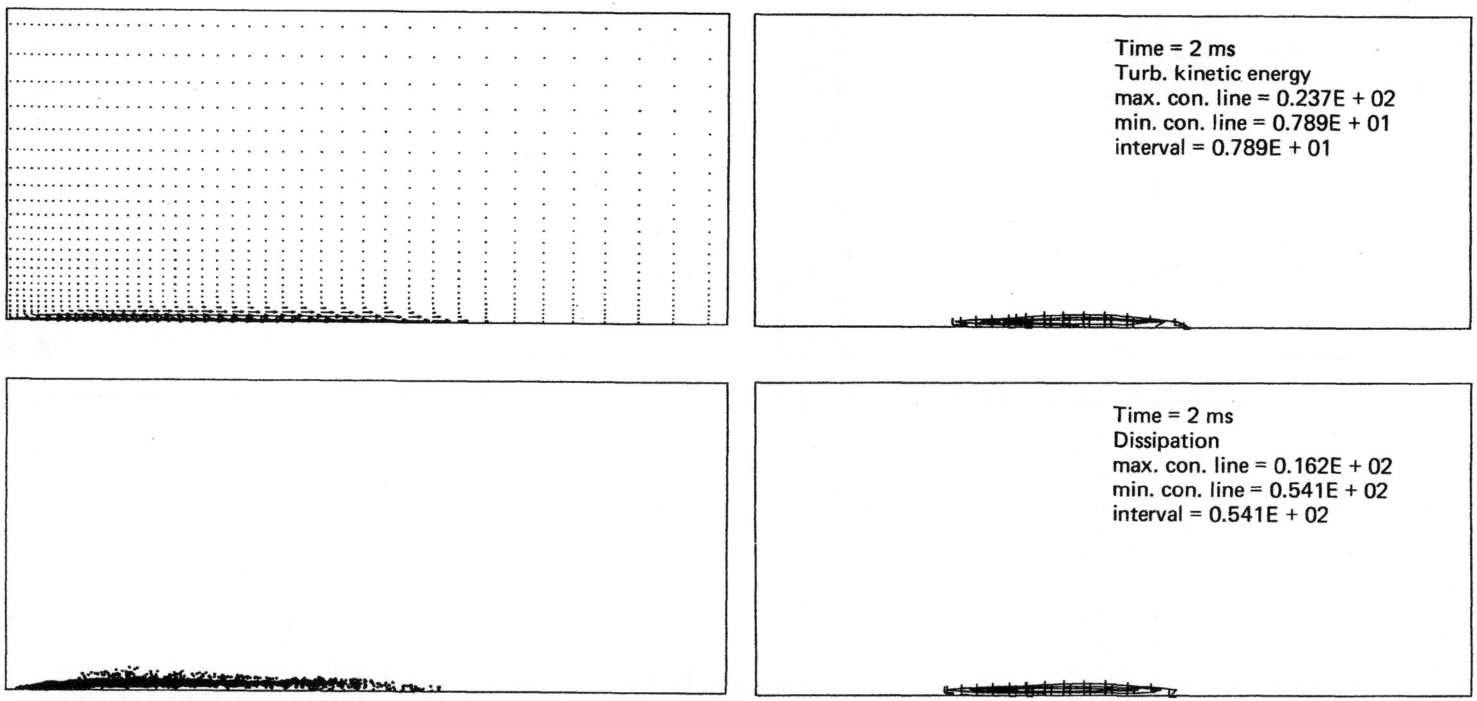

FIGURE 13.5. Vector mean gas velocities, particle distribution, turbulence kinetic energy, and dissipation contour lines at time 2 ms (case A).

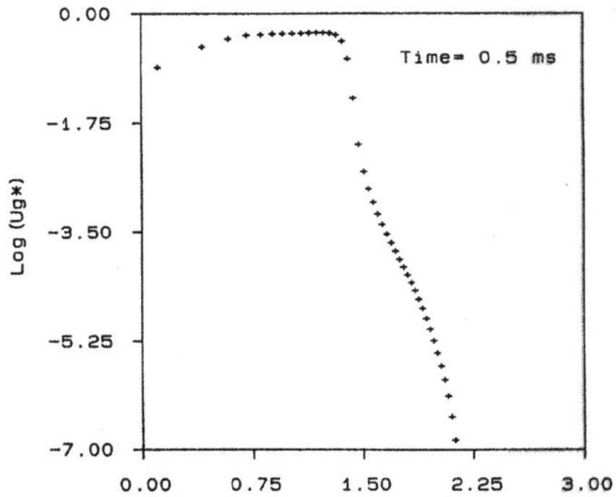

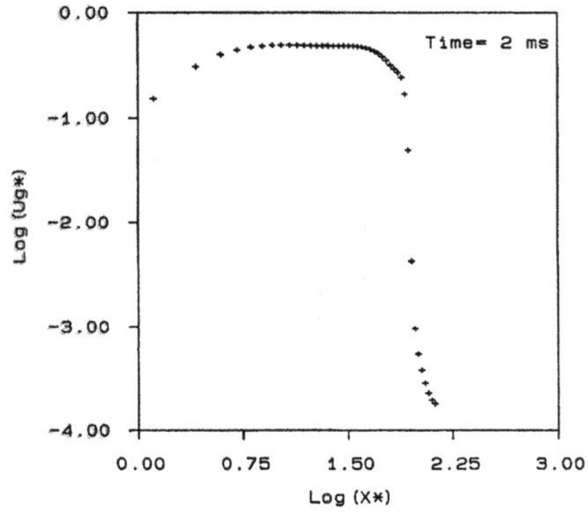

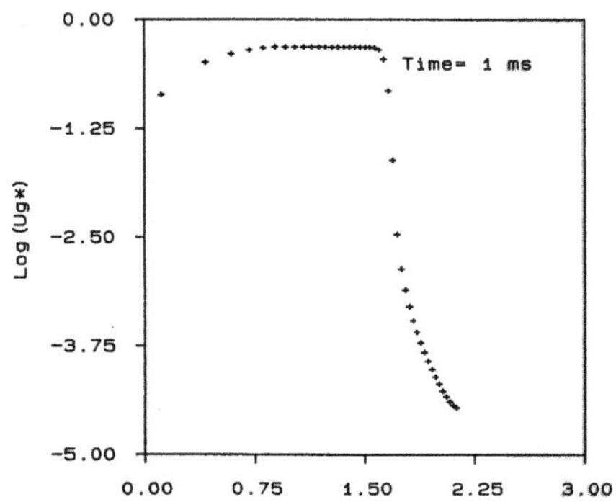

FIGURE 13.6. Mean gas velocity versus axial position at time 0.5, 1, and 2 ms (case A).

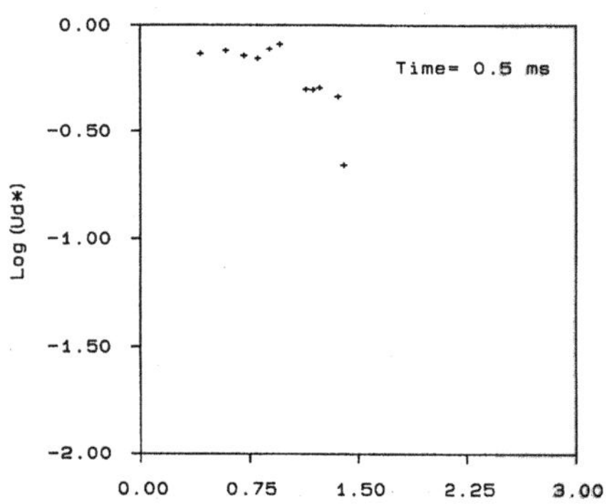

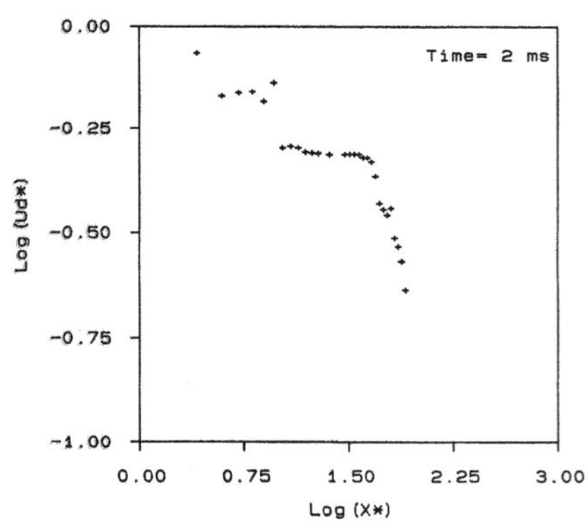

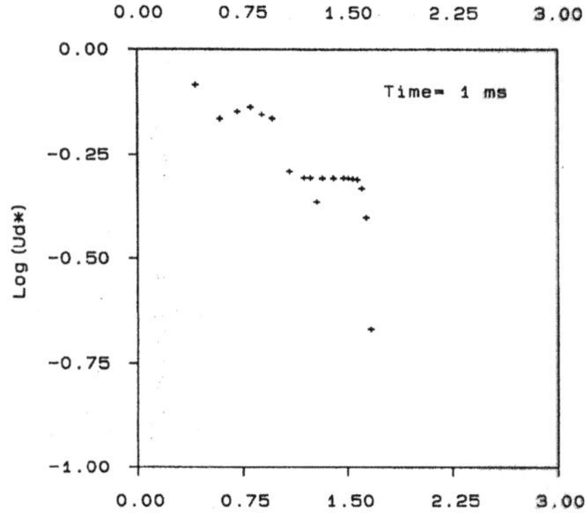

FIGURE 13.7. Droplet velocity versus axial position at time 0.5, 1, and 2 ms (case A).

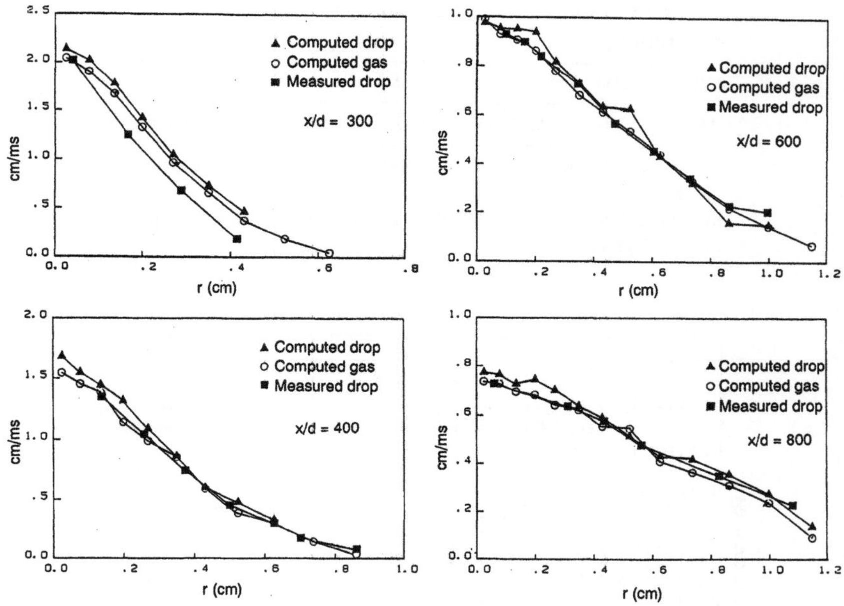

FIGURE 13.8. Steady computed and measured mean axial droplet velocity and computed gas velocity at different axial locations (case A).

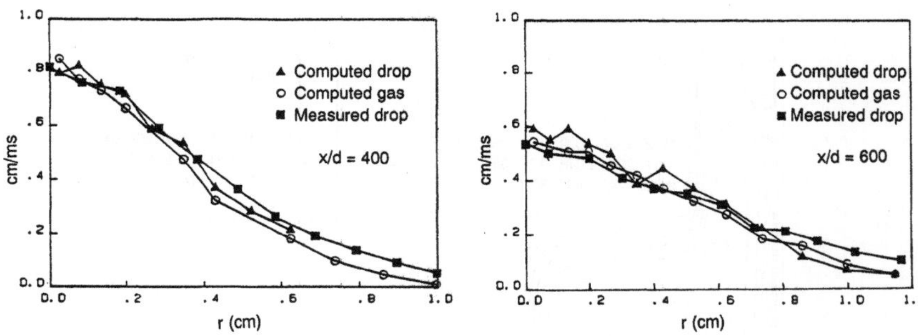

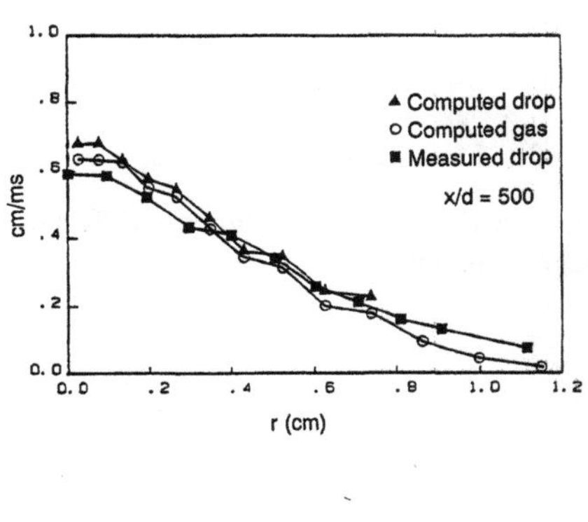

FIGURE 13.9. Steady computed and measured mean axial droplet velocity and computed gas velocity at different axial locations (case B).

TABLE 13.2. Spray in confined volume (diesel-type conditions).

CR	RPM	B	S	d	V_{inj}
18	1200	9.14 cm	12.7 cm	150 μm	100 m/s

liquid: hexadecane; gas: air; initial gas and droplet temperature: 350 K
initial crank angle of injection 30 degrees BTDC; duration of injection 10 degrees.

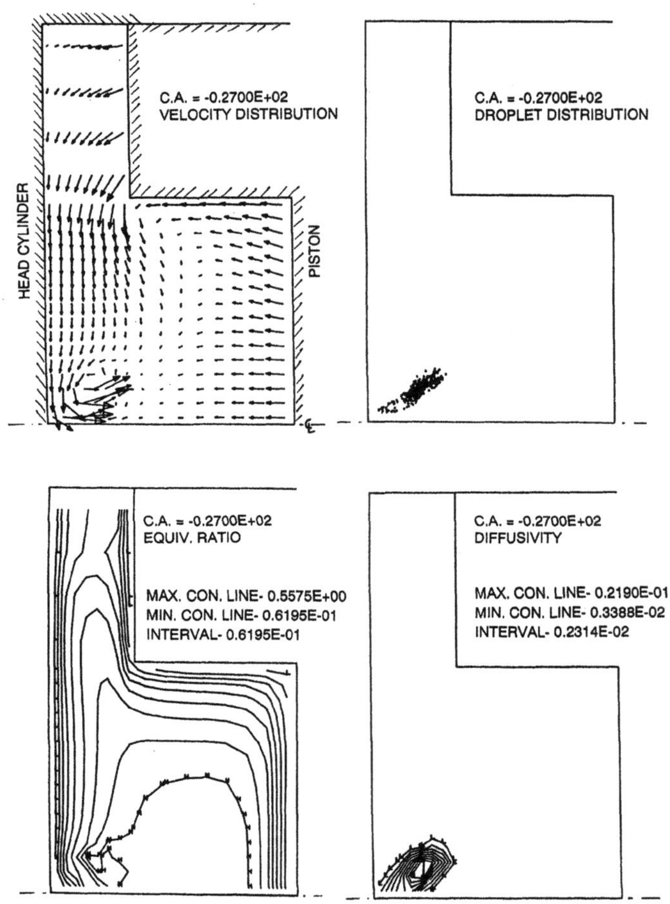

FIGURE 13.10. Vector mean gas velocities, particle distribution, turbulence diffusivity, and equivalence ratio contour lines at 27 degrees BTDC.

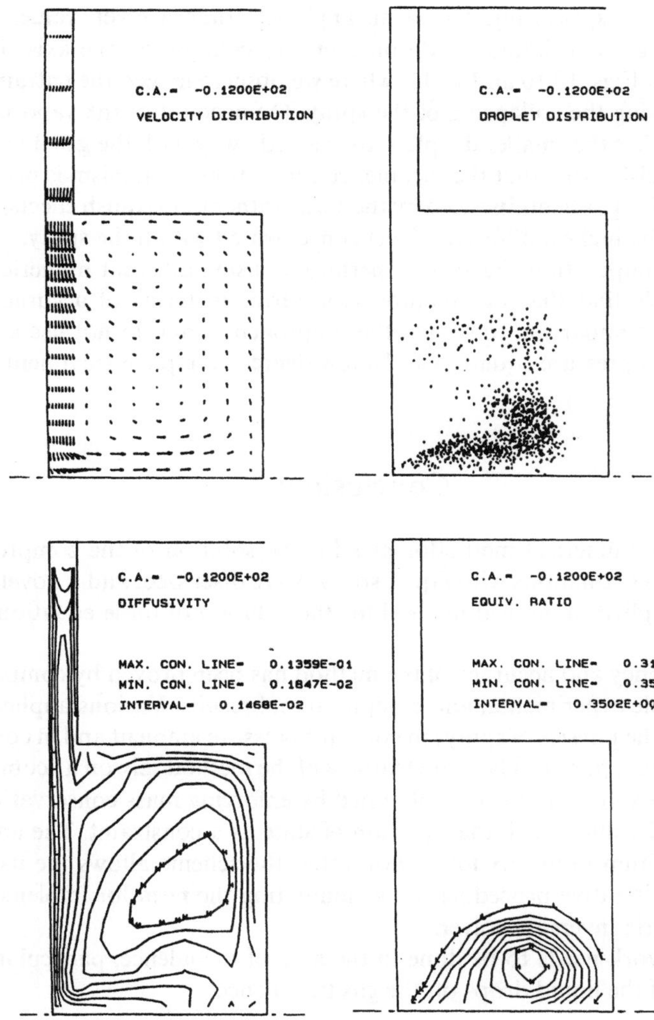

FIGURE 13.11. Vector mean gas velocities, particle distribution, turbulence diffusivity, and equivalence ratio contour lines at 12 degrees BTDC.

The conditions for the test case corresponding to the evolution of a vaporizing hollow-cone spray in confined volume are reported in Table 13.2. The geometry resembles an axisymmetric combustion chamber typical of a direct injection diesel engine with a cylindrical cup in piston. The spray is injected at an angle of 30 degrees with respect to the axis when the piston is at an angular position of 30 degrees before top dead center (*BTDC*). In Fig. 13.10 and 13.11 the droplet distribution, the mean gas vector plot, the turbulent diffusivity and the equivalence ratio contour lines are shown at two different crank-angle positions.

Hollow-cone sprays injected at an angle less than a given value, which depends on gas and liquid conditions, tend to collapse on the axis. This is evident from Figs. 13.10 and 13.11, where we observe indeed the entrainment process causing the collapsing of the spray. On account of the vaporization we also see that the smaller droplets are carried away with the gas due to the effects of squish. Note that the turbulence generation mechanism is primarily due to velocity gradients induced by the spray rather than squish effects; this is evident by the highest diffusivity level concentrated within the spray.

Similar computations have been performed using different numerical approaches. We find that our method is superior in terms of accuracy and efficiency in comparison to some other approaches that do not use a rigorously derived pressure equation and a low degree of implicit treatment of the equations.

Conclusions

In this work numerical methodologies for the solution of the compressible Navier-Stokes equations with liquid sprays were discussed, and a novel iterative fully implicit numerical method for the solution of these equations was presented.

The efficiency and accuracy of the method has been proven by comparison with available experimental and computational results. Various applications to simulate the physics of sprays in constant pressure ambient and in confined volumes were reported. The main features of the method rely on a "compressible pressure solver" rigorously obtained by enforcing mass conservation for each control volume with the equation of state as a constraint. The analysis of the algorithm in matrix form shows that the scheme allows the use of a Jacobi-type iterative procedure, thus eliminating the need for implementing a direct matrix inversion solver.

Further work needs to be done in the area of turbulence; particularly for the effects of the liquid drops on the gas turbulence.

Acknowledgments. The authors wish to thank Prof. F.V. Bracco for his support, suggestions, and help throughout the years. Some of the results have been presented at the International Symposium on Computational Fluid Dynamics held in Sydney, Australia, August 24–27, 1987, and at ATA Seminars held in Turin, Italy, at Flat Research Center, on December 4, 1987. This work was partly supported by CNR-PFE2, MPI, and some computations were performed at Princeton University, Engine Laboratory Group.

Nomenclature

A_θ = constant of the initial spray angle.
B_d = constant of the initial drop size equation.

B	= drop transfer number; bore.
c_{p_g}	= average specific heat at constant pressure of gas.
c_l	= specific heat of liquid.
C_c	= constant of the intact core length equation.
C_1, C_2, C_3	= constants of the turbulence model ($C_1 = 1.44$; $C_2 = 1.92$; $C_3 = 1$).
C_D	= drop drag coefficient.
CR	= compression ratio.
d_j	= intact jet diameter at nozzle exit.
d, d_n	= nozzle diameter.
$\bar{d}_{p,0}$	= diameter of injected drops.
D	= species diffusion coefficient.
$\tilde{E}_g$	= specific total energy.
f	= drop number distribution, $f(\underline{x}_p, \underline{u}_p, r_p, T_p, t)$.
h_g	= gas-specific enthalpy.
h_k	= specific enthalpy of species k in gas.
I	= unit tensor.
k	= turbulence kinetic energy.
p	= static pressure.
r_p	= drop radius.
$r_{0.5}$	= jet half-width at half the average axial centerline velocity.
R	= gas constant.
S	= stroke.
t	= time.
T	= temperature.
T_p	= drop temperature.
T_s	= drop surface temperature.
$\underline{u}$	= gas velocity.
$\underline{u}_p$, $\underline{u}_d$	= drop velocity.
V_{inj}	= mass-mean injection velocity.
V_T	= relative liquid-gas velocity in Taylor's theory of drop formation.
W^k	= molecular weight of species k.
We	= Weber number.
x, r	= cylindrical coordinates.
Y^k	= mass fraction of species k.
Y_v	= mass fraction of volatile species.
Y_{vs}	= mass fraction of volatile species at the drop surface.

Greek symbols

ε	= rate of dissipation of turbulence kinetic energy.
η_{ab}	= collision efficiency of drops (or particles) a, b.
θ	= spray angle; void fraction.
λ_g	= thermal conductivity of gas.
λ_l	= thermal conductivity of liquid.

μ	= gas viscosity.
$\nu_{a,b}$	= collision frequency of drops a, b.
ρ_g	= gas density.
ρ_l	= liquid density.
σ_l	= liquid surface tension.
$\underline{\underline{\sigma}}$	= stress tensor.
$\sigma_k, \sigma_\varepsilon$	= constants of the turbulence model ($\sigma_k = 1$, $\sigma_\varepsilon = 1.3$).
$\tilde{\gamma}$	= ratio of specific heat coefficients.
γ	= constant ($\gamma = 1$ for volatile components; $\gamma = 0$ for nonvolatile components).

References

Aggarwal, S.K., Tong, A.Y., and Sirignano, W.A., 1984. "A Comparison of Vaporization Models in Spray Calculations," *AIAA J.*, *22*, 10, 1448–1457.

Amsden, A.A., Ramshaw, J.D., O'Rourke, P.J., and Dukowicz, J.K., 1985, "*KIVA*: A Computer Program for Two- and Three-Dimensional Fluid Flows with Chemical Reactions and Fuel Sprays," *Report no. LA-10245-MS*, Los Alamos Scientific Laboratory.

Andrews, M.J., and Bracco, F.V., 1989, "On the Structure of Turbulent Dense Spray Jets," *Encyclopedia of Fluid Mechanics*, *8*, N.P. Cheremisinoff, (Ed.).

Bracco, F.V., 1985, "Modeling of Engine Sprays," *SAE Paper* 850394.

Butler, T.D., Cloutman, L.D., Dukowicz, J.K., and Ramshaw, J.D., 1979, "*CONCHAS*: An Arbitrary Lagrangian-Eulerian Computer Code for Multicomponent Chemically Reactive Fluid Flow at All Speeds," *Report no. LA-8129-MS*, Los Alamos Scientific Laboratory.

Casulli, V., and Greenspan, D., 1984, "Pressure Method for the Numerical Solution of Transient, Compressible Fluid Flows," *Int. J. for Numerical Methods in Fluids*, *4*, 1001–1012.

Chatwani, A.U., and Bracco, F.V., 1985, "Computation of Dense Spray Jets," *ICLASS-85*, London, U.K.

Chorin, A.J., 1966, "Numerical Study of Thermal Convection in a Fluid Layer Heated from Below," *AEC Report NYO-1480-61*.

Dukowicz, J.K., 1980, "A Particle-Fluid Numerical Model for Liquid Sprays," *J. of Computational Physics*, *35*, 229–253.

Elghobashi, S.E., and Abou-Arab, T.W., 1983, "A Two-Equation Turbulence Model for Two-Phase Flows," *Physics of Fluids*, *26*, 4.

El-Tahry, S., 1985, "Application of a Reynolds Stress Model to Engine-Like Flow Calculations," *Transactions of the ASME*, *107*, 444–450.

Grasso, F., 1981, "On Flows in Internal Combustion Engines," *Ph.D. Thesis 1537-T*, Princeton University, Dept. of Mechanical and Aerospace Engineering.

Grasso, F., and Magi, V., 1985, "A Predictor Corrector Semi-Implicit Pressure Solver for Compressible Two-Phase Flows," *Lecture Notes in Physics*, *245*, Springer-Verlag.

Gupta, H.C., and Bracco, F.V., 1978, "Numerical Computations of Two-Dimensional Unsteady Sprays for Applications to Engines," *AIAA J. 16*, 10, 1053–1061.

Haselman, L.C., 1980, "*TDC*—A Computer Program for Calculating Chemically Reacting Hydrodynamic Flows in Two-Dimensions," *Report no. UCRL-52931*, Lawrence Livermore Laboratory.

Haselman, L.C., and Westbrook, C.K., 1978, "A Theoretical Model for Two-Phase Fuel Injection in Stratified Charge Engines," *SAE Paper 780318*.

Hirt, C.W., and Cook, J.L., 1972, "Calculating Three-Dimensional Flows Around Structures and Over Rough Terrain," *J. of Computational Physics, 10*, 324–340.

Hotchkiss, R.S., and Hirt, C.W., 1972, "Particulate Transport in Highly Distorted Three-Dimensional Flow Fields," *Proc. of the 1972 Summer Simulation Conference*, SHARE, San Diego, Calif., June 14–16.

Issa, R.I., 1985, "Solution of the Implicitly Discretised Fluid Flow Equations by Operator-Splitting," *J. of Computational Physics, 62*, 40–65.

Launder, B.E., and Spalding, D.B., 1974, "The Numerical Computations of Turbulent Flows," *Computer Methods in Applied Mechanics and Engineering, 3*, 269–289.

Magi, V., 1987, "REC-87 A New 3-D Code for Flows, Sprays and Combustion in Reciprocating and Rotary Engines," *Report no. 1793*, Princeton University, Dept. of Mechanical and Aerospace Engineering, Oct.

Magi, V., and Grasso, F., 1985, "A Computer Program for Two-Dimensional Axisymmetric Flows with Sprays and Combustion (REC-2DA-FSC-85)," *Report no. 1766*, Princeton University, Dept. of Mechanical and Aerospace Engineering, Nov.

Margolin, L.G., 1978, "Turbulent Diffusion of Small Particles," *Report no. LA-7040-T*, Los Alamos Scientific Laboratory.

Martinelli, L., Reitz, R.D., and Bracco, F.V., 1984, "Comparisons of Computed and Measured Dense Spray Jets," *AIAA Progress in Astronautics and Aeronautics, 95*.

Nichols, J., 1972, "Stream and Droplet Breakup by Shock Waves," *NASA Sp. 194*, D.T. Harrje and F.H. Reardon, Eds.

O'Rourke, P.J., 1981, "Collective Drop Effects on Vaporizing Liquid Sprays," *Ph.D. Thesis 1532-T*, Princeton University, Dept. of Mechanical and Aerospace Engineering.

O'Rourke, P.J., and Amsden, A.A., 1987, "The TAB Method for Numerical Calculation of Spray Droplet Breakup," *SAE Paper 872089*.

Patankar, S.V., 1980, *Numerical Heat Transfer and Fluid Flow*, McGraw-Hill, New York.

Patankar, S.V., and Spalding, D.B., 1972, *Int. J. of Heat Mass Transfer, 15*.

Reitz, R.D., and Diwakar, R., 1986, "Effect of Drop Breakup on Fuel Sprays," *SAE Paper 860469*.

Rivard, W.C., Farmer, O.A., and Butler, T.D., 1975, "*RICE*: A Computer Program for Multicomponent Chemically Reactive Fluid Flow at All Speeds," *Report no. LA-5812*, Los Alamos Scientific Laboratory.

Stewart, H.B., and Wendroff, B., 1984, "Two-Phase Flow: Models and Methods," *J. of Computational Physics, 56*, 363–409.

Stone, H.L., 1968, "Iterative Solution of Implicit Approximations of Multidimensional Partial Differential Equations," *SIAM J. Numerical Analysis, 5*, 3.

Watkins, A.P., Gosman, A.D., and Tabrizi, B.S., 1986, "Calculation of Three-Dimensional Spray Motion in Engines," *SAE Paper 860468*.

Williams, F.A., 1965, *Combustion Theory*, Addison-Wesley, Reading, Mass.

Wu, K.-J., Santavicca, D.A., Bracco, F.V., and Coghe, A., 1984, "LDV Measurements of Drop Velocity in Diesel-Type Sprays," *AIAA J. 22*, 9, 1263–1270.

Wygnanski, I., and Fiedler, H., 1969, "Some Measurements in the Self Preserving Jet," *J. of Fluid Mechanics*, 38, 577–612.

IV
Turbomachinery and Power Cycles

14
Convective Heat Transfer with Film Cooling Around a Rotor Blade

T. ARTS

ABSTRACT: This paper deals with an experimental convective heat transfer investigation around a high pressure gas turbine film cooled rotor blade. The measurements were performed in the von Karman Institute short duration isentropic light piston compression tube facility allowing a correct simulation of Mach and Reynolds number as well as free stream to wall and free stream to coolant temperature ratios. The airfoil was mounted in a linear stationary cascade environment and heat transfer measurements were obtained by using platinum thin film gages painted on a blade made of machinable glass ceramic.

The coolant flow was ejected simultaneously through the leading edge (3 rows of holes), the suction side (2 rows of holes), and the pressure side (1 row of holes). The coolant hydrodynamic behavior is described and the effects of overall coolant to free stream mass weight ratio, coolant to free stream temperature ratio, and free stream turbulence intensity on the convective heat transfer distribution are successively described.

Introduction

A classical way to improve the thermal efficiency of a Joule/Brayton cycle is to increase the turbine entry temperature and pressure ratio. As a result, specific fuel consumption, size, and weight of aeroengines have been significantly reduced during the two last decades. A 25:1 pressure ratio and a 1800 K TET are presently typical values encountered in high performance jet engines. However, the latter are limited by material properties and an efficient internal and/or external cooling is most often required to overcome the high temperature operation problems. An accurate knowledge of the airfoil and endwall temperature and heat flux is therefore an important, even essential, part of the design in order to perform any detailed heat conduction or thermal stress analysis and to guarantee the lifetime of the different components.

The numerous parameters to be investigated in this field concern both the main (or free stream) and the secondary (or coolant) flow; they can be listed in a nonexhaustive way as follows:

- Airfoil geometry: curvature distribution, coolant emission location.
- Coolant emission geometry: hole shape, diameter and spacing, inclination and/or sweep angle of the hole, number of rows of holes.
- Blade loading: transition location, boundary layer status, shock/boundary layer interaction.
- Free stream Reynolds number.
- Free stream turbulence intensity.
- Blowing ratio or coolant to free stream mass weight ratio.
- Coolant to free stream temperature ratio.

From a numerical point of view, the accurate heat transfer pattern determination in a turbine, with or without any cooling scheme, remains an extremely difficult problem. The flow is highly three-dimensional, viscous, rotational, transonic, and unsteady. Reynolds numbers of the order of 5×10^5 to 3×10^6 (based on the true chord and the downstream conditions) are most often encountered. Also the free stream turbulence levels are quite high. In such an environment, the solution of the full three-dimensional Navier-Stokes equations requires an enormous computational effort (numerical method, CPU time, and computer memory) and will still remain a challenge in the coming years. In order to help the designer, simplified approaches have been considered: boundary layer codes, two-dimensional Navier-Stokes equations programs (parabolized or partially parabolized solutions), etc. (Crawford and Kays 1976; Dodge, 1976; Hah 1984; Lawerenz 1984; Lucking 1982; Moore and Moore 1981). These codes depend anyway on some empirical or experimental input (Reynolds stress modeling, boundary layer transition criterion, intermittency behavior, etc.).

From an experimental point of view, some of the representative available measurements on film cooled turbine cascade models were presented by Lander et al. (1972), Nicolas and Le Meur (1974) Ito et al. (1978), Daniels (1979), Dring et al. (1980), Horton et al. (1985) Camci and Arts (1985a, 1985b), and Arts and Bourguignon (1989). A large number of these heat transfer data are, however, difficult to use as such for modern cooled gas turbine design. Some of these are presented in terms of adiabatic efficiency and, as a matter of fact, in the severe environment of a film cooled turbine blade, the large temperature differences existing between the mainstream and the blade surface induce a wall temperature pattern quite different from an adiabatic distribution. Although adiabatic wall temperatures have been widely used for good reasons in predicting temperatures on film cooled turbine blades over the last 15 or 20 years, it appears that, considering the important spatial temperature variations and strongly varying heat flux distributions downstream of a film cooling row of holes, the most representative heat transfer parameter seems to be the convective heat transfer coefficient h.

The aim of the present contribution is to present detailed heat transfer data measured around a film cooled rotor blade. The latter was mounted in a

6-profile, stationary, linear cascade arrangement and was subjected to correctly simulated flow conditions, i.e., Mach and Reynolds numbers as well as free stream wall/coolant temperature ratios. The mainstream flow was generated in the von Karman Institute isentropic light piston compression tube facility and the coolant flow was independently ejected through the leading edge, the suction side, and the pressure side. The effects of both external and internal flows are considered in terms of Mach and Reynolds number, free stream turbulence intensity, blowing ratio and coolant to free stream temperature ratio.

Experimental Apparatus

The experimental techniques currently used to investigate heat transfer on gas turbine components can be divided into two categories: steady state techniques and short duration or transient techniques. The first approach has been intensively used in order to investigate the basic principles of convective heat transfer (e.g., Ko et al. 1986; Schwarz and Goldstein 1988). Their disadvantage is that, most often, they are not able to simultaneously provide the correct free stream Mach and Reynolds numbers, turbulence intensity, and free stream wall/coolant temperature ratios. Only engine test rigs provide a full similarity but their construction, maintenance, and operating costs prohibit their use in university or nonindustrial research laboratories. The use of short duration facilities allows running at full scale (mean) engine conditions but in a transient mode so that, although all the flow parameters are correctly duplicated, the total energy consumption is tremendously reduced. Three categories of short-duration testing facilities are principally used to investigate heat transfer and aerodynamic phenomena in turbine components: shock tunnels (e.g., Dunn and Chupp 1988; Louis 1977), blowdown cascades (e.g., Guenette et al. 1988), and light piston isentropic compression tubes (e.g., Arts and Graham 1985; La Graff et al. 1988). This last type of wind tunnel was used for the present investigation.

Description of the Facility

The experimental investigation was carried out in the von Karman Institute light piston isentropic compression tube facility (Fig. 14.1). The operating principles of this kind of wind tunnel were developed by Schultz & Jones (Jones and Schultz 1973; Schultz and Jones 1978) about 15 years ago. The VKI CT-2 facility, constructed in 1978, consists of a 5-m long, 1-m diameter cylinder containing a lightweight piston driven by the air of a high pressure reservoir. This cylinder is isolated from the test section by a fast-opening slide valve. As the piston moves, the gas in front of it is nearly isentropically compressed until it reaches the pressure, and hence the temperature levels defined by the operator. The fast-opening valve is then actuated, allowing this

FIGURE 14.1. VKI isentropic compression tube facility.

pressurized air to flow through the test section. Constant free stream conditions are maintained in the test section until the piston completes its stroke. The typical test duration is about 500 ms. The free stream conditions can be varied between 300 and 600 K and between 0.5 and 7 bar.

Description of the Model

All measurements were carried out on the two-dimensional rotor blade section already tested without any film cooling present by Consigny and Richards (1982). The blade and cascade geometry are illustrated in Fig. 14.2. The profile coordinates are listed in Table 14.1 and the main cascade dimensions are summarized as follows:

Chord length = 80.0 mm.
Blade span = 100.0 mm.
Stagger angle = 38.5 degrees.
Pitch-to-chord ratio = 0.670.
Gauging angle = 21.0 degrees.
Design inlet flow angle = 30.0 degrees (referred to axial direction).
Leading edge diameter = 6.25 mm.
Trailing edge diameter = 3.0 mm.

The cascade consists of 1 ceramic and 5 aluminium airfoils.

The film cooling configuration applied on the ceramic blade is represented in Fig. 14.2. Three rows of cylindrical holes ($d = 0.8$ mm; $s/c = -0.031$, 0., 0.031) are located around the leading edge (rows LP, LM, LS). The row and hole spacing are both 2.5 mm. These holes are spanwise angled at 30 degrees from the tangential direction and drilled in a plane formed by the blade span and the local perpendicular to the blade surface. Two staggered rows of conical holes ($d = 0.8$ mm; $s/c = 0.206$, 0.237) are located on the suction side (rows S). The row and hole spacing are, respectively, 2.5 and 2.6 mm. The holes

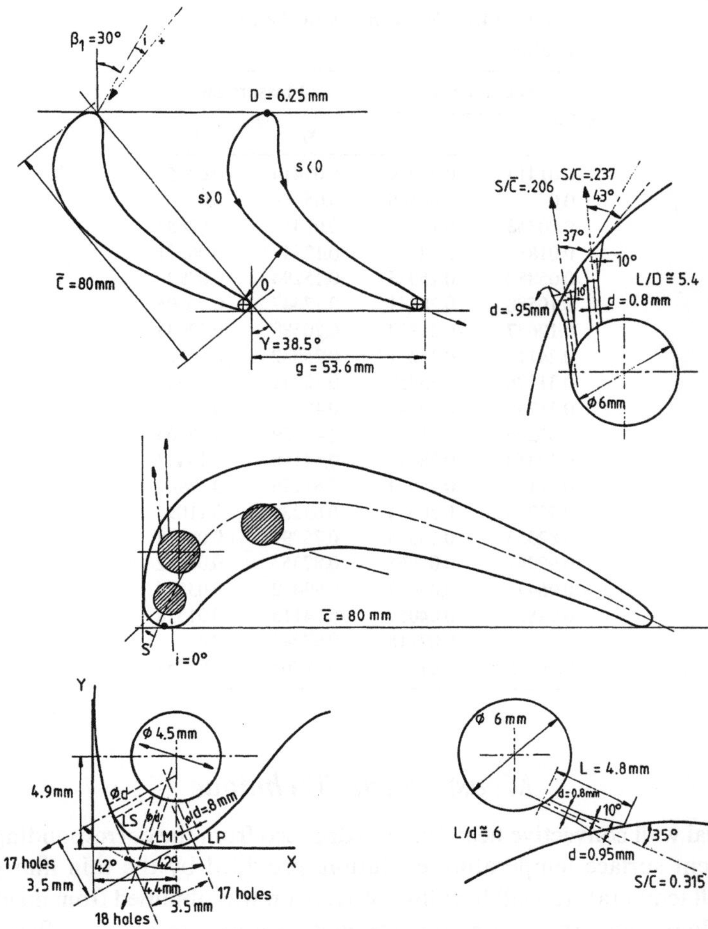

FIGURE 14.2. Blade and cascade geometry.

are inclined at 37.0 degrees and 43.0 degrees with respect to the local blade surface and drilled in a plane perpendicular to the blade span. One row of conical holes ($d = 0.8$ mm; $s/c = -0.315$) is located along the pressure side (row P). The hole spacing is 2.6 mm. These holes are inclined at 35.0 degrees with respect to the local blade surface and drilled in a plane perpendicular to the blade span.

The blade instrumented for heat flux measurements was milled from MACOR glass ceramic and 45 platinum thin films were applied on its surface (Fig. 14.3). Three independent cavities were drilled along the blade span to act as plenum chambers. The coolant flow was supplied by a regenerative-type cryogenic heat exchanger.

TABLE 14.1. Coordinates of the blade profile.

Suction Side		Pressure Side	
x/c	y/c	x/c	y/c
0.01414	0.02059	0.01414	0.02059
0.0	0.08588	0.05059	0.0
0.00353	0.13294	0.09412	0.02682
0.01882	0.18588	0.12941	0.06871
0.05882	0.25035	0.15294	0.08976
0.10588	0.28741	0.17647	0.10506
0.17647	0.31824	0.20588	0.12035
0.24118	0.33271	0.25882	0.14129
0.31176	0.33458	0.32941	0.15647
0.38235	0.32647	0.40	0.16012
0.45296	0.31	0.47059	0.15565
0.52353	0.28647	0.54118	0.14447
0.61176	0.25094	0.61176	0.12941
0.71765	0.20024	0.68235	0.11094
0.82353	0.14318	0.75294	0.08918
0.92941	0.07965	0.82353	0.06412
0.96471	0.05647	0.89412	0.03576
0.98824	0.04059	0.94118	0.01518
1.0	0.02118	0.97941	0.0
0.99765	0.01059	0.99765	0.01059

Measurement Technique

The local wall convective heat flux was deduced from the corresponding time dependent surface temperature evolution, provided by the thin film gages. The wall temperature/wall heat flux conversion was obtained from an electrical analogy, simulating a one-dimensional semi-infinite body configuration; the latter can be represented by the following equation and boundary conditions (Fig. 14.4):

$$\frac{\partial^2\theta}{\partial x^2} = \frac{\rho c}{k}\frac{\partial\theta}{\partial t} \qquad [\theta = T(x,t) - T(\infty)]$$

$$x = 0 \; : \dot{q}_w = -k\frac{\partial\theta}{\partial x}$$

$$x = \infty : \; \theta = 0$$

$$t = 0 \; : \; \theta = 0$$

A detailed description of this transient technique was presented by Schultz and Jones (1973). The convective heat transfer coefficient is defined as the ratio of the measured wall heat flux and the difference between the free stream recovery and the local wall temperatures. A recovery factor equal to 0.896 was

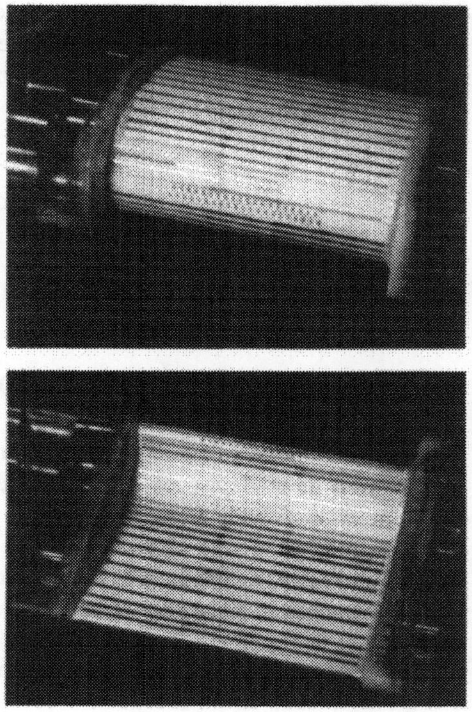

FIGURE 14.3. Blade instrumented for heat transfer measurements.

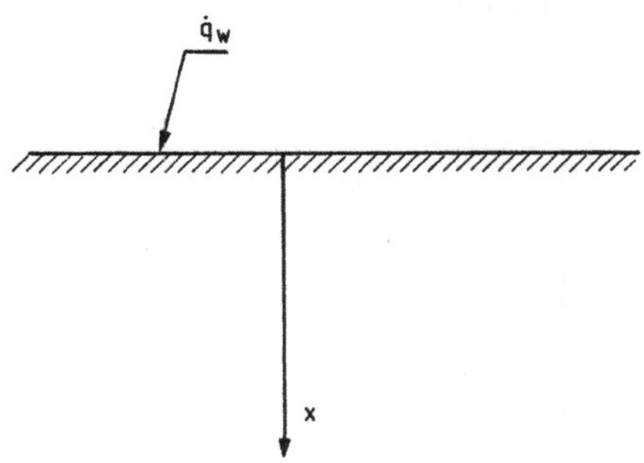

FIGURE 14.4. One-dimensional semi-infinite body configuration for heat transfer measurements.

used, as if the boundary layer on the blade surface was turbulent everywhere. The coolant mass flow was measured by means of choked orifices and miniature total pressure and total temperature probes continuously provided the coolant characteristics at the inlet and exit of the plenum cavities.

The free stream turbulence was generated by a grid of spanwise oriented cylindrical bars, displaced upstream of the model. Its intensity was measured by means of a constant temperature hot wire probe. The sampling rate was set at 1 kHz for heat transfer, pressure, and temperature measurements and at

25 kHz for turbulence intensity measurements. The free stream total temperature was selected to be 415 K. The uncertainty on the different measured quantities has been estimated as follows, based on a 20:1 confidence interval (Kline and McClintock 1953):

$$h = 1000\,\text{W/m}^2\text{K} \pm 50\,\text{W/m}^2\text{K}$$

$$p = 10^5\,\text{N/m}^2 \pm 750\,\text{N/m}^2$$

$$T = 100\,\text{K} \pm 1\,\text{K}$$

$$\dot{m}_c = 0.020\,\text{kg/s} \pm 0.0005\,\text{kg/s}$$

Blade Velocity Distributions

Blade surface pressure measurements were obtained from 31 static pressure taps located at midspan on one aluminum profile. Local isentropic Mach numbers were defined from these static pressures and from the total inlet pressure, measured one chord upstream of the cascade. The downstream Mach number was obtained from 17 wall static pressure taps located one chord downstream in the flow direction in a plane parallel to the blade trailing edges.

The isentropic Mach number distributions measured along the blade profile at zero incidence and without film cooling are shown in Fig. 14.5 for two isentropic outlet Mach numbers (0.62, 1.15). Along the pressure side, a velocity peak was predicted at $s/c = -0.08$. Because of the small leading edge radius, detailed measurements could unfortunately not be carried out around the

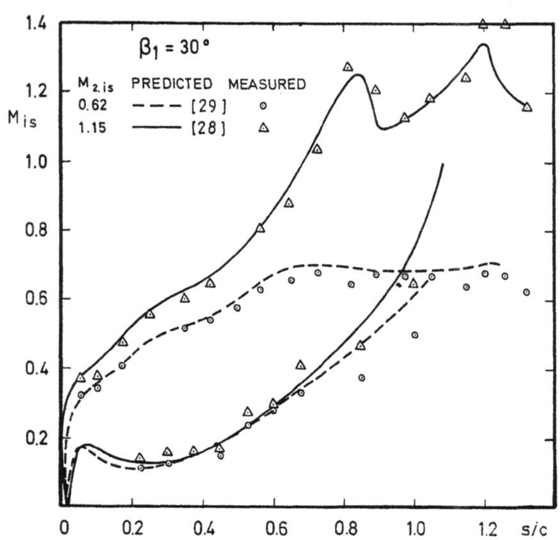

FIGURE 14.5. Blade velocity distributions.

stagnation point position. A two-dimensional inviscid time marching program (Arts 1982) provided a valuable prediction of the transonic blade velocity distribution, whereas a good approximation of the low exit Mach number velocity distribution was obtained by means of a singularity method (Van den Braembussche 1973). Both calculation methods were inviscid and no attempt was made to simulate the effect of film cooling.

The singularity method was also used to accurately determine the stagnation point position. The latter was calculated to be at $s/c = -0.019$. This result suggested that the suction side boundary layer would be affected by rows LM and LS, whereas the pressure side boundary layer would only be affected by row LP. These calculations were confirmed by detailed heat transfer measurements conducted around the leading edge by Camci and Arts (1985a). These measurements were also performed for different incidence angles.

Blade Convective Heat Transfer Distributions Without Film Cooling

The convective heat transfer coefficient distributions measured at zero incidence, and without any coolant emission, are shown in Fig. 14.6 for three different free stream Reynolds numbers (based on chord length and upstream conditions). A possible free stream air recirculation among the leading edge and/or suction side rows of holes was avoided by filling the three plenum chambers with flexible inserts. In the absence of the inserts, as was demonstrated by oil-flow visualizations (Camci 1985), free stream air entered into the leading egde plenum through row LM and was ejected through rows LS and LP, influencing the local heat transfer rates. The same phenomenon was observed across the two suction side rows (S).

The highest wall heating rates were measured in the leading edge region. Figure 14.7 demonstrates a definite influence of the existence of rows LM and

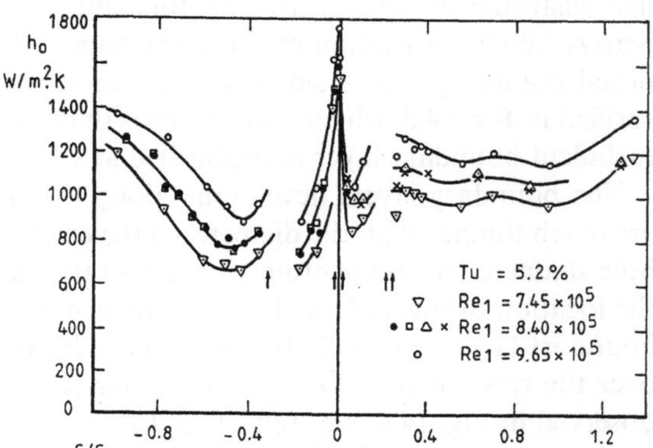

FIGURE 14.6. Blade convective heat transfer coefficient distributions (effect of Reynolds number).

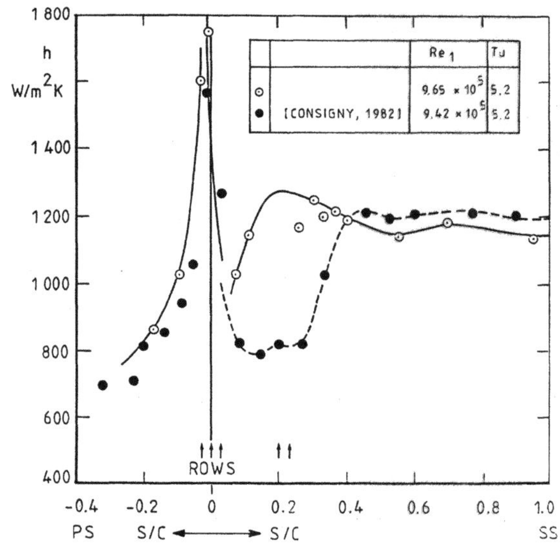

FIGURE 14.7. Blade convective heat transfer coefficient distribution (transition of suction side boundary layer).

LS on the suction side boundary layer behavior between $s/c = 0.0$ and 0.22. The results along the present model (open symbols) were obtained without any coolant flow emission, the plenum cavities being filled with the inserts previously mentioned; the free stream turbulence intensity was equal to 5.2 percent. The comparison between the present data and those obtained by Consigny and Richards (1982; closed symbols) for the same turbulence intensity around an identical but smooth undrilled blade, indeed reveals an earlier transition. Along the pressure side, an eventual tripping effect of row LP is not as obvious: similar heat transfer distributions were measured along the present blade and the one of Consigny and Richards. As a matter of fact, the early pressure side boundary layer transition is principally due to the existence of the velocity peak and the curvature inversion.

The influence of free stream turbulence is shown in Fig. 14.8. As already demonstrated by several investigators (Buyuktur et al. 1964; Junkhan and Serovy 1967), a variation of this parameter only affects laminar and transitional boundary layers developing along curved surfaces. This behavior is verified in Fig. 14.8, which also confirms that along the suction surface a fully turbulent boundary layer is established at $s/c = 0.25$.

The boundary layers developing along the suction and pressure surfaces are much thinner than the diameter of the emission holes. A local value of the hole diameter to momentum thickness ratio equal to 43 was evaluated at the location of row S. One direct consequence of this situation is that such a boundary layer most probably undergoes a local separation and reattachment over the rows of emission. This behavior is exemplified by the data scatter observed in Fig. 14.6 over rows S and P.

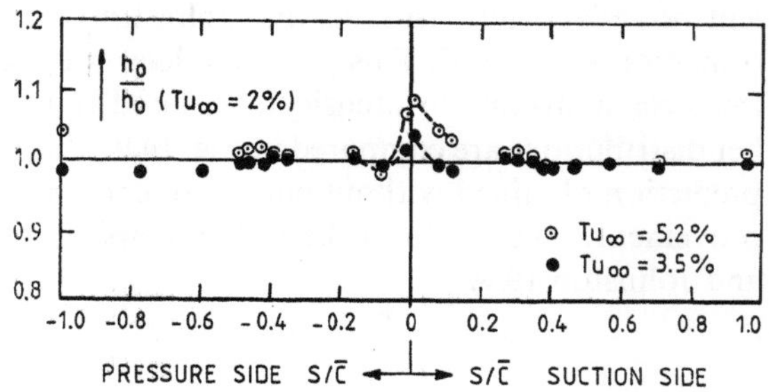

FIGURE 14.8. Blade convective heat transfer coefficient distribution (effect of free stream turbulence).

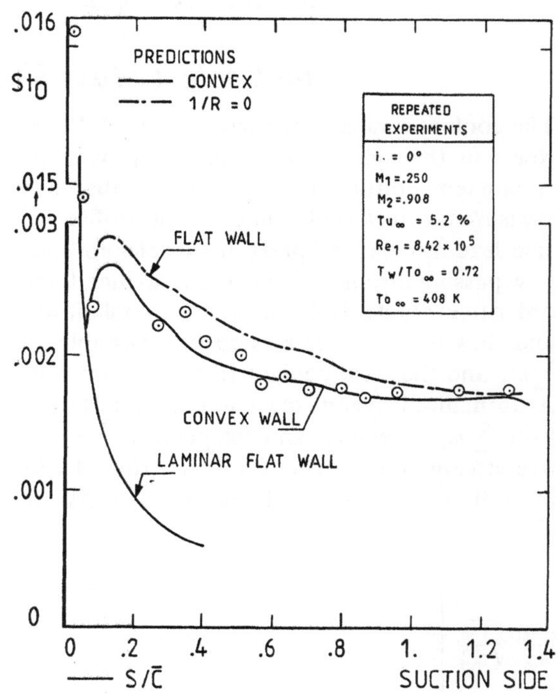

FIGURE 14.9. Blade convective heat transfer coefficient distribution (comparison with 2-D boundary layer calculations).

A numerical prediction of the convective heat transfer distribution without cooling along the suction side was obtained from a two-dimensional finite-difference boundary layer code (STAN5) developed by Crawford and Kays (1976) at Stanford University. This program is based on the classical Spalding-Patankar approach (Patankar and Spalding 1967) to compute boundary layer flows. It uses a finite difference technique to solve, through a streamwise space marching procedure, the simplified two-dimensional boundary layer equations as applied to flows developing, e.g., along a flat wall or in an axisymmetric tube. The streamwise curvature effects were taken into account by implementing a mixing length modification in the outer region of a turbulent boundary layer (Adams and Johnston 1984). An early boundary layer transition was forced. This program does not take into account, as such, the free stream turbulence intensity. Measured and computed local Stanton number distributions are compared in Fig. 14.9. This result clearly shows the overprediction obtained without curvature correction and demonstrates the improvement obtained from the rather simple correction presented in (Adams and Johnston 1984).

Coolant Flow Characteristics

Total Coolant Mass Flow Rate

The coolant flow across rows S, L (S, M, P) and P originated from a single reservoir through a heat exchanger providing the required coolant to free stream temperature ratios. This implies that the total coolant mass flow $\sum m_{ci}$, measured by means of a unique sonic orifice, was shared between the suction side, leading edge, and pressure side plenum chambers. The amount of coolant flow passing through each of these cavities, therefore, had to be determined in order to evaluate the local values of coolant to free stream mass weight ratio and blowing rate. The first step was to establish a unique dependency between $\sum m_{ci}$ and the local coolant to free stream static pressure ratio (Fig. 14.10). A normalized overall mass weight ratio $(\sum m_{ci}/m_\infty)(T_{oc}/T_{\text{ref}})^{1/2}$ was defined from $\sum m_{ci}$, pressure, and temperature measurements in each plenum and free stream static pressure at each row of emission; m_∞ is the free stream mass flow through one blade passage. All the measurements were taken

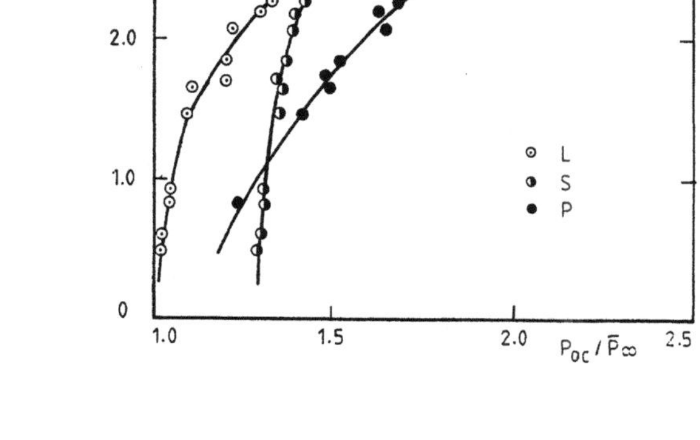

FIGURE 14.10. Normalized overall mass weight ratio evolution.

for three different values of the coolant to free stream temperature ratio ($T_{oc}/T_{0_\infty} = 0.7, 0.6, 0.5$).

Discharge Coefficient

The second step was to determine local values of the discharge coefficient, defined as the ratio between the real and isentropic mass flow rates through a row of film cooling holes. Averaged values of this parameter were measured (Fig. 14.11) at the location of rows S, L (S, M, P), and P from independent investigations. Significant losses were observed across the leading edge holes compared to the two other emission sites. These values showed, nevertheless, qualitative agreement with data presented by Tillman et al. (1984), obtained in incompressible flow (a water tunnel). The relatively low C_D values measured in the leading edge region were expected to occur because of the highly complicated nature of the coolant flow, with compound angle emission. Across the pressure and suction side rows, C_D values varied between 0.4 and 0.5.

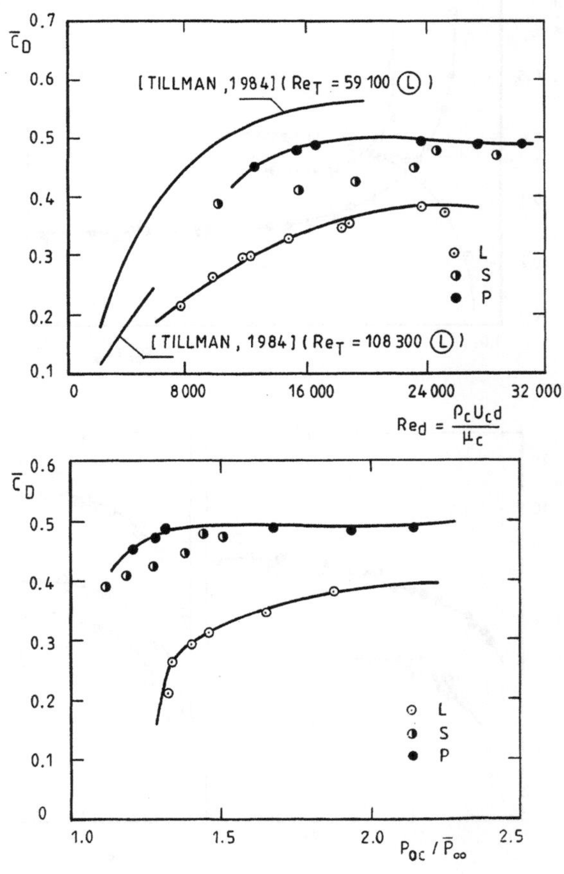

FIGURE 14.11. Discharge coefficient evolution for rows L, S, and P.

Coolant Flow Distribution

The third and final step was the quantitative determination of the coolant mass flow rate through each ejection site. It was obtained by combining data from Figs. 14.10 and 14.11. The coolant to free stream static pressure ratio for each plenum was obtained from Fig. 14.10, knowing the measured total mass flow rate $\sum m_{ci}$. The isentropic mass flow rate across the corresponding film-cooling rows was then calculated. The application of the corresponding discharge coefficient values (Fig. 14.11) finally provided the three actual mass flow rates, characterized by the ratio $m_{ci}/\sum m_{ci}$ (Fig. 14.12). For very low-pressure ratios, the flow conditions were not well defined in the leading edge plenum and very low C_D values were responsible for quite low local mass flow rates. The cooling rows S and P then performed the largest percentage of the emission. However, for a typical value of the overall mass weight ratio

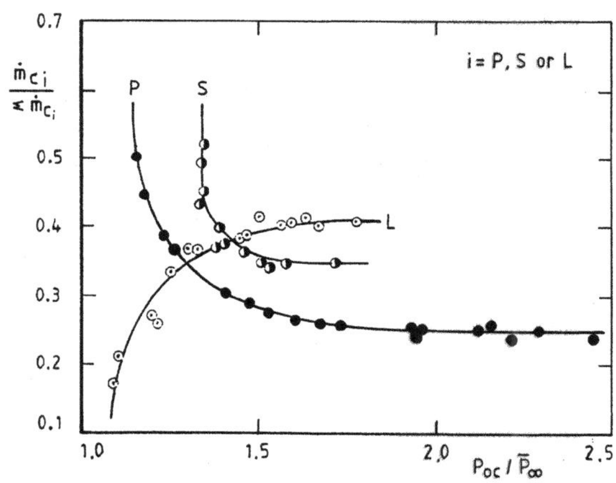

FIGURE 14.12. Coolant flow distribution among rows L, S, and P.

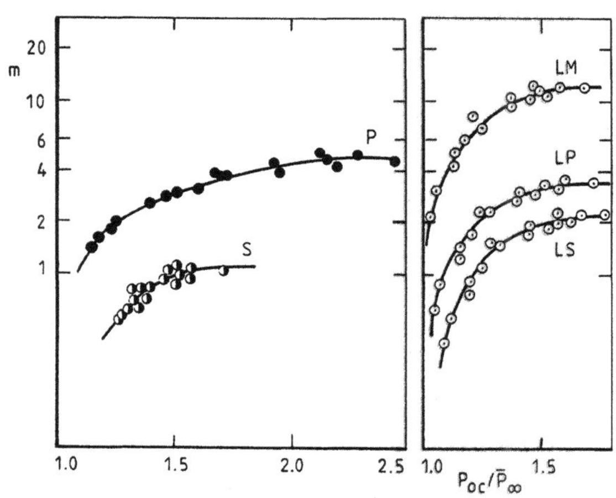

FIGURE 14.13. Blowing rate distribution among rows L, S, and P.

($\sum m_{ci}/m_\infty = 2$ to 3 percent), the coolant split was found to be 40 percent-35 percent-25 percent respectively through the leading edge, suction, and pressure side film cooling rows. The local blowing rate variations (Fig. 14.13) were determined from the three emission mass flow rate distributions (Fig. 14.12), the corresponding emission surfaces (Fig. 14.2), and the local free stream conditions (Fig. 14.5). The data of Fig. 14.13 are representative for three different coolant to free stream total temperature ratios ($T_{oc}/T_{0_\infty} = 0.7$, 0.6, and 0.5).

Blade Convective Heat Transfer Distributions with Film Cooling

The convective heat transfer measurements with film cooling were performed for constant values of the downstream Reynolds (2.32×10^6, based on chord length and downstream isentropic conditions) and isentropic Mach (0.925) numbers. The overall mass weight ratio ($\sum m_{ci}/m_\infty$) was varied between 0.5 and 3.3 percent and coolant to free stream temperature ratios (T_{oc}/T_{0_∞}) ranging between 0.51 and 0.70 were considered.

Effect of Overall Mass Weight Ratio

The results presented in this paragraph were obtained for a coolant to free stream temperature ratio equal to 0.7. The effect of overall mass weight ratio is presented in Fig. 14.14. Downstream of row LS, the heat transfer coefficient distribution was observed to be quite smooth for low values of $\sum m_{ci}/m_\infty$ (0.50 to 0.93 percent). As expected, increasing values of the mass weight ratio resulted in decreasing wall convective heat fluxes. For higher values (2.07 to 3.09 percent) however, a continuously increasing heat transfer coefficient was measured around $s/c = 0.08$. This behavior was explained by the high blowing rate values (up to 2.05) observed along this highly curved surface. The effect of the coolant film was more to augment the local free stream turbulence, and hence heat transfer, in this not yet fully turbulent region, than to efficiently protect the surface.

Downstream of row LP, a similar behavior was observed for the different values of the overall mass weight ratio. An additional effect was due to the existence of the pressure side curvature inversion; the latter induced the reattachment of the coolant layers ($s/c = -0.16$), separated from the wall at the high blowing rate values.

Downstream of row S, different phenomena were identified. For low values of $\sum m_{ci}/m_\infty$ (0.5 to 0.93 percent), a significant difference was observed in heat flux levels, although the blowing rate variation was rather limited (0.59 to 0.62). This behavior was explained by a cumulative effect, due to the additional contribution of the leading edge films (rows LM + LS), more effective for the

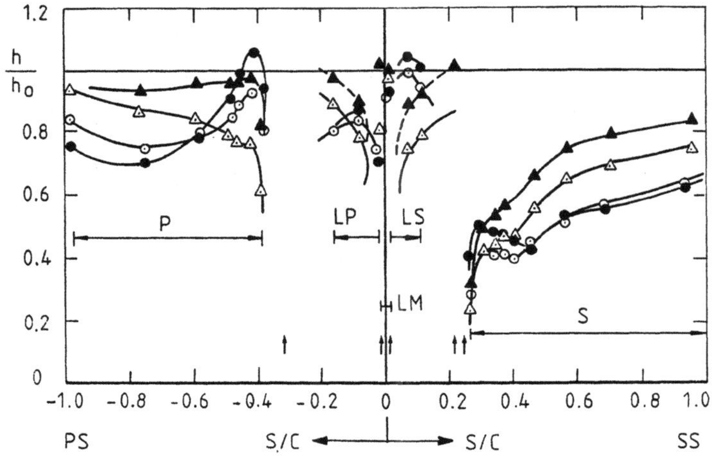

	$\Sigma\dot{m}_{ci}/\dot{m}_{\infty}$	$T_{oc}/T_{o\infty}$	m_P	m_{LP} / P^*	m_{LM} / P^*	m_{LS} / P^*	m_S
▲	0.50%	0.71	1.42	/ 1.004	/ 0.984	/ 1.038	0.59
△	0.93%	0.71	1.94	/ 1.022	/ 1.002	/ 1.056	0.62
⊙	2.07%	0.70	3.03	1.74 / 1.21	5.98 / 1.19	1.11 / 1.25	0.72
●	3.09%	0.70	4.17	3.20 / 1.58	11.0 / 1.55	2.05 / 1.63	0.93

$P^{\times} = P_{oc}/\bar{P}_{\infty}$

FIGURE 14.14. Heat transfer distribution with film cooling (effect of overall mass-weight ratio).

highest value of the overall mass weight ratio (0.93 percent). For even higher values (2.07 to 3.09 percent), a more significant heat transfer coefficient increase was measured downstream of row S. This phenomenon, due to the strong jet/mainstream interaction, was spread over a length equal to almost 30 film cooling hole diameters. No significant differences were observed farther downstream.

Downstream of row P, two regions were considered. Just downstream of the cooling holes, the heat transfer coefficient first decreased with increasing overall mass weight ratio (0.5 to 0.93 percent). For higher values (2.07 to 3.09 percent) the classical heat transfer augmentation took place. Farther downstream, on the contrary, the lowest heat transfer coefficients were observed for the highest values of $\sum m_{ci}/m_{\infty}$. This was most probably due to the concave nature of the wall; the free stream pulled the separated jets back to the wall.

Effect of Coolant to Free Stream Temperature Ratio

Values of the coolant to free stream temperature ratio as low as 0.5 are most usually observed in advanced aeroengines. In order to verify the effect of this important parameter, measurements were taken for three different values (0.7, 0.56, and 0.51), while $\sum m_{ci}/m_\infty$ was maintained at a constant value of about 3.1 percent. The results of Fig. 14.15 were obtained.

As obviously expected, significant heat transfer coefficient reductions were obtained when lowering the coolant temperature. The only exception was observed along the suction side front part, just downstream of row LS, where the heat transfer reduction is less important. This phenomenon was explained by the very high local blowing rate value, causing a separation of the coolant film. Although local blowing rates were maintained at constant values, just downstream of rows P, LP, and S, the importance of the wall heat flux augmentation identified in the preceding paragraph decreased with T_{oc}. As a matter of fact, a decrease of the temperature ratio (T_{oc}/T_{o_∞}) results in an

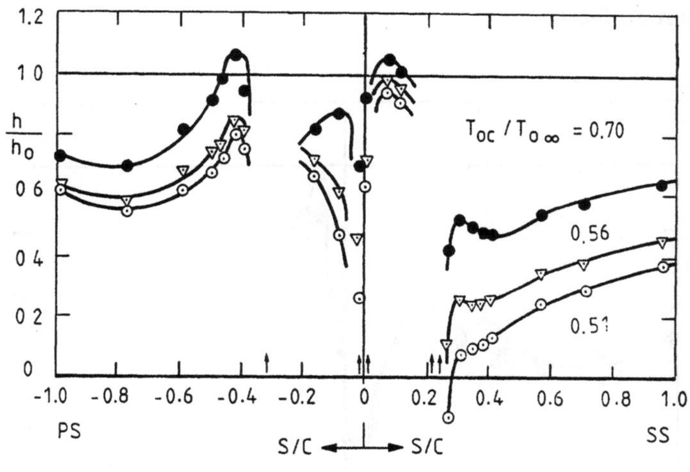

	$\Sigma\dot{m}_{ci}/\dot{m}_\infty$	$T_{oc}/T_{o\infty}$	m_P	m_{LP} / p^*	m_{LM} / p^*	m_{LM} / p^*	m_S
●	3.09%	0.70	4.17	3.20 / 1.58	11.0 / 1.55	2.05 / 1.63	0.93
▽	3.32%	0.56	4.67	3.36 / 1.53	11.5 / 4.50	2.15 / 1.58	0.99
⊙	3.12%	0.51	4.37	3.02 / 1.41	10.4 / 1.38	1.93 / 1.46	0.98

$p^* = P_{oc}/\bar{P}_\infty$

FIGURE 14.15. Heat transfer distribution with film cooling (effect of temperature ratio).

increase of the density ratio (ρ_c/ρ_∞) and, hence, for a constant blowing rate, in a decrease of the velocity ratio (u_c/u_∞) and moreover of the momentum ratio ($\rho_c u_c^2/\rho_\infty u_\infty^2$). The penetration of a "colder" film in the boundary layer is thus less severe, or in other words, the turbulence augmentation just downstream of these film cooling rows is less important.

Effect of Free Stream Turbulence Intensity

The effect of free stream turbulence on heat transfer with film cooling was finally investigated (Fig. 14.16). The turbulence level was varied from 0.8 percent to 5.2 percent while maintaining at almost constant values $\sum m_{ci}/m_\infty$ (2.5 percent) and T_{oc}/T_{0_∞} (0.5 percent). The values of h and h_0 plotted in figure 14.16 were obtained for equal values of the turbulence intensity. No significant changes in the wall heat flux were observed when varying Tu. As a matter of fact, this behavior was expected because of the dominant effect of the coolant flow (almost a new boundary layer) just downstream of the emission rows and of the turbulent nature of the boundary layers developing along the suction and pressure surfaces.

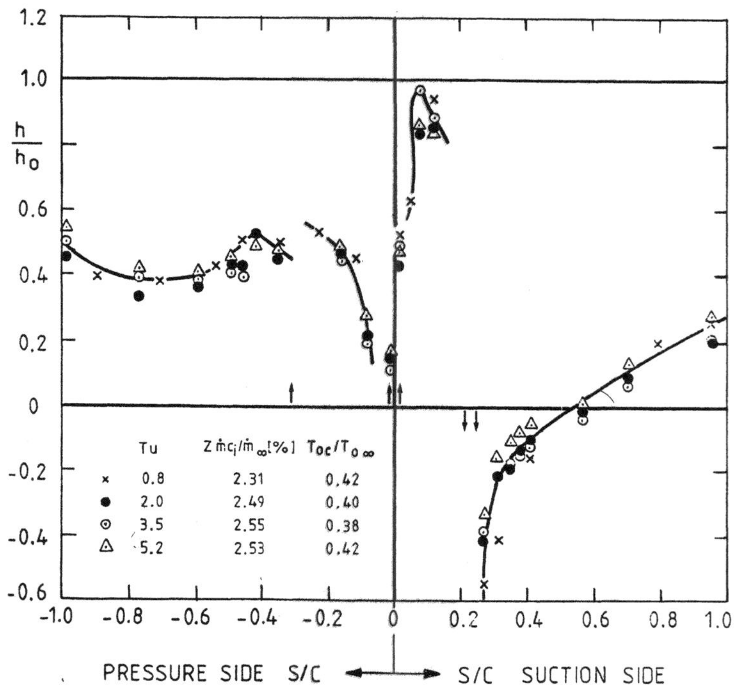

FIGURE 14.16. Heat transfer distribution with film cooling (effect of free stream turbulence).

Conclusions

Detailed convective heat transfer measurements were obtained for a high pressure, film cooled rotor blade mounted in a stationary cascade arrangement and submitted to correctly simulated aeroengine conditions (Mach and Reynolds number as well as temperature ratios).

The main conclusions of this investigation can be listed as follows:

- In the absence of coolant emission, the suction side boundary layer was dominated by the existence of the leading edge film cooling holes while the pressure side boundary layer behavior was dominated by the free stream pressure gradient. Moreover, the influence of free stream turbulence on the heat transfer coefficient distribution was extremely limited.
- Film cooling around the leading edge was verified to be quite effective for moderate overall mass weight ratio values. The same conclusion was drawn downstream of rows S and P. For higher values of $\sum m_{ci}/m_{\infty}$, increases in the local heat transfer were measured just downstream of the different cooling rows.
- The convective heat transfer coefficient distribution was proved to be strongly dependent on the coolant to free stream temperature ratio.
- No significant effect of free stream turbulence on convective heat transfer with film cooling was identified.

Nomenclature

c	= blade chord.
$\bar{C}_D$	= discharge coefficient.
d	= film cooling hole diameter.
h	= convective heat transfer coefficient.
LP, LM, LS	= leading edge film cooling rows of holes.
M	= Mach number.
m	= blowing rate.
$\dot{m}$	= mass flow rate.
P	= pressure side film cooling row of holes.
p	= pressure.
Re	= Reynolds number.
S	= suction side film cooling rows of holes.
s	= curvilinear coordinate (+ along suction surface − along pressure surface).
T	= temperature.
T_{ref}	= reference temperature (290 K).
Tu	= free stream turbulence intensity.

u = velocity.
ρ = density.

Subscripts

c = coolant.
∞ = free stream.
o = total condition.
1 = upstream of the cascade.
is = isentropic.

References

Adams, E.W., and Johnston, J.P., 1984, "A Mixing Length Model for the Prediction of Convex Curvature Effects on Turbulent Boundary Layers," *J. Engrg for Gas Turbines & Power, 106*, 1, Jan., 142–148.

Arts, T., 1982, "Cascade Flow Calculation Using a Finite Volume Method," "*Numerical Methods for Flows in Turbomachinery*, VKI LS 1982-05.

Arts, T., and Bourguignon, A.E., 1989, "Behaviour of a Two Rows of Holes Coolant Film Along the Pressure Side of a High Pressure Nozzle Guide Vane," ASME 34th Int. Gas Turbine and Aero-Engine Congress and Exposition, Toronto, Canada, June.

Arts, T., and Graham, C.G., 1985, "External Heat Transfer Study on a HP Turbine Rotor Blade," *Heat Transfer and Cooling in Gas Turbines*, AGARD CP 390, Paper 5.

Buyuktur, A.R., Kestin, J., and Maeder, P.F., 1964, "Influence of Combined Pressure Gradient and Turbulence on the Transfer of Heat From a Plate," *Int. J. of Heat & Mass Transfer*, 7, 11, Nov., 1175–1186.

Camci, C., 1985, "An Experimental and Theoretical Heat Transfer Investigation of Film Cooling on a High Pressure Gas Turbine Blade," *Ph.D. Thesis*, Katholieke University Leuven, Belgium.

Camci, C., and Arts, T., 1985a, "Experimental Heat Transfer Investigation Around the Film Cooled Leading Edge of a High Pressure Gas Turbine Rotor Blade. *J. Engrg for Gas Turbines & Power, 107*, 4, Oct., 1016–1021.

Camci, C. and Arts, T.: 1985b, Short duration measurements and numerical simulation of heat transfer along the suction side of a film cooled turbine blade. *J. Engrg for Gas Turbines & Power* 107, 4, Oct., 991–997.

Crawford, M.E., and Kays, W.M., 1976, "STAN5—A Program for Numerical Computation of Two Dimensional Internal/External Boundary Layer Flows," *NASA CR 2742.*

Consigny, H., and Richards, B.E., 1982, "Short Duration Measurements of Heat Transfer Rate to a Gas Turbine Rotor Blade," *J. Engrg for Power, 104*, 3, July, 542–551.

Daniels, L.C., 1979 "Film Cooling of Gas Turbine Blades," *Ph.D. Thesis*, University of Oxford.

Dodge, P., 1976, "A Numerical Method for Two and Three Dimensional Viscous Flow," *AIAA Paper 76-425.*

Dring, R.P., Blair, M.F., and Joslyn, H.D., 1980, "An Experimental Investigation of Film Cooling on a Turbine Rotor Blade," *J. Engrg for Power, 102*, 1, Jan., 81–87.

Dunn, M.G., and Chupp, R.E., 1988, "Time Averaged Heat Flux Distributions and Comparison with Prediction for the Teledyne 702 HP Turbine Stage," *J. Turbomachinery, 110*, 1, Jan., 51–56.

Guenette, G.R., Epstein, A.H., Giles, M.B., Haimes, R., Norton, R.J.G., 1988, "Fully Scaled Transonic Turbine Rotor Heat Transfer Measurements," *ASME Paper 88 GT 171.*

Hah, C., 1984, "A Navier-Stokes Analysis of Three Dimensional Turbulent Flows inside Turbine Blade Rows at Design and Off-Design Conditions," *J. Engrg for Gas Turbines & Power, 106*, 2, April, 421–429.

Horton, F.G., Schultz, D.L., and Forest, A.E., 1985, "Heat Transfer Measurements with Film Cooling on a Turbine Blade Profile in Cascade," *ASME Paper 85 GT 117.*

Ito, S., Goldstein, R.J., and Eckert, E.R.G., 1978, "Film Cooling of a Gas Turbine Blade," *J. Engrg for Power, 100*, 3, July, 476–489.

Jones, T.V., Schultz, D.L., and Hendley, A.D., 1973, "On the Flow in an Isentropic Free Piston Tunnel," *ARC R&M 3731.*

Junkhan, G.H., and Serovy, G.K., 1967, "Effects of Free Stream Turbulence and Pressure Gradient on Flat Plate Boundary Layer Velocity Profiles and Heat Transfer," *J. Heat Transfer, 69*, 2, May, 169–176.

Kline, S.J., and McClintock, F.A., 1953, "Describing Uncertainties in Single Sample Experiments," *J. Mechanical Engrg., 75*, 1, Jan., 3–8.

Ko, S.Y., Yao, Y.Q., Xia, B., and Tsou, F.K., 1986, "Discrete Hole Film Cooling Characteristics Over Concave and Convex Surfaces," *8th Int. Heat Transfer Conf., 3*, Washington, Hemisphere Publ. Corp., 1297–1301.

La Graff, J.E., Ashworth, D.A., and Schultz, D.L., 1988, "Measurement and Modeling of the Gas Turbine Blade Transiton Process as Disturbed by Wakes," *ASME Paper 88 GT 232.*

Lander, R.D., Fish, R.W., and Suo, M., 1972, "External Heat Transfer Distribution on Film Cooled Turbine Vanes," *J. Aircraft, 9*, 10, Oct., 707–714.

Lawerenz, M., 1984, "Calculation of the Three Dimensional Viscous Flow in Annular Cascades Using Parabolized Navier-Stokes Equations," "Secondary Flows and Endwall Boundary Layers in Axial Turbomachines," VK1 LS 1984–85.

Louis, J.F., 1977, "Systematic studies of Heat Transfer and Film Cooling Effectiveness," *High Temperature Problems in Gas Turbine Engines*, Paper 28, AGARD CP 229.

Lucking, P., 1982, "Numerische Berechnung des Dreidimensionalen Reibungsbfreien und Reibungsbehafteten Strömung Durch Turbomaschinen," *Ph.D. Thesis*, Aachen.

Moore, J., and Moore, J.G., 1981, "Calculations of the Three Dimensional Viscous Flow and Wake Development in a Centrifugal Impeller," *J. Engrg for Power, 103*, 2, April, 367–372.

Nicolas, J., and Le Meur, A., 1974, "Curvature Effects on a Turbine Blade Cooling Film," *ASME Paper 74 GT 156.*

Patankar, S.V., and Spalding, D.B., 1967, *Heat and Mass Transfer in Boundary Layers*, Morgan & Grampian, London.

Schultz, D.L., and Jones, T.V., 1973, "Heat Transfer Measurements in Short Duration Hypersonic Facilities," *AGARDograph 165.*

Schultz, D.L., Jones, T.V., Oldfield, M.L.G., and Daniels, L.C., 1978, "New Transient Facility for the Measurement of Heat Transfer Rates," *High Temperature Problems in Gas Turbine Engines*, AGARD CP 229, Paper 31.

Schwarz, S.G., and Goldstein, R.J., 1988, "The Two Dimensional Behavior of Film Cooling Jets on Concave Surfaces," *ASME Paper 88 GT 161.*

Tillman, E.S., Hartel, E.L., and Jen, H.F., 1984, "The Prediction of Flow Through Leading Edge holes in a Film Cooled Airfoil with and Without Inserts," *ASME Paper 84 GT 4.*

Van den Braembussche, R.A., 1973, "Calculation of Compressible Subsonic Flow in Cascades with Varying Blade Height," *J. Engrg for Power, 95*, 4, Oct., 345–351.

15
Unsteady Flow in Axial Flow Compressors

F.A.E. Breugelmans

Abstract: A review is made of the experimental rotating stall work on a low-speed compressor stage. The instantaneous flow field inside the cell is explored using multiple hot wires in the absolute frame of reference. Rotor blade stall and radial drift of the boundary layer is investigated by on-rotor blade instrumentation. The large flow variations suggest a strong unsteady behavior of the blades, which is demonstrated by the unsteady loss-incidence curve as measured with fast response instrumentation in the relative motion.

Introduction

The aerodynamic and mechanical performances of single- and multistage compressors are seriously influenced by the occurrence of rotating stall. As we move along the compressor characteristic toward small flow coefficients, a rotor subsynchronous phenomenon occurs which covers the whole or a part of the blade span and extends in the circumferential direction. This pattern is organized in one or more periodic events, called rotating stall cells. It is important to avoid or delay this event by adapting the design, to detect and control it in order to make full use of the operating range.

The understanding of the physics of flow in rotating stall cells of axial flow compressors has improved in recent years through the results of detailed measurements of the unsteady flow field. Having a physical picture of the flow is necessary for developing theoretical models for rotating stall prediction. Although considerable new information has been gained, there are still questions to be answered about the structure of the flow.

The first extensive detailed measurements of velocity magnitude and direction in stall cells of single- and multistage compressors have been reported by Day and Cumpsty (1978). Their results dispelled ideas about the cells being "dead wake" regions and an active flow structure was proposed. A systematic examination of the influence of design parameters on stall features was pursued and a correlation for stalled compressor performance was developed on the basis of their experimental results (Day et al. 1978). Information about flow in stall cells has also been provided by Das and Jiang (1983, 1984).

It should be noticed that these investigations have been referred to as flow in "big" stall cells. The distinction between two types of rotating stall has been reported by the authors in Breugelmans et al. (1983). While "big" (or "deep") stall is connected to large velocity fluctuations and the appearance of a recirculation through the ejection of fluid upstream of rotors, "small" stall occurs with the appearance of small fluctuations around a mean flow, which keeps moving in the through-flow direction. The development of this type of stall has been investigated and reported in Mathioudakis and Breugelmans (1985).

The three-dimensional character of the flow in big stall cells has been noticed in the previously mentioned experimental studies. The three dimensionality is exhibited by a considerable change of the flow pattern at different radial positions. The radial component of the velocity has not been measured, however, in any of these studies, and suggested meridional flow patterns were deduced from two-dimensional results at different radial positions. The present work provides a three-dimensional description of the flow, obtained by measurements of the three velocity components within stall cells.

The understanding of the flow phenomena near the surface of a compressor rotor blade is important for the design and analysis of this type of machine. The boundary layer development and flow separation on.the blades can lead to the occurrence of rotating stall in a compressor.

The first experiments on an oscillating airfoil of a helicopter model rotor were performed by W.J. McCroskey et al., using thc hot-film technique (McCroskey et al. 1976; McCroskey and Fisher 1972). The boundary layer development on a stator is analyzed by R.L. Evans (1977) and G.J. Walker (1977). Measurements in a rotor passage are performed by A.K. Anand and Lakshminarayana (1978). The limiting streamline angle is detected using an ammonia visualization technique. The analysis of the boundary layer evolution is performed in Lakshminarayana and Govindan (1981), Lakshminarayana et al. (1982), and Thompkins and Usab (1982) for the stable part of the compressor characteristic. The first attempt to detect flow separation and/or reversal on a rotor blade was reported by B. Gyles et al. (1982). A double or triple parallel wire arrangement is used, applying the thermal tuft technique. Further development of this idea conducted to the V-shaped wires for flow vector measurements near the blade surface under all low-speed operating conditions. These results are discussed herein.

The prediction of the onset of rotating stall requires knowledge of the blade charcteristics. The question can be asked whether the steady-state characteristic is applicable during the stall cell appearance. A series of experiments is organized to measure the unsteady loss characteristic of the rotor blade during rotating stall.

Small perturbation theories and nonlinear approaches investigated the onset of rotating stall. The compressor is modeled as a cascade or as axisymmetric surfaces. The flow field itself is not predicted, but some success was obtained in the prediction of number of cells, propagating velocity, and onset of stall (Montgomery and Braun 1957; Rannie and Marble 1957). A correlation for a stalled compressor performance, assuming a simplified flow model combined with experimental results, has been developed (Stenning et al. 1959).

A three-dimensional description of the perturbations is used. Three-dimensional perturbations to predict rotating stall in a rectangular cascade have been used in Yeh (1959), while in Dixon (1961) a similar approach was applied to an annular cascade. The complete solution for small perturbations in an annulus with free vortex mean flow was derived in Dunham (1965) and applied to study rotating stall. A theoretical model for rotating stall in a straight cascade with shear mean flow has been proposed recently in Sasaki and Takata.

In the present approach the solution of Dunham (1965) for three-dimensional perturbations is used to formulate transfer relations for the propagation of a disturbance through a compressor. A matrix formulation allows the derivation of an occurrence criterion of rotating stall and the study of its development characteristics. Compressors with any number of rows, rotors or stators, can be studied (the corresponding solution of Dunham (1965) was restricted to rotating stall of a single stator). The performance data of the compressor are considered in the form of the individual blade row characteristics and not as the overall performance map. A similar type of approach for two-dimensional perturbations has been developed in Ferrand and Chauvin (1982) where an analogy to linear servomechanisms was used to predict rotating stall.

In the present contribution, the single stall cell pattern was chosen to illustrate the detailed and averaged properties of the flow field in the absolute and relative frame of reference.

Experimental Facility

The single-stage low-speed compressor used for these experiments is the VKI R-l facility in Fig. 15.1. It is an open loop, constant annular section facility with an inner diffuser and throttle valve. The rotor is driven by a 55-KW dc motor with a speed control from 0 to 1500 rpm. The outer diameter is 0.704 m (constant) and the hub-tip ratio equals 0.78. The number of blades is 39 (IGV), 25 (rotor), and 25 (stator) with a midspan solidity of 1, 1.015, and 1, respectively. The blades are of the NACA 65 series.

The distance between IGV trailing edge and rotor leading edge is 0.094 m and between rotor and stator 0.069 m, at midradius. The geometric data for IGV and rotor blades are given in Table 15.1. The stator blades are 25 constant thickness, nontwisted, thin-metal airfoils of 35-degree camber and 34-degree stagger, constant chord of 0.080 m and $t/c = 0.025$.

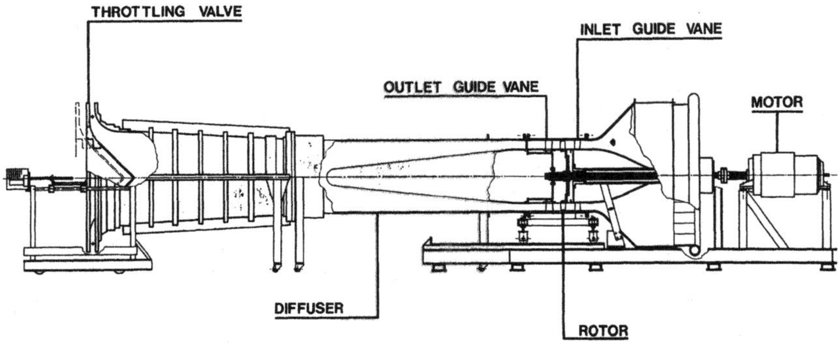

FIGURE 15.1. Low speed axial flow compressor.

TABLE 15.1. Geometry of inlet guide vane and rotor blading.

Blade Shape: NACA 65 (Cl_0A10)
Inlet Shape: NACA 65 (Cl_0A10)

h	$\varphi(°)$	$\left(\frac{t}{c}\right)_{max}$	$\gamma(°)$	c(mm)
0	47.01	0.06	21.69	44.5
0.25	44.76	0.06	20.65	47.5
0.50	42.55	0.06	19.63	50.5
0.75	40.73	0.06	18.79	53.5
1	38.86	0.06	17.73	56.5

Rotor, 25 blades

$\varphi(°)$	$\left(\frac{t}{c}\right)_{max}$	$\gamma(°)$	c(mm)
22.34	0.06	24.56	70.4
18.34	0.06	31.54	75.2
14.83	0.06	37.33	80.0
11.74	0.06	42.11	84.8
9.23	0.06	46.03	89.6

Instrumentation

The stage performance is obtained using directional pressure probes of the NACA short prism type (Erwin 1964), positioned 10 mm in front of the IGV and 25 mm downstream of the stator vanes.

The unsteady measurements in the stationary frame are performed using two crossed hot wire probes placed 40 percent of rotor blade chord upstream and 33 percent of a chord downstream of the rotor. The hot wires are 9-μm platinum-plated tungsten wires with $L/D = 200$. Their calibration curves are derived in the form of the ratios

$$\frac{Ve_1}{Ve_2}, \quad \frac{Ve_1 - Ve_2}{V}$$

as a function of the angle between the first wire and the velocity vector.

For measurements near the rotor blade surface, two sensors are used, namely, a thermal tuft and V-shaped hot wire combination. Both are placed at the midchord and midspan position on the rotor blade during successive experiments (Fig. 15.2). The thermal tuft consists of two parallel wires, operating at different overheat ratios. Under these circumstances, the ratio of the two effective cooling velocities can give an indication whether the flow is attached to the blade surface and normal to the wires and its value is 0.2. When the velocity vector becomes more parallel to the wires or small amounts of separation occur the ratio lies between 0.2 and 1.0. Finally, when flow reversal occurs the value is 1.20 (Ligrani et al. 1983). The V-shaped sensor is operated on the same principle as the crossed hot wire probes. It is used to measure the velocity magnitude and angle with respect to a reference direction near the blade surface.

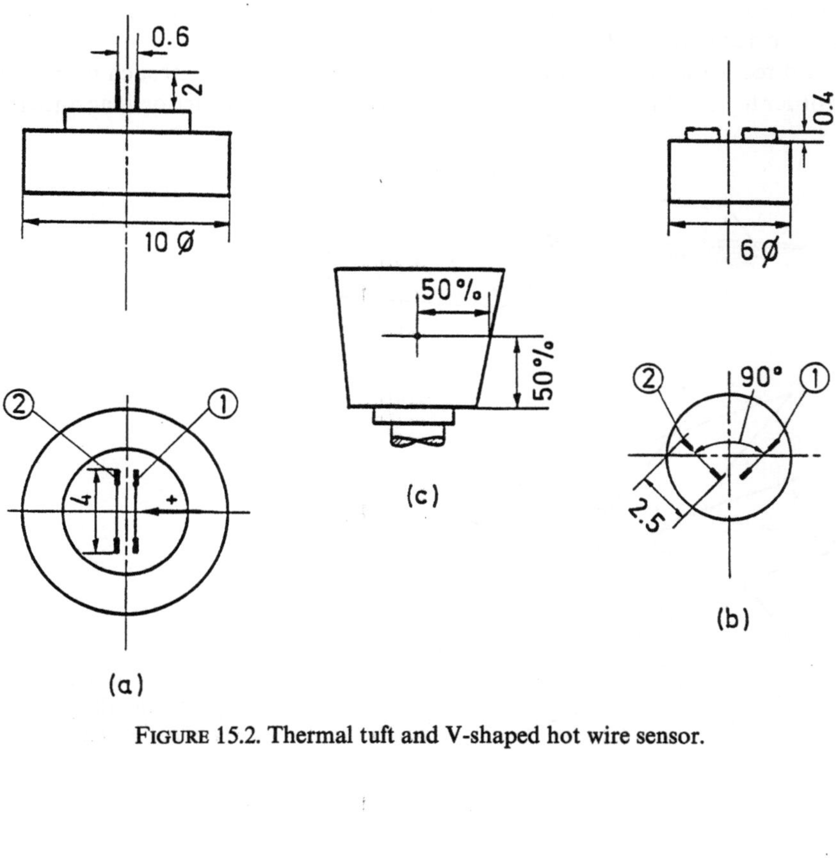

FIGURE 15.2. Thermal tuft and V-shaped hot wire sensor.

All the hot wires are operated at a constant temperature using VKI anemometer bridges. A low-noise mercury slip ring ($< 50\,\mu$V) is used to transmit the signal from the hot wires to the bridges. The anemometer bridges are connected to a 16-channel data-acquisition system. It contains an amplifier and a variable antialiasing filter per channel followed by an analog to digital converter/multiplexer with maximum sampling rate of 50 kHz. Data are sent as 12-bit words to a PDP 11/34 via a serial line transmitter of 1 MB/s for later processing.

In order to perform unsteady 3-D velocity measurements in rotating stall, a measurement technique has been developed, employing a triple hot wire probe designed for this purpose. Existing triple hot wire techniques have a limited range of flow direction that can be measured, which make them unsuitable for rotating stall measurement. A detailed description of the present method can be found in Mathioudakis and Breugelmans (1985). A brief description is given here.

A schematic of the probe is shown in Fig. 15.3. The wires are platinum-plated tungsten wires of $d = 9\,\mu$m and $l/d = 222$. Thev are designed to form an orthogonal system if they are accurately manufactured. A rectangular coordinate system with respect to the wires is defined as shown in Fig. 15.3. Two octants (an octant is the solid angle defined in space by the direction of three orthogonal axes) are used for the measurements; the one defined by the positive x, y, and z axes and the one defined by positive y, z, and negative x of the orthogonal system.

A direct calibration of the probe is done. This type of approach is chosen in order to account for probe interference effects and construction inaccuracies

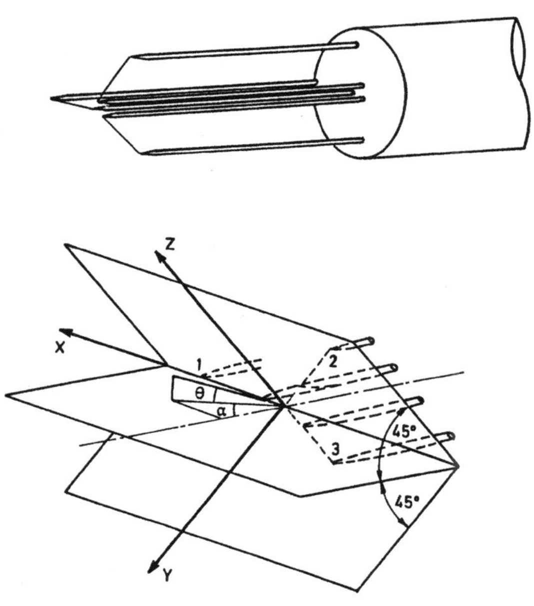

FIGURE 15.3. Schematic of the probe and the related orthogonal system.

and to provide a wide range of operation of the probe. Three parameters Y, P, and Q are defined as functions of the cooling velocities Ve_1, Ve_2, and Ve_3 of the three wires. They are chosen in such a way that Y is primarily sensitive to yaw angle changes, P to pitch angle changes, and Q to velocity magnitude. The formulas for their derivation are given in Mathioudakis and Breugelmans (1985).

The measurements in the relative frame of reference are performed with a hot wire and a total pressure transducer. A single hot wire, placed at midspan in the rotor blade leading edge plane, is used as the trigger for the data acquisition of the measurements in the rotating frame. This hot wire is located at an available position 8 blade pitches tangentially separated from the pressure sensor. The total pressure is measured with a fast response probe mounted at midblade height in a plane 4.5 mm downstream of the rotor trailing edges. The tangential traversing is done in steps of 1 degree. The sensor is a Kulite XCS-093 with an outside diameter of 2.4 mm and fitted with a protective screen of the type "*B*." The probe transducer is placed with the silicon diaphragm plane in the direction of the centrifugal forces in order to minimize this influence. The triggering hot wire and Kulite signal are taken out by means of a low-noise mercury slip ring mounted on the shaft extremity.

Performance and Rotating Stall Cell Description

The Performance Map

The stage performance map (Fig. 15.4) is obtained by radial flow surveys of the complete stage at 1000 rpm and a complete characteristic, with hysteresis loop is measured. The operating points are indicated $A \rightarrow H$, where A = clean flow; B = maximum rotor efficiency of 90 percent; C = maximum loading C, D, E = multicell rotating stall; F = one cell; G = two cells and H = one cell on the return branch of the characteristic, as summarized in Table 15.2.

Small-scale unsteady effects, such as IGV wake-rotor interactions are observed in points A and B. The rotor wake thickens as the aerodynamic loading increases from $B \rightarrow C$ when flow separation is already felt on the rotor blade. The inlet flow field gets organized in an eight-cell pattern, point D, whereafter a not-so-well-defined perturbation of 7 to 10 cell structure occurs before transiting to F where one large cell is observed. Further throttling, point G, produces a two-cell pattern, which disappears into a single cell at the opening cycle of the throttle and a hysteresis loop is complete in point $H \rightarrow B$. No stable operating points can be obtained between $E \rightarrow F$ and $H \rightarrow B$. Small perturbations are not observed during the throttle valve opening.

The correlation from Greitzer (1980) indicates the occurence of rotating stall and no surge based on the parameter B smaller than 0.70. Full-span

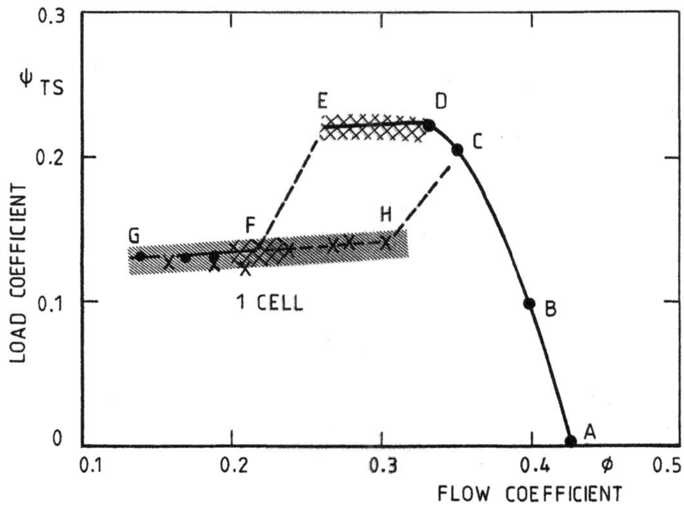

FIGURE 15.4. Performance map of the compressor.

TABLE 15.2

Point	N (degree) cells	U_c/U (%)	Passing frequency (Hz)	Type	Stall
A	—		Clean flow		
D	8	84	113.0	Full span	Small perturbation
E	7– 10			Full span	Small perturbation
F	1	31	5.2	Full span	Deep span
G	2	32	10.7	Full span	Deep span
H	1	30	5.0	Full span	Deep span

rotating stall is possible when the blockage due to the cells is more than 30 percent of the annulus. This is confirmed in those experiments on a compressor model with $B = 0.12$. The local (radial) and overall (annular) blockage factor λ is evaluated from the velocity measurements. Table 15.3 summarizes values λ for the single- (point *F*)- and double-cell pattern (point *G*), upstream of the rotor at five radial positions *h*. The average over the total annulus is $\lambda_1 = 0.42$, and $\lambda_2 = 0.48$, which is well above the limit value of 0.30. The total-to-static load coefficient is preferred in order to allow a comparison with correlations for stalled and unstalled predictions (Greiter 1980). The ψ_{TS} in stall is close to 0.11, while ϕ_H equals 0.70 times the ϕ value on the stable charateristics at the same local coefficient $\psi_{TS-(H)}$.

TABLE 15.3

h	$F(\lambda_1)$	$G(\lambda_2)$
0.90	0.72	0.76
0.70	0.52	0.55
0.50	0.41	0.34
0.30	0.35	0.26
0.10	0.34	0.25

The Power Spectra and Autocorrelation

This analysis is performed in order to study the energy content in the fluctuating components at the different frequencies, the shape of the cell, the preservation with time, and correspondence from cell to cell.

The power spectra are calculated up to half the sampling frequency using the velocity signals and a fast Fourier transform. The $\log|E(f)|$ is presented as a function of frequency.

In the unstalled condition, $A \rightarrow C$, only the rotor blade passing frequency is observed in the two measuring planes and has been filtered in all following measurements. The power spectra for point F is shown in Fig. 15.5.

The one-cell pattern—point F—has the peak at 5 Hz with rapidly decaying harmonics due to the nonsinusoidal shape of the stall cell. The expansion of the scale in the low-frequency domain allows the exact definition of the period for the event.

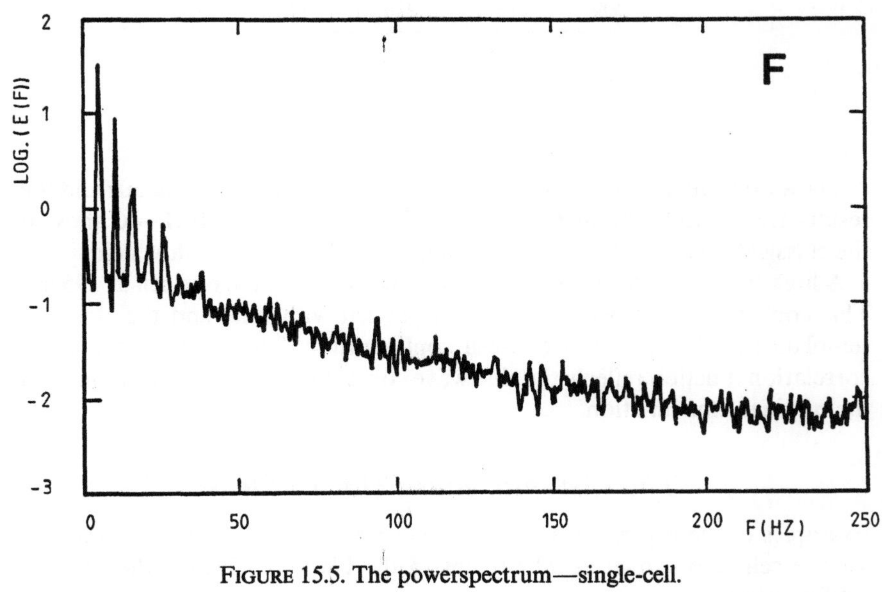

FIGURE 15.5. The powerspectrum—single-cell.

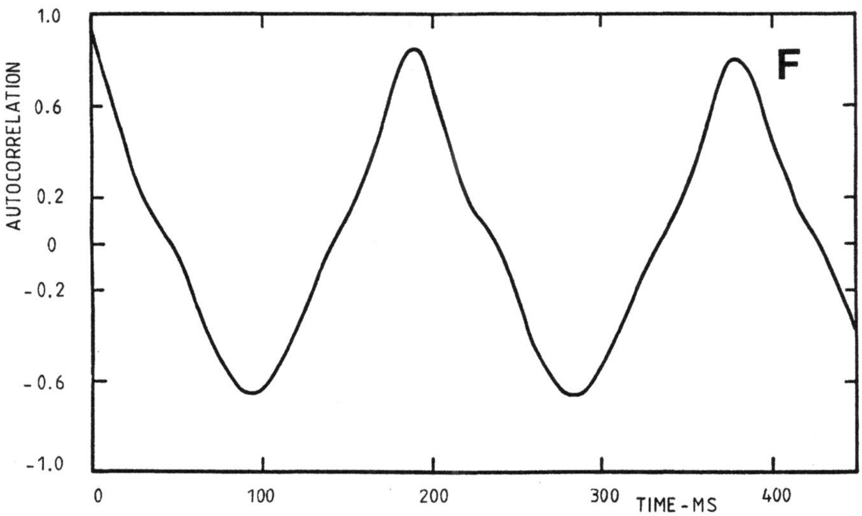

FIGURE 15.6. The autocorrelation—single-cell.

The cross-spectra calculations, using the velocity (and angle-average spectra, do show a phase difference for the angular fluctuations with respect to the velocity. These results show a delay for the velocity variations with respect to the angle for the eight cells and opposite for the other patterns.

Cells	Phase	$\Delta\tau$(ms)	f_c(Hz)
8	−55	−1.34	113
1	47.7	26.7	5.2
2	54	13.6	10.7

The autocorrelations are identical for the velocity and angle signals and the results are shown for point *F* (Fig. 15.6). The operating points $A \rightarrow C$ showed the classical turbulent flow autocorrelation for the upstream flow field.

A high level, ~0.85, exists for the autocorrelation of the one cell at 195 ms. This confirms an almost-invariant cell pattern with time and the close resemblance of the cells in a two-cell configuration. The shape of the autocorrelation function reflects the sine wave content and the wide-band noise of the original time function.

The Instantaneous Velocity Field

Examples of the usptream instantaneous velocity and flow angle variations for the cells are shown at 90 percent of the blade height, for the one-cell configuration.

The one cell, point F, occupies more than half the circumference and the flow direction is 36 percent of the time past 90 degrees from axial, indicating reverse flow in that area. In the same region, the flow velocity equals or surpasses the local rotor speed. The same behavior is seen in each cell for the case of the two cells as well as for the one cell in the return branch.

The propagating cell moves at a fraction of the rotor speed and a fixed instrumentation will capture the unstalled flow first, then the leading front of the cell, followed by the central part, and finally the trailing end as shown in the velocity-time diagram. The rotor, on the contrary, approaches the cell from the trailing end due to the larger peripheral velocity. Therefore, the leading edge of the rotating stall cell corresponds to the unstalling process of the compressor rotor blading, while stalling of the rotor occurs at the trailing end of the cell.

Sinusoidal variations of the local flow direction are observed when the leading edge of the propagating stall cell passes a fixed instrumentation. This phenomenon is seen in the case of a single cell. It is interpreted as a vortex forming the leading part of the cell. This leading vortex is estimated to occupy a tangential extension of one-and-a-half blade passages. The trailing end of the cells also has an indication of these vortices but the steep gradient masks the effect. This phenomenon occupies 10 percent of the cell extension and is observed on all of the large samples that have been made, but it will be attenuated in the averaging procedures due to the fluctuations in the cell extension. Some theoretical approaches (Kriebel et al. 1960) used vortices in the analysis of the rotating stall, where the airfoils lose or gain their circulation when entering or leaving stall by shedding vortices in the downstream flow field. During the reverse flow the "downstream" field is in front of the rotor and the previous mentioned observations might be the corresponding shed vortices.

Structure of the Rotating Stall Cell

The structure of a big rotating stall cell is best described using the instantaneous measurements of the single and double cell at a radius where reverse flow is present. Such a trace can be found at 90 percent and 70 percent of the blade height for the operating points F and G. The different zones that can be distinguished are found by combining velocity and flow angle and relating the results to the rotor blade. It is useful to follow the trace in opposite direction to the timescale, $A \rightarrow D$. Six zones are found (Fig. 15.7).

- *Zone A*: where the velocity is constant and the absolute flow angle is still adjusting itself to the clean flow condition.
- *Zone B*: the velocity decrease and angle increase result in a build up of the rotor blade incidence angle. An instantaneous incidence angle of 25 degrees can be obtained during this unsteady condition.
- *Zone C*: the absolute velocity remains at a very low level and the inlet flow angle suddenly increases to 90 degrees and permanent stall has occurred.

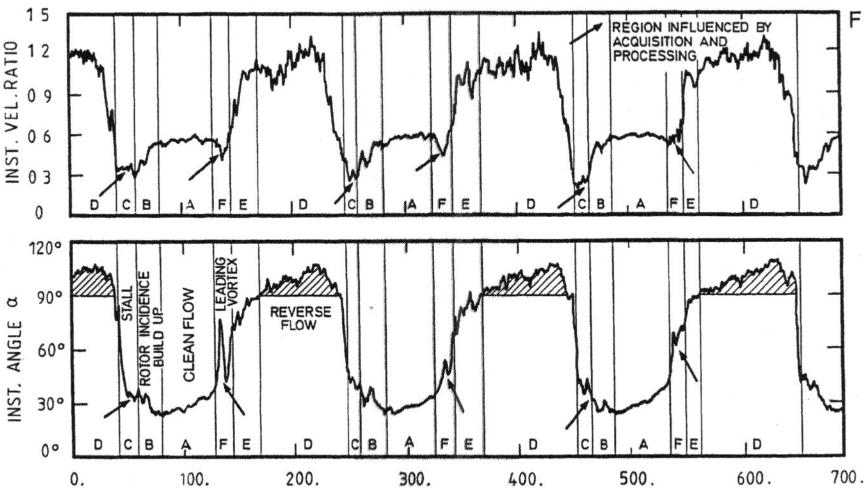

FIGURE 15.7. The instantaneous absolute velocity and angle variation upstream of rotor blade—90 % height.

- *Zone D*: the tangential flow from the rotor starts entraining the fluid in the IGV-rotor space at a velocity slightly above the peripheral speed and a large zone of reverse flow appears. This reverse flow region is centered toward the back end of the stall cell. In fact, it is fed by the rotor.
- *Zone E*: as we move toward the cell leading edge a decay of the rotor return flow occurs and the flow angle becomes smaller than 90 degrees; the velocity follows a similar trend.
- *Zone F*: this is the front or "head" of the rotating stall cell. The combination of velocity and angle variations suggests a vortex type of motion by which flow around the rotor blades is restored to the clean flow condition.

The rotating stall cell can be defined by the zones $[C, D, E, F]$. Some of these zones are not always present at all radial positions; others will be strongly attenuated or suppressed completely due to the sampling technique in the acquisition phase or the averaging technique in the data processing. According to this definition, the rotating stall cell presented in this figure occupies 63 percent of the circumference.

Phase Locked Average Results

The phase-locked averaged results for the velocity and angle evolutlon during one stall event ($0 \rightarrow 2\pi$) are presented for five radial positions h. The velocities are nondimensionalized with respect to mean rotor speed. The condition F is given. Since the measurements at the different radial positions are not performed at the same time, some synchronization of the different traces is

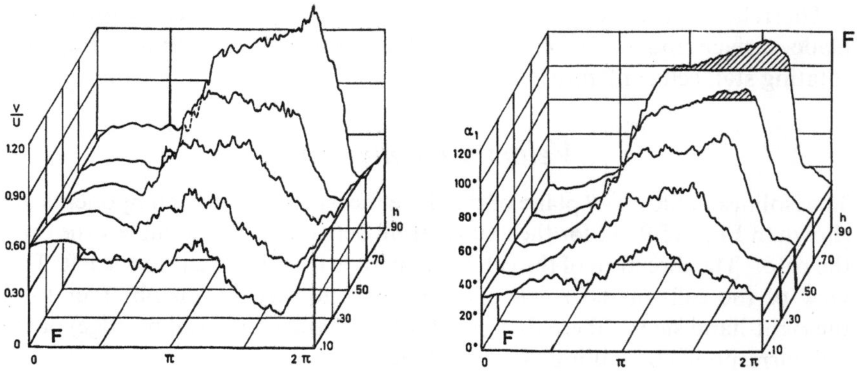

FIGURE 15.8. The phase averaged absolute velocity and angle variation upstream of rotorblade.

needed. The alignment is done by using distinct features of the stall cell in the velocity or angle traces and centering them for the presentation. A sudden increase of angle at the leading and trailing edge of the cell, the center of the cell, and the clean flow zone have been used (Fig. 15.8).

The stall cell region is characterized by high velocities and large flow angles in the one-cell case, point F. Reverse flow is present at the outer radii; these zones are indicated at 70 and 90 percent of the blade height as well as their tangential extension. The absolute velocity in the stall cell equals or exeeds the rotor peripheral speed. Near the hub, at 30 and 10 percent height, the flow angle reaches maximum values of 70 and 60 degrees, respectively, and a small axial component is present. This cell can be viewed as propagating in the tangential direction, being fed at the outer radius by fluid returning from the rotor. This return flow executes a helicoidal motion and is found back at the inner radius, where it contributes to a small mass flow influx into the rotor blades during the occurrence of rotating stall.

The fluctuating components of the velocity outside of the stall cell are 2 percent with respect to the average velocity, while at the edges of the cell the fluctuations increase to 25 percent at all radii. Inside the cell a variation from 6 percent (tip) to 20 percent (hub) is observed. These levels are very similar for the two-cell configuration. The downstream fluctuations are much larger.

Flow Field: Experimental Results

The results of the 3-D measurements are now presented. Representative results have been selected from the total amount of the data obtained for the presentation. Three mutually orthogonal planes are used for the presentation of the velocity components in the absolute reference frame.

The relative flow quantities are illustrated by the velocity vector on the blade surface and the total pressure loss coefficient variation during one rotating stall cell configuration.

Velocity Components

The isolines in the $r - \theta$ plane of the three measured velocity components are shown in Figs. 15.9–11 for the axial station between the inlet guide vanes and the rotor. The direction of the cell movement is from the right to the left. This view of the cell on the $r - \theta$ plane is the one seen by an observer from the right-hand side in these figures. The right-hand edge (trailing edge) of the cell corresponds to stalling of rotor blades.

Axial Velocity

The isolines of the measured axial velocity components are shown in Fig. 15.9. These isolines upstream of the rotor (station *C*) show that the stall cell is characterized by a core of reversed flow covering the upper central part of the

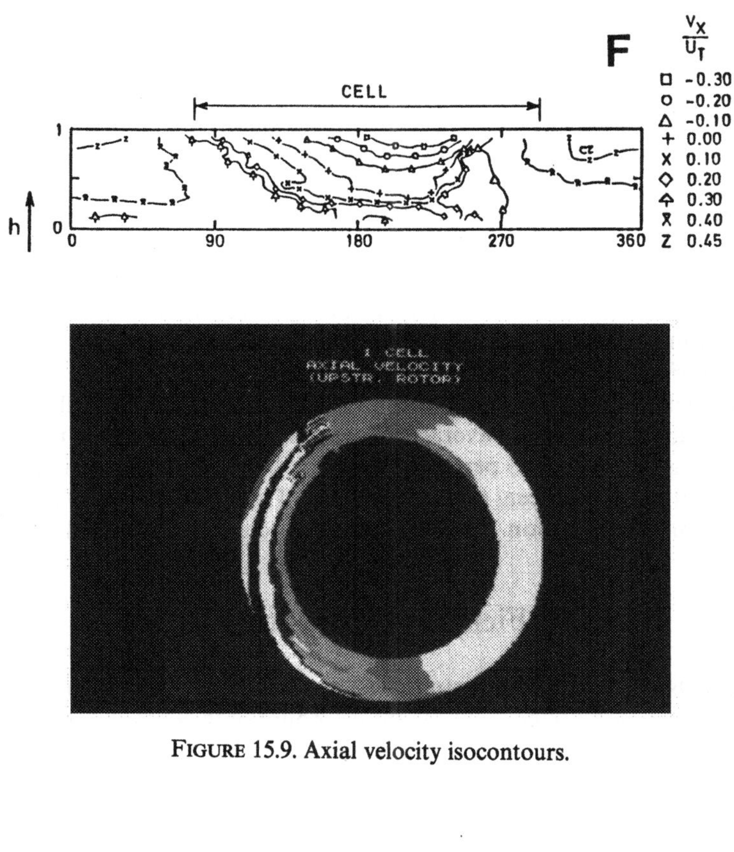

FIGURE 15.9. Axial velocity isocontours.

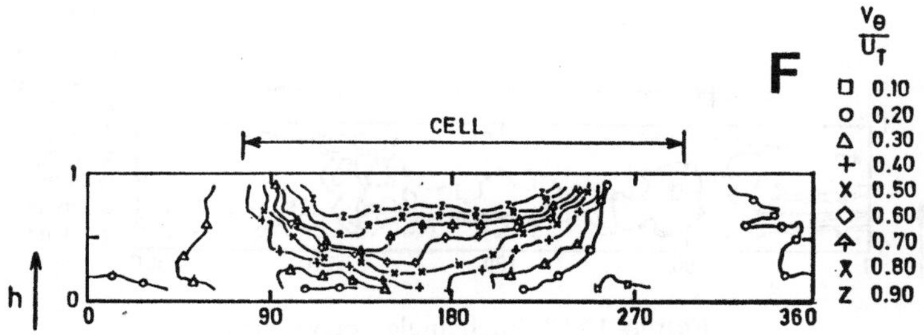

FIGURE 15.10. Tangential velocity isocontours.

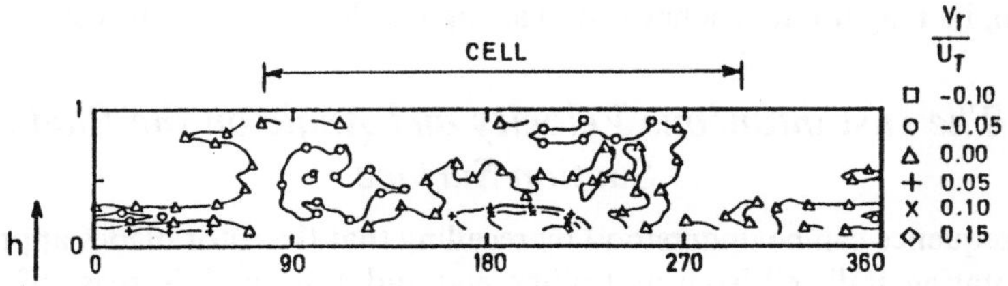

FIGURE 15.11. Radial velocity isocontours.

cell. The largest negative velocities are observed near the tip. Large gradients occur near the edges of the cell while the clean flow is rather uniform.

Tangential Velocity

The isolines of the measured tangential velocity components are shown in Fig. 15.10. Upstream of the rotor the tangential velocity is very high inside the cell, the maximum values correspond to the maximum reversed flow of Fig. 15.9. They decrease toward the edges of the cell to much lower values inside the clean flow region. The opposite trend occurs between the rotor and the stator, where the lowest tangential velocities happen in the reversed flow region of the cell.

Radial Velocity

The isolines of the radial velocity components are shown in Fig. 15.11. At the upstream station, the radial velocity is negative in the upper and front part of the cell, while an area of positive radial velocity is indicated at the lower half of the cell.

Reference to the radial velocity alone does not give a picture of the direction of the flow because the magnitude of the pitch angle, formed by the velocity vector and the $\theta - x$ plane, depends on the axial and circumferential components also. The isocontours of pitch angle have therefore been calculated and

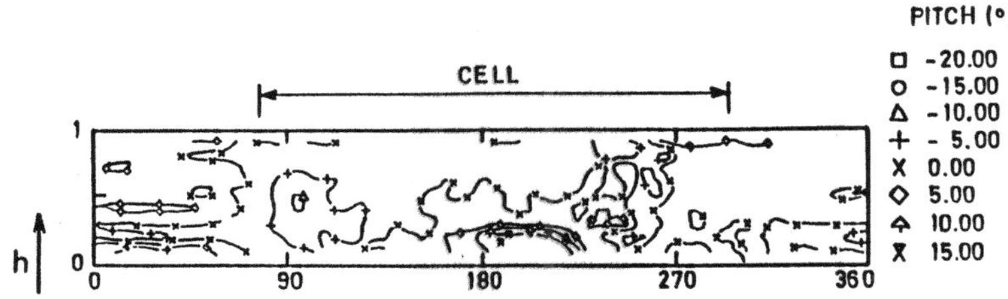

FIGURE 15.12. Pitch angle isocontours.

are presented in Fig. 15.12. While the pitch angle is small in the clean flow region, its magnitude reaches values as high as 20 degrees in the cell.

The Instantaneous Velocity and Angle on the Blade Suction Surface

This sequence can be understood by recalling that the rotor blades approach the rotating stall cell from its trailing end and a gradual decrease of blade velocity and flow separation occur. The large radial drift and the return flow in the tip region create a recirculation in the rotor blades, forcing the entering flow to have a negative meridional flow angle at the midblade height in the rotor blade leading edge plane. This negative angle is detected by sensor $S-1$ and grows to -30 degrees during the flow reversal. The sensor $S-2$ is already influenced by the boundary layer drift and instantaneous angles of $+60$ degrees are seen. The third sensor $S-3$ indicates a permanent stall or separation towards the blade trailing edge. The sensor $S-1$, at 25 percent chord, is selected as an example for the averaged results during the single-cell pattern.

The Phase-Averaged Results

In the unstalled flow, (compressor operating points A to D) a constant flow angle of -7 degrees from horizontal is observed and the local velocity decreases gradually from 27 m/s to 12 m/s, caused by the steeper velocity gradient on the suction surface. An increased unsteadiness is measured at this forward location when the maximum efficiency (point C) and maximum loading (point D) are reached.

The one-cell (F) pattern is recognizable in Fig. 15.13 and causes the local flow angle to deviate to a minimum value of -30 degrees, when at the same time a maximum velocity is indicated. Similar phase-averaged results are obtained for sensor $S-2$, while the $S-3$ sensor indicated serious flow separation and phase-locked averaging cannot be performed.

The flow angle increases between sensors $S-1$ and $S-2$, indicating the boundary layer shift along the suction surface. The sensor location is much deeper in the boundary layer at $S-2$ and therefore detects the centrifugation

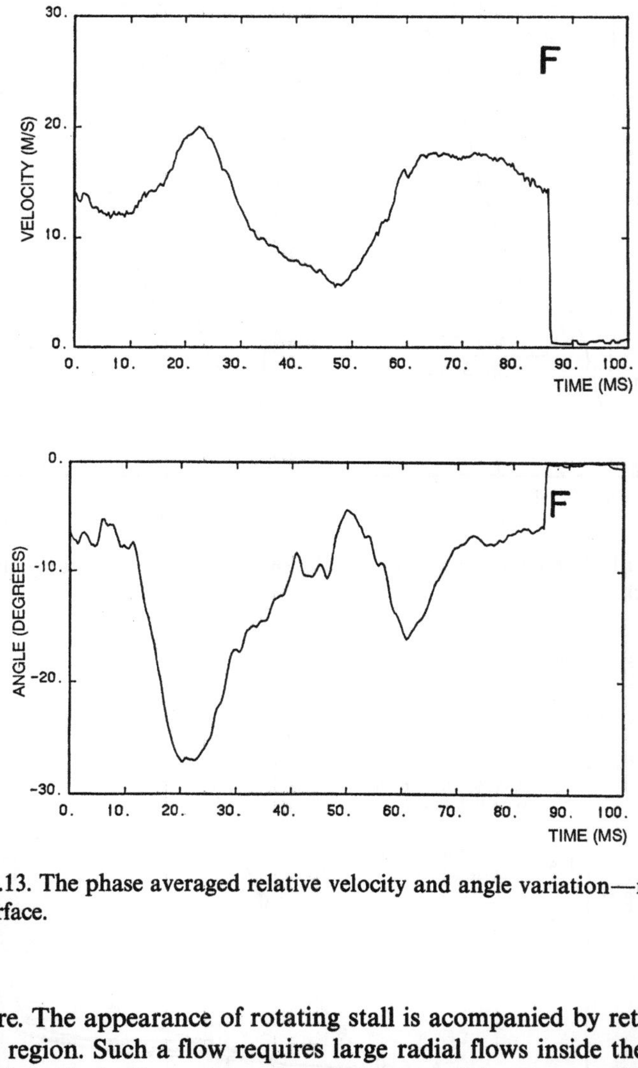

FIGURE 15.13. The phase averaged relative velocity and angle variation—rotorblade suction surface.

effect more. The appearance of rotating stall is acompanied by return flows in the tip region. Such a flow requires large radial flows inside the blading, downwards in the leading edge region and outwards in the second half of the passage (Breugelmans et al. 1983).

Phase-Averaged Total Pressure

The total pressure in the relative frame of reference is measured by a fast-response transducer, which can be positioned at different locations in the blade-to-blade plane. A full pitch is covered in the trailing edge plane and the equivalent of a cascade test can be performed for what concerns the total pressure loss coefficient. The unstalled and the rotating stall cell points can be analyzed and the loss evaluation determined (Breugelmans et al. 1989).

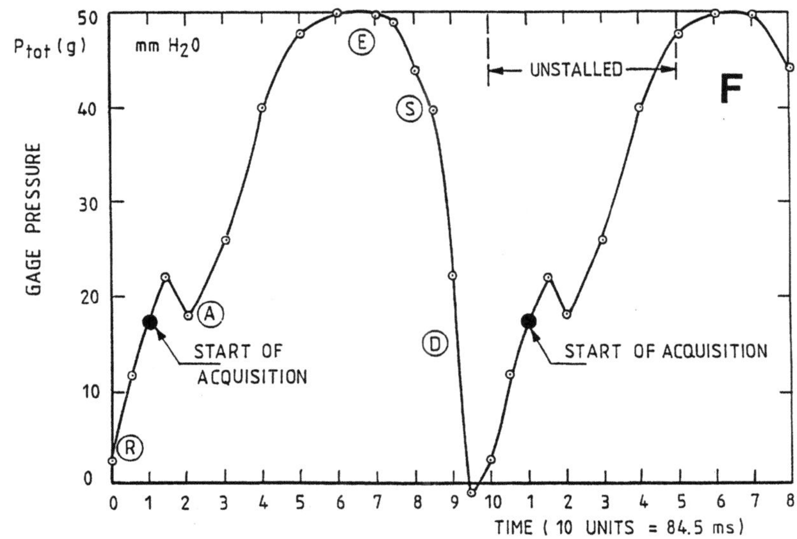

FIGURE 15.14. Corrected total pressure evolution—relative motion, tangential position at mid pitch.

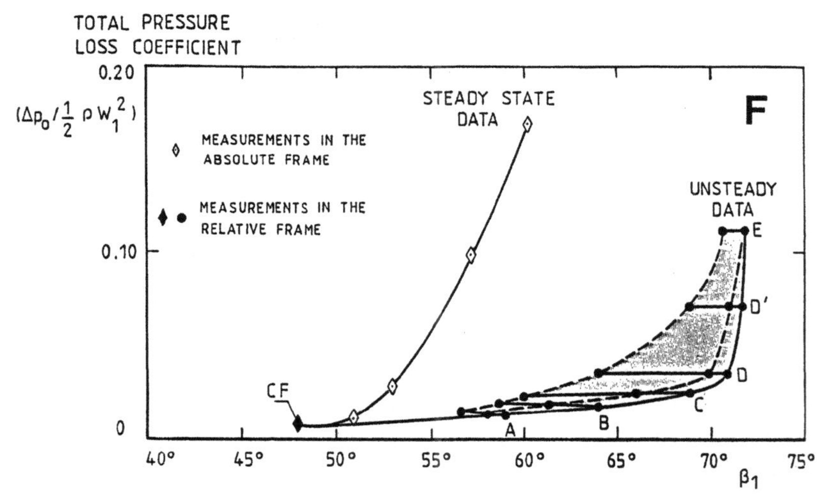

FIGURE 15.15. Mid-height section characteristic.

An example of the relative total pressure evolution during a single stall cell is shown in Fig. 15.14 as measured at the midpitch position. The incidence angle increases from $A \rightarrow E$, stall occurs in point S and reverse flow in D. Fifteen tangential positions (every degree) are being investigated. The phase-averaged results are recombined at selected instants during the inception of stall for the derivation of the loss coefficient at the different incidence angles. The unstalled operation is also measured with the rotating probe and serves as a checking point.

The results are shown in Fig. 15.15 with a comparison between the steady-state loss characteristic and the unsteady results. Much higher incidence angles are accepted by the rotor blade midsection before a sharp rise of the losses occurs. This illustrates the unsteady response of the rotor blade to the unsteady motion generated by the rotating stall cell.

Conclusions

A single rotating stall cell pattern has been used to illustrate the investigations performed in the absolute and relative frame of reference.

The stall cell has a strong harmonic content and autocorrelates well.

Special techniques for triple hot wire measurements and on-blade surface investigation have been developed.

The complicated three-dimensional flow field radial drift of the flow along the rotating blades and the unsteady response are demonstrated.

Nomenclature

$A \rightarrow H$ = operating points of compressor (Fig. 15.4).
$A \rightarrow F$ = phases during stall velocity trace (Fig. 15.7).
$A \rightarrow F$ = phases during stall pressure trace (Fig. 15.14).
B = Greitzer's parameter.
C = chord.
C_{l0} = isolated airfoil lift coefficient.
h = percentage height of channel.
IGV = inlet guide vanes.
t = thickness.
V = velocity.
U = peripheral velocity.
W = relative velocity.

Greek Symbols

β = relative flow angle.
γ = blade stagger angle.
λ = cell blockage factor.
ΔP_0 = pressure drop.
ρ = density.
ψ = compressor load coefficient.
φ = compressor flow coefficient.

Subscripts

A = axial component.
C = stall cell related.
e_1, e_2 = effective cooling velocity wire 1, 2.
T = tip.
TS = total static.
R = radial component.
θ = peripheral component.
o = stagnation condition.

References

Anand, A.K., and Lakshminarayana, B., 1978, "An Experimental Study of Three Dimensional Turbulent Boundary Layer and Turbulence Characteristics Inside a Turbomachinery Rotor Passage." *ASME Paper 78 GT 114.*

Breugelmans, F.A.E., Huang, L., Larosiliere, L., and Andrew, P., 1989, "Unsteady Loss in a Low Speed Axial Flow Compressor During Rotating Stall," *9th ISABE Symp.*, Athens, Sept., Paper 4.7.

Breugelmans, F.A.E., Lambropoulos, L., and Mathioudakis, K., 1983, "Measurement of the Radial Flow Along a Low Speed Compressor Blading During Unstalled and Stalled Operation," *Int. Gas Turbine Congress 83-Tokyo-IGTC-76*, Tokyo.

Breugelmans, F.A.E., Mathioudakis, K., and Casalini, F., 1983, "Flow in Rotating Stall Cells of a Low Speed Axial Flow Compressor," *6th ISABE*, Paris, June, Paper 83–7073; also VKI Preprint 1982–27.

Das, D.K., and Jiang, H.K., 1983, "Flow Measurements Within Rotating Stall Cells in Single and Multistage Axial Flow Compressors," *6th ISABE*, Paris, June, Paper 83–7072.

Das, D.K., and Jiang, H.K., 1984, "An Experimental Study of Rotating Stall in a Multistage Axial Flow Compressor," *J. Engrg for Gas Turbines & Power*, *106*, 3, July, 542–551.

Day, I.J., and Cumpsty, N.A., 1978, "The Measurements and Interpretation of Flow Within Rotating Stall Cells in Axial Compressors," *J. Mech. Engrg. Sciences*, *20*, 2, 101–114.

Day, I.J., Greitzer, E., and Cumpsty, N.A., 1978, "Prediction of Compressor Performance in Rotating Stall. *J. Engrg. for Power*, *100*, 1, Jan. 1–14.

Dixon, S.L., (1961), Some Three Dimensional Effects of Rotating Stall," *ARC CP 609*, May.

Dunham, J., 1965, "Non-Axisymmetric Flows in Axial Compressors," *Mech. Engrg. Science, Monograph No. 3*, Oct.

Erwin, J.R., 1964, *Experimental Techniques. Section D of Aerodynamics of Turbines and Compressors*, Princeton University Press.

Evans, R.L., 1977, "Boundary Layer Development on an Axial Flow Compressor Stator Blade," *ASME Paper 77 GT 11.*

Ferrand, P., and Chauvin, J., 1982, "Theoretical Study of Flow Instabilities and Distortions in Axial Compressors," *J. Engrg for Power*, *104*, 3, July, 715–721.

Greitzer, E.M., 1980, "Review—Axial Compressor Stall Phenomena," *J. Fluids Engrg.*, *102*, 2, June, 134–151.

Gyles, B., Ligrani, P., and Breugelmans, F.A.E., 1982, "Rotating Stall in an Axial Flow Single Stage Compressor," *On-Blade Velocity Measurements.* AFOSR-80-01198, April.

Kriebel, A.R., Seidel, B.S., and Schwind, R.G., 1960, "Stall Propagation in a Cascade of Airfoils," *NASA TR R 61.*

Lakshminarayana, B., and Govindan, T.R., 1981, "Analysis of Turbulent Boundary Layer on Cascade and Rotor Blades of Turbomachinery," *AIAA J.*, *19*, 10, Oct., 1333–1341.

Lakshminarayana, B., Hah, C., and Govindan, T., 1982, "Three Dimensional Turbulent Boundary Layer Development on a Fan Rotor Blade," *AIAA Paper 82-1007.*

Ligrani, P.M., Gyles, B.R., Mathioudakis, K., and Breugelmans, F.A.E., 1983, "Sensor for Flow Measurements Near the Surface of a Compressor Blade," *J. Scientific Instruments, 16*, 5, May.

Mathioudakis, K., and Breugelmans, F.A.E., 1985, "Use of Triple Hot Wires to Measure Unsteady Flows with Large Direction Changes," *J. Scientific Instruments, 18*, 5, 414–419.

Mathioudakis, K. and Breugelmans, F.A.E., 1985, "Development of Small Rotating Stall in Axial Compressors," *ASME Paper 85 GT 227*; also VKI Preprint 1984–25.

McCroskey, W.J., Carr, L.W., and McAlister, K.W., 1976, "Dynamic Stall Experiments on Oscillating Airfoils," *AIAA J.*, *14*, 1, 57–63.

McCroskey, W.J., and Fisher, R.K., 1972, "Detailed Aerodynamic Measurements on a Model Rotor in the Blade Stall Regime," *J. American Helicopter Society*, *17*, 1, 20–30.

Montgomery, S.R., and Braun, I.J., 1957, "Investigation of Rotating Stall in a Single Stage Axial Compressor," *NACA TN 3823*, Jan.

Rannie, W.D., and Marble, F.E., 1957, "Unsteady Flows in Axial Turbomachines," *ONERA Comptes Rendus des Journées Internationales des Sciences Aéronautiques*, Paris.

Sasaki, I., and Takata, H., 1984, "Rotating Stall in Blade Rows Operating in Shear Flow (2d Report)," *Bull. JSME*, *27*, 225, March, 411–418.

Sekido, T., Sasaki, I., and Takata, H., 1981, "Rotating Stall in Blade Rows Operating in Shear Flow (1st Report)," *Bull. JSME*, *24*, 198, Dec., 2074–2081.

Stenning, A.H., Seidel, B.S., and Senoo, Y., 1959, "Effect of Cascade Parameters on Rotating Stall," *NASA Memo 3-16-59W*, April.

Thompkins, W.T., and Usab, W.J., 1982, "A Quasi–Three Dimensional Blade Surface Boundary Layer Analysis for Rotating Blade Rows," *J. Engrg. for Power*, *104*, 2, April, 439–449.

Walker, G.J., 1974, "The Unsteady Nature of Boundary Layer Transition on an Axial Flow Compressor Blade." *ASME Paper 74 GT 135.*

Yeh, H., 1959, "An Actuator Disk Analysis of Inlet Distortion and Rotating Stall in Axial Flow Turbomachines," *J. of Aerosp. Sciences*, Nov., 739–753.

16

Organic Working Fluid Optimization for Space Power Cycles

G. Angelino, C. Invernizzi and E. Macchi

Abstract: The merits of organic fluid space power cycles are surveyed and compared with those of alternate options. Selection of an optimum working fluid is recognized as an important tool to improve system performance. The main characteristics of organic power cycles are shown to be predictable with a good level of accuracy through a general method, which requests the knowledge of a limited information about the fluid properties: specific heat in the ideal gas state, a portion of the saturation curve, and the critical parameters. On the ground of such a theory the adoption of fluids with a relatively complex molecular structure and condensation at the lowest practically admissible reduced temperature allow a better efficiency than achievable with the use of toluene, which is taken as a reference fluid. The influence of turbine efficiency actually achievable in real machines on cycle performance is then addressed; performance diagrams of optimized turbines in the power range of interest for space cycles are calculated and presented. It is shown that only the combined optimization of thermal and fluid dynamic variables leads to the definition of an optimum working fluid and power cycle. A class of fluids is examined, that of the methyl-substituted benzenes, offering a wide variation of thermal properties. A thorough optimization that considers a wide range of power outputs, one- and two-stage turbines, saturated and superheated cycles is performed. For a power output of about 30 kW trimethylbenzene is found to offer the best overall efficiency, a moderate maximum pressure, reasonable turbine dimensions, and rotating speed. A thermodynamic conversion efficiency in excess of 30 percent seems achievable at a maximum temperature of 360°C for a condensation temperature of 60°C. Such energy performance suggests that ORC systems could represent a viable multifuel prime mover option also for terrestrial power generation. Thermal stability of the proposed fluid is experimentally investigated and found to be similar to that of toluene, but its definite evaluation is shown to require further testing.

Introduction

Electrical power needed in space missions is steadily increasing and will reach an unprecedented value of about 100 kW in connection with the construction of the first orbiting manned space station (van Landingham 1988). The only proven general-purpose power technology presently available for these applications is the solar photovoltaic, which will supply power for the station buildup and contingency power in case the pointing capability of future solar dynamic generators is lost. Hybrid systems, relying on both photovoltaic (PV) and solar dynamic (SD) modules, seem to offer the best technical and economical solution, in the 20- to 40-kW power range, at least in the near future (Teren 1987). The main merits of SD in comparison to PV systems are: (1) a higher overall conversion efficiency (20 to 30 percent against about 14 percent, according to Nored and Bernatowicz (1986), which implies a smaller collecting surface, a reduced drag in low orbit flight, and hence, a reduced fuel consumption for maintaining the orbit original parameters; (2) a slower performance degradation due to aging; and (3) a potentially lighter and higher-efficiency (90 percent against 70 to 80 percent) energy-storage system relying on thermal rather than chemical energy, provided the newly developed equipment succeeds in achieving an adequate reliability and operating life.

Several thermodynamic cycles, powered by focused solar radiation, each performed by means of a specific thermal engine, have been considered for development: the closed Brayton cycle (CBC), the organic Rankine cycle (ORC), the Stirling cycle, the metal vapor, and binary Rankine cycle. CBC and ORC have long been intended as alternative options for the first flight application, the other systems represent promising solutions, presently in a lower development stage. In CBC peak temperatures of 750° to 800°C allow a thermal-to-mechanical conversion efficiency[1] of about 35 percent, whereas in ORC working fluid degradation problems limit the top temperature to about 400°C, and conversion efficiency to about 28 percent for an effective radiator temperature similar to that of CBC (Chandoir et al. 1985; Pietsch and Trimble 1985).

Besides a better potential efficiency, CBC is not negatively influenced by a micrograyitational environment, owing to its single-phase gaseous working fluid. Its high-temperature heat input, on the other hand, poses severe problems in concentrator and receiver design and fabrication (Trudell et al. 1988; Valade 1988), in structural material qualification, and in heat storage material selection. ORC systems, on the contrary, potentially allow a lightweight and conservative design for the concentrator, which can achieve a high energy performance at a concentration ratio as low as 500 (Heidenreich et al. 1985), with realistic allowance for mirror quality and pointing errors and a wider selection of heat-storage materials with stronger driving forces in the charging/

[1] Here and in the following the quoted efficiencies do not take into account mechanical and electrical losses.

discharging cycle (Downing and Parekh 1985; Faget et al. 1985; Phillips and Stearns 1985). In case energy is stored as sensible rather than latent heat, using, for example, liquid lithium (Jin Song and Louis 1988), which could solve many containment and material problems, low-temperature operation directly contributes to achieving a lightweight solution for the heat storage system by making a wider temperature fluctuation feasible. On the debit side of the ORC concept are the problems connected with the two-phase flow handling in a microgravitational environment, which, however, could be solved by introducing proper additional equipment in the plant, such as the rotary fluid management device described in Chandoir et al. (1985).

From a strictly thermodynamic point of view, notwithstanding the exceptionally high projected performance of both Brayton (Boyle et al. 1988) and Stirling (West 1988) engines, organic Rankine systems offer the best potential quality for the conversion cycle; this suggests that ORC engines could be used as high-quality components in the moderate-temperature range, to build complex high-efficiency heat engines capable of managing the whole temperature range, which is technically controllable with existing materials (binary and ternary cycles). Furthermore, the comparatively simple and inexpensive materials required by the ORC moderate temperature represent a guarantee that the transfer to terrestrial applications of advanced and successful space systems will not be barred by an excessively sophisticated and costly material technology, as could be the case for the alternate options.

Preliminary Working Fluid Optimization

Although the destination of an ORC power plant to the space environment predetermines some of its characters, a number of important variables still survive (power level; cycle arrangement; specific technical characteristics of concentrator, storage, and radiator) that request a thermodynamic cycle optimization within a potentially wide bunch of options. The nature of the working fluid represents the main parameter for achieving the best plant performance (Casci and Angelino 1969).

Following a general method, similar to the one reported in Angelino and Invernizzi (1988) for heat pump cycles, it could be shown that the cycle performance is basically determined by a limited number of parameters:

1. An index σ, which accounts for the fluid molecular complexity; it is basically a function of the number and mass of the atoms forming the molecule and controls the shape of the saturation curve in the $T - S$ plane:

$$\sigma = \frac{10^{-3} T_{cr}^2}{R} \left(\frac{\partial S}{\partial T}\right)_{\text{sat}, Tr=0.7} \tag{1}$$

2. The reduced condensation temperature T_{rC}, giving the location of the conversion cycle within the fluid state diagram.

3. The maximum-to-minimum cycle temperature ratio $\tau = T_{max}/T_{min}$ (or the ideal efficiency $\eta_{id} = 1 - \tau^{-1}$), giving the temperature extension of the cycle.

Referring for sake of simplicity to saturated cycles, the basic influence of these parameters can be inferred from Fig. 16.1.

In Fig. 16.1 (a) and (b) cycles having a typical simple ($\sigma = 0$ K) and complex ($\sigma = 15$ K) fluid are compared for $\eta_{id} = 0.2$ and for $T_{rC} = 0.5$ and 0.75, respectively. Assuming a nonregenerative arrangement, the cycle quality worsens at increasing T_{rC} and σ (as visually shown by the increasing departure of the cycle configuration from that of the Carnot cycle). In the regenerative arrangement (as for cycle *C* or *D* of Fig. 16.1 (b), in which most of the heat of superheating at the turbine exhaust is transferred to the compressed liquid) an acceptable level of quality is regained, since regeneration removes either the highest temperature fraction of the rejected waste heat and the lowest temperature fraction of the primary heat. Similarly, Fig. 16.1 (c) and (d) show a loss of quality caused, in the nonregenerative arrangement, by the increased temperature extension of the cycle (from $\eta_{id} = 0.20$ to $\eta_{id} = 0.40$), both for simple-molecule fluids (owing to the increase of the fraction of the primary heat, which is introduced at lower than top temperature) and for complex-molecule fluids (owing also to the increased fraction of the waste heat to be rejected at higher than minimum temperatures).

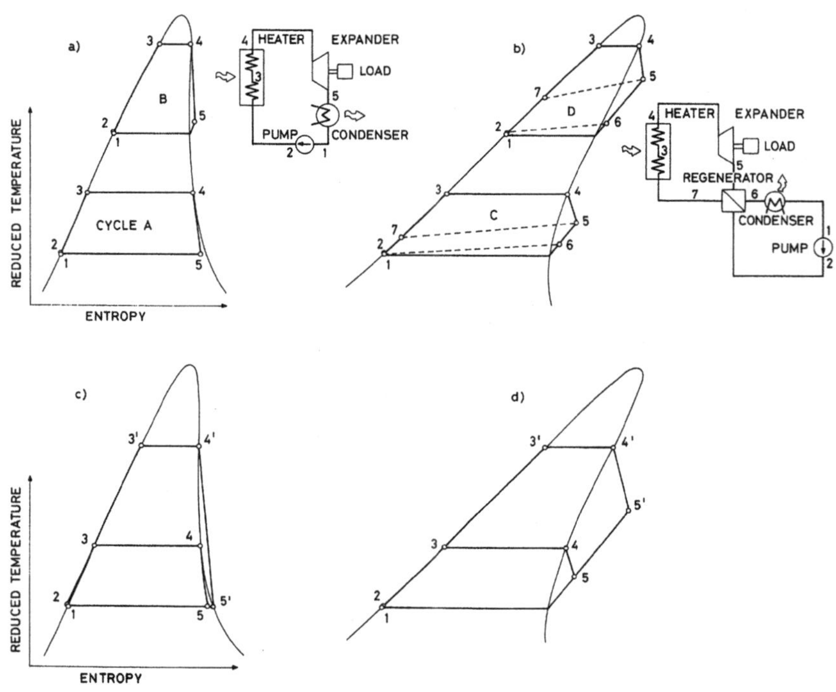

FIGURE 16.1. Influence of the main thermodynamic variables on saturated cycles configuration.

In order to obtain quantitative information on the performance trend just outlined, a systematic analysis of organic fluid thermodynamic cycles was performed. A computer program was developed (Angelino and Invernizzi 1988; Portinari 1988) capable of evaluating the thermodynamic properties of working fluids from a small number of data (molecular structure, critical pressure and critical temperature, acentric factor ω). The program was validated with reference to a number of fluids (water, carbon dioxide, various hydrocarbons, and refrigerants) well known from the literature and found to be accurate at least for evidencing general trends (departures in cycle efficiency of less than 2 to 5 percent were found in the subcritical region). The following classes of fluids were considered: linear, cyclic, and aromatic hydrocarbons; linear perfluorocarbons; linear and cyclic siloxanes.

The results of calculation are summarized in Fig. 16.2 in terms of cycle quality factor QF:

$$QF = \frac{\eta}{\eta_{id}} = \frac{\eta}{(1 - \tau^{-1})} \tag{2}$$

For graphical clearness, nonregenerative cycles, which are comparatively unimportant, are illustrated only through averaged curves.

A turbine isentropic efficiency of 0.75, a pump efficiency of 0.5, and a minimum temperature difference in the regenerator (whenever employed) of 15°C were assumed in the calculation.

Two values of η_{id} of particular interest for space power systems (0.3 and 0.4) and two T_{rC} were examined, the lowest one producing a definitely subcritical evaporative heat input, the highest implying a just critical top temperature. Inspection of Fig. 16.2 suggests the following observations:

1. Molecular complexity has a negative effect on nonregenerative cycles, markedly for condensation at high T_r (for $\eta_{id} = 0.3$, $T_{rC} = 0.7$, QF drops from 0.43 to 0.35 by changing a from 0 to 15 K); on the other hand, the efficiency of regenerative cycles is positively influenced by σ, either slightly (for low T_{rC}) or strongly (for high T_{rC}).
2. Condensation at high reduced temperatures has a negative effect on efficiency, mainly when this implies evaporation in the vicinity of the critical point (for $\sigma = 15$ K, $\eta_{id} = 0.40$, QF drops from 0.67 down to 0.58 by changing T_{rC} from 0.5 to 0.6). The main obstacle to operation at low T_{rC} is represented by the minimum cycle pressure, which becomes unpractically low at T_r below 0.4 to 0.5. As shown in Fig. 16.3, for a given condensation pressure, simple-molecule fluids allow condensation at the lowest T_r, cyclic and aromatic hydrocarbons being favored among the classes of fluids investigated (for example, for a σ of about 15 K and for $p_C = 1$ kPa, methyl-benzenes condense at $T_{rC} = 0.47$, linear perfluorocarbons at $T_{rC} = 0.52$).
3. For a given T_{rC}, a wide-cycle temperature extension $\Delta T/T_{max}$ has a negative effect on QF in the nonregenerative configuration, while it affects very slightly the efficiency of regenerative cycles. For a given T_{rE}, on the contrary

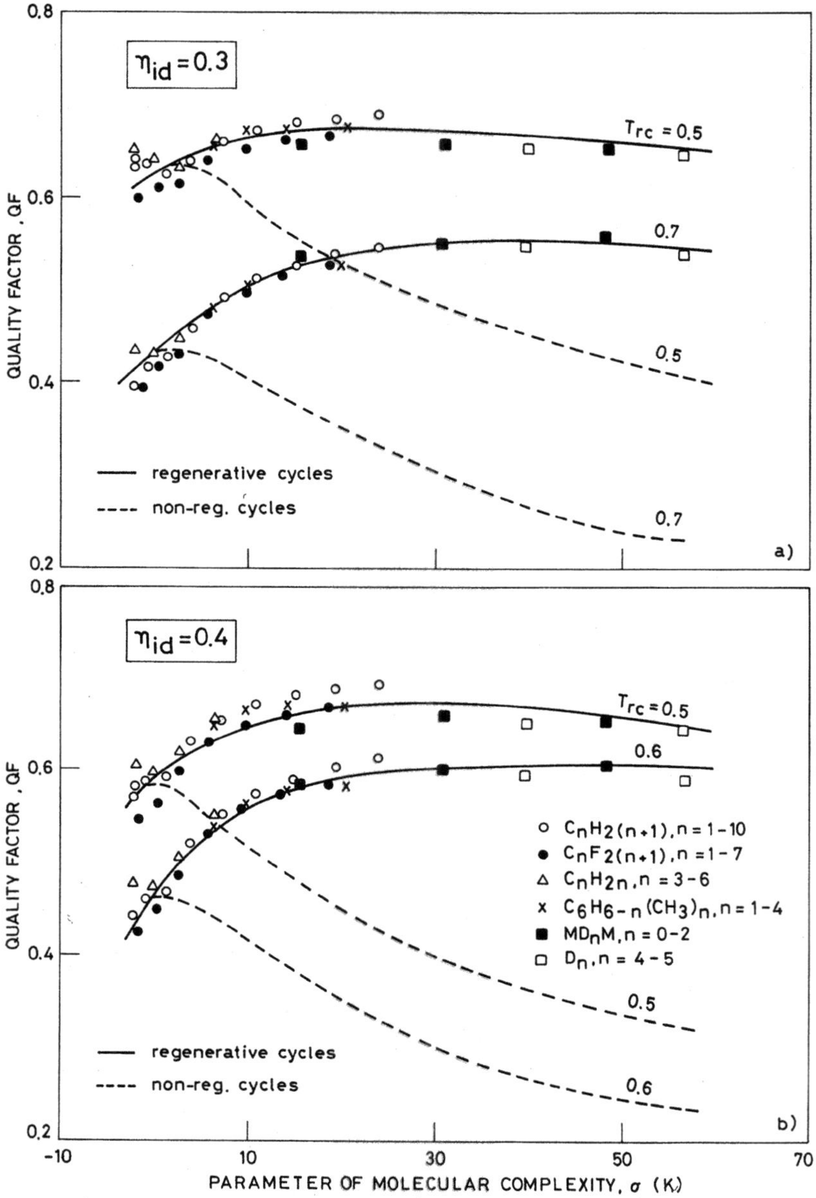

FIGURE 16.2. Quality factor QF for saturated cycles as a function of the parameter of molecular complexity.

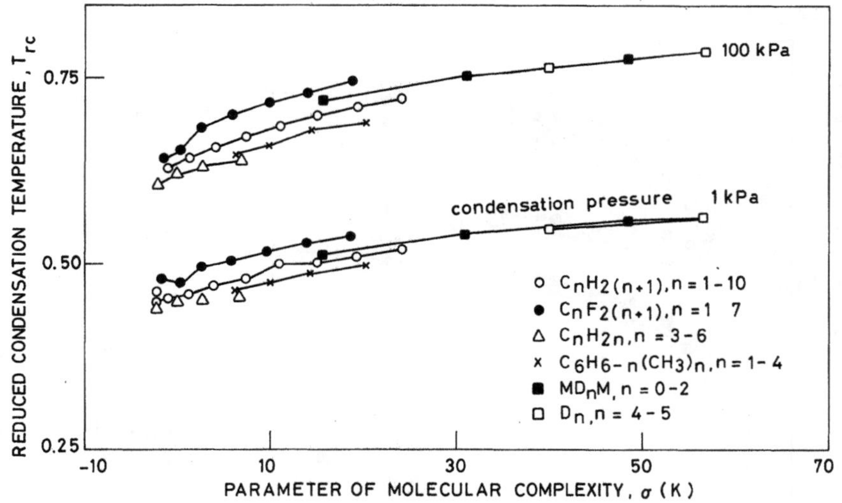

FIGURE 16.3. Admissible reduced condensation temperature for various classes of fluids.

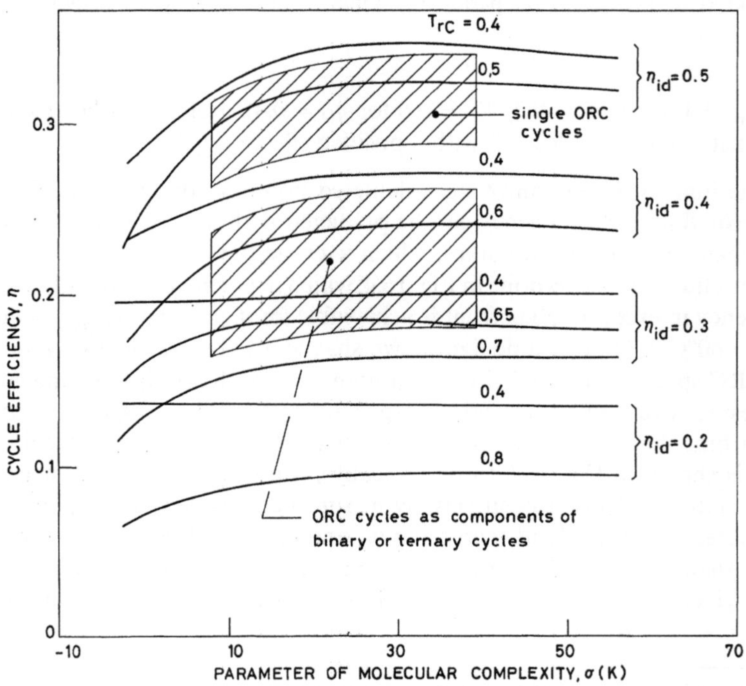

FIGURE 16.4. Achievable cycle efficiency as a function of the parameter of molecular complexity.

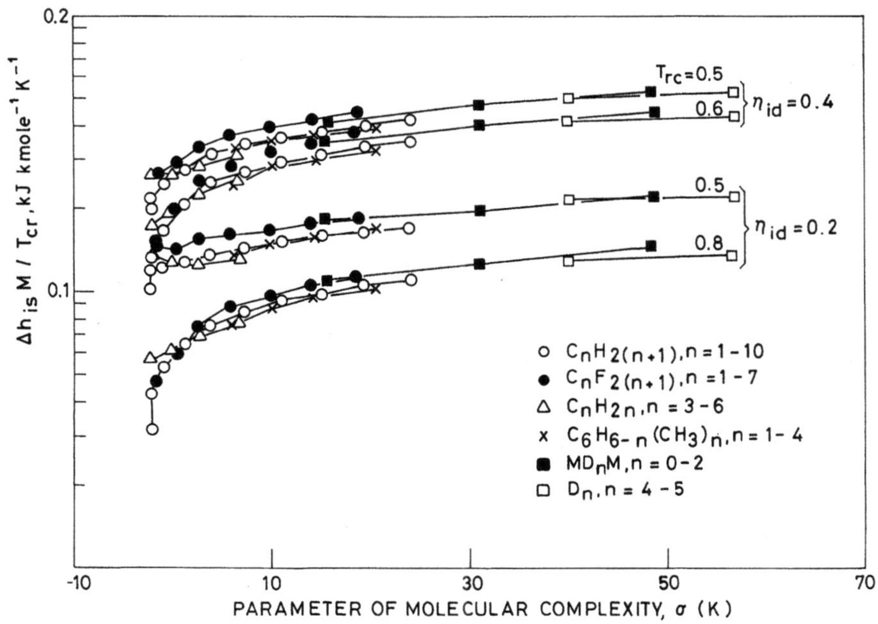

FIGURE 16.5. Influence of the parameter of molecular complexity on turbine enthalpy drop.

($T_{rC} = 0.7$ for $\eta_{id} = 0.3$; $T_{rC} = 0.6$ for $\eta_{id} = 0.4$), cycles with large $\Delta T/T_{max}$ must condense at low T_r, with a positive effect on QF.

The limited dispersion of the calculated points in the graph of Fig. 16.2 suggests that cycle efficiency can be predicted with a good level of accuracy by means of generalized curves, which average the results obtained for the various fluid classes, as done in Fig. 16.4, which in particular shows that a cycle efficiency in excess of 30 percent is achievable for $\eta_{id} = 0.5$, (say $t_{max} = 400°C$, $t_{min} = 60°C$[2]). On the same figure, two shaded areas show the likely location of ORC space cycles used either autonomously or in multiple arrangement.

Besides cycle efficiency, other important parameters of special relevance to turbine design lend themselves to a generalized evaluation: specific work Δh_{is} is shown in Fig. 16.5 to be basically proportional to T_{cr}/M, negatively influenced by T_{rC} and positively by σ; turbine expansion ratio $\beta_t = p_{in}/p_{out}$ increases dramatically at increasing η_{id} and σ, while it decreases as T_{rC} rises (Fig. 16.6); specific heat transfer surface requirement for regeneration, which, for a given heat transfer coefficient can be shown to be proportional to

[2] This condensation temperature is extensively stipulated in the following. No attempt was made to optimize radiator temperature, which, depending on the mission characteristics, could be somewhat different from the assumed value.

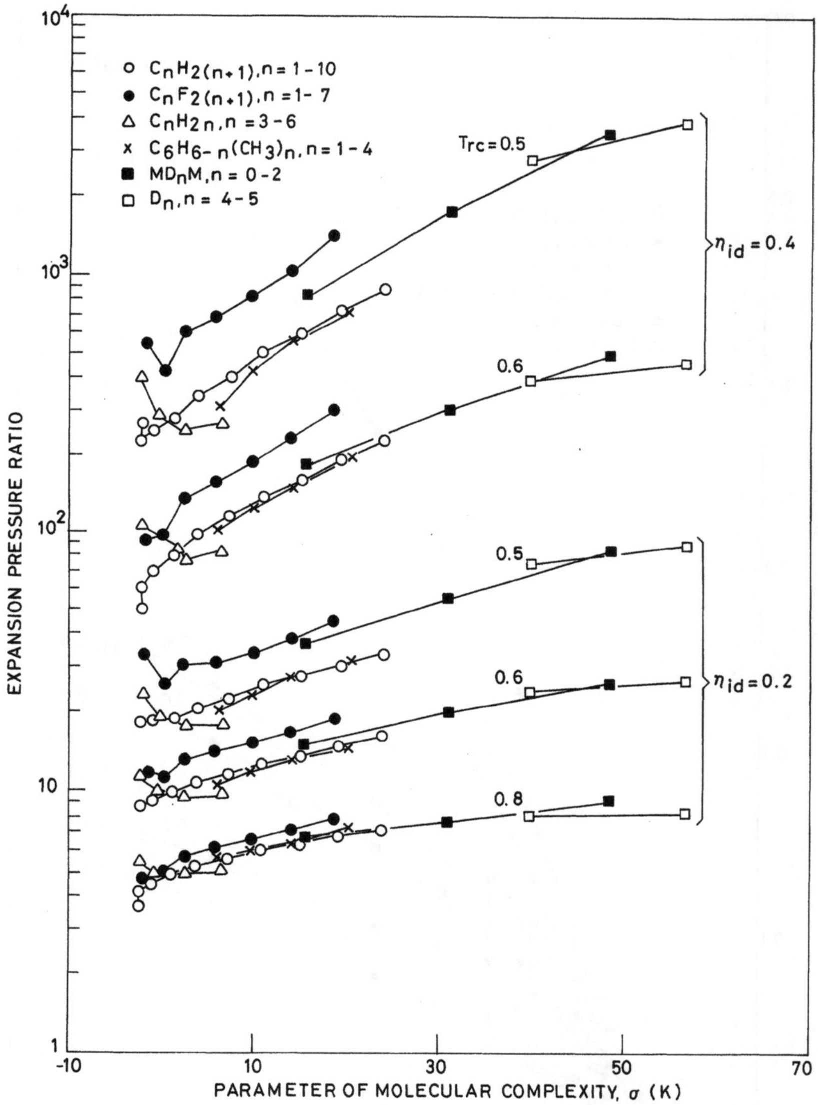

FIGURE 16.6. Turbine expansion ratio as a function of the parameter of molecular complexity.

$$S^* = \frac{\Delta h_r}{W \wedge} \quad (3)$$

(Δh_r is the regenerated heat, W the net work, and $\wedge$ the regenerator log-mean temperature difference), increases severely at increasing σ and T_{rC} but is only marginally influenced by η_{id} (Fig. 16.7).

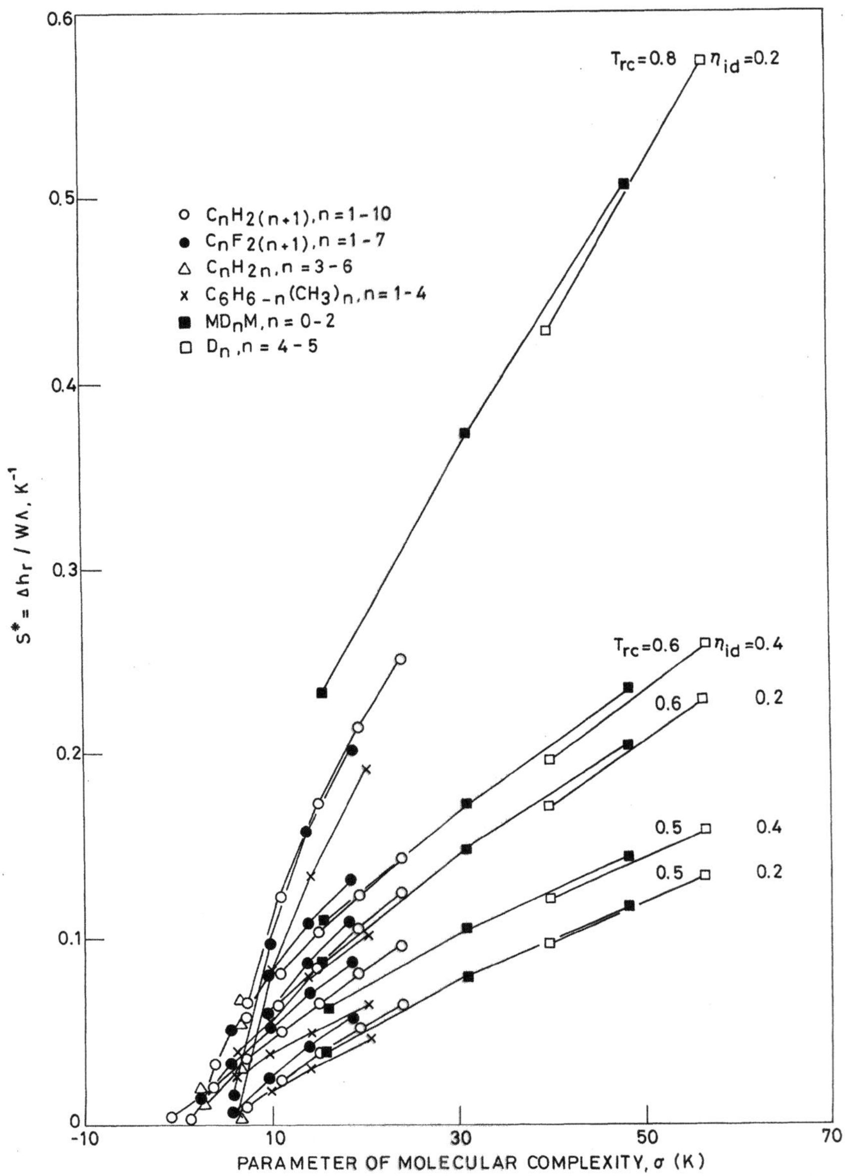

FIGURE 16.7. Heat transfer surface requirement for regeneration, in terms of parameters S^*, as a function of the fluid molecular complexity.

Novel Working Fluids

In agreement with the results of the previous section, organic Rankine cycles should operate at the minimum admissible reduced condensation temperature, i.e., use a fluid with a very high critical temperature, provided the corresponding pressure is not too low, and have a moderately complex molecule; very complicated molecular structures would achieve an acceptable efficiency level only by making use of a very large regenerator.

With reference to the question of how to establish the minimum tolerable pressure, a definitive answer can only come from the detailed analysis of volume flows and other relevant parameters in the low-pressure section of the plant. However, the fluid dynamics and mechanics of small turbines are favorably affected by increasing volume flows. Furthermore, by designing the regenerator close to or within the laminar range, heat transfer coefficients can be kept sufficiently high even at very low pressures. Remembering that large steam turbines operate at condensing pressures of a few kPa in a nonsealed system, it seems that pressures of 1 to 5 kPa would be acceptable in space cycles, provided this is beneficial to turbine design.

Toluene, which is presently the favorite candidate fluid for the proposed application, does not meet either of the concomitant requirements of high critical temperature and low condensation pressure at optimum condensation temperature. In the search of fluids of similar molecular structure, hopefully having a similar thermal stability, but with higher critical temperatures, the whole class of methyl-substituted benzenes, in which toluene is the simplest constituent, was investigated both theoretically and experimentally.

The chief thermodynamic parameters of the considered fluids are reported in Table 16.1. The progressive substitution of one hydrogen atom with a

TABLE 16.1. Some thermodynamic parameters of the Methyl-substituted benzenes considered in this work.

Type of fluid*	①	②	③	④
Structural formula	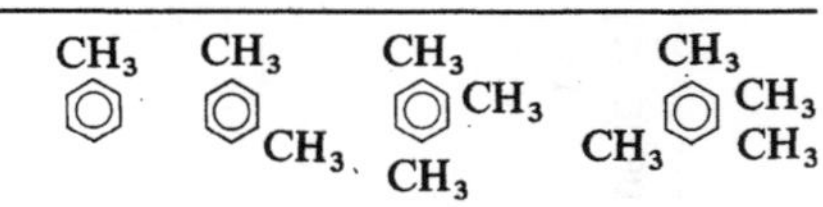			
Critical temperature, t_{cr}, °C	318.65	343.95	376.05	405.85
Critical pressure, p_{cr}, MPa	4.1	3.54	3.23	3.216
Molecular mass, M, kg kmole^{-1}	92.14	106.17	120.19	134.21
Ideal molar specific heat capacity, J mole^{-1} K^{-1}	106.97	129.65	156.23	187.9
Acentric factor, ω	0.263	0.325	0.376	0.4032
Parameter of molecular complexity, σ, K	6.16	9.8	14.15	20.3
Condensation temperature at 1 kPa, °C	1.66	23.02	46.23	68.77
Condensation temperature at 5 kPa, °C	31.2	53.95	79.08	103.28
Vapor pressure at 60°C, kPa	17.92	6.59	2.05	0.623
Heat of vaporization at 60°C, kJ kg^{-1}	392.24	384.3	381.28	383.4

*① methylbenzene (toluene), ② 1,3-dimethylbenzene (m-xylene), ③ 1,2,4-trimethylbenzene, ④ 1,2,3,5-tetramethylbenzene.

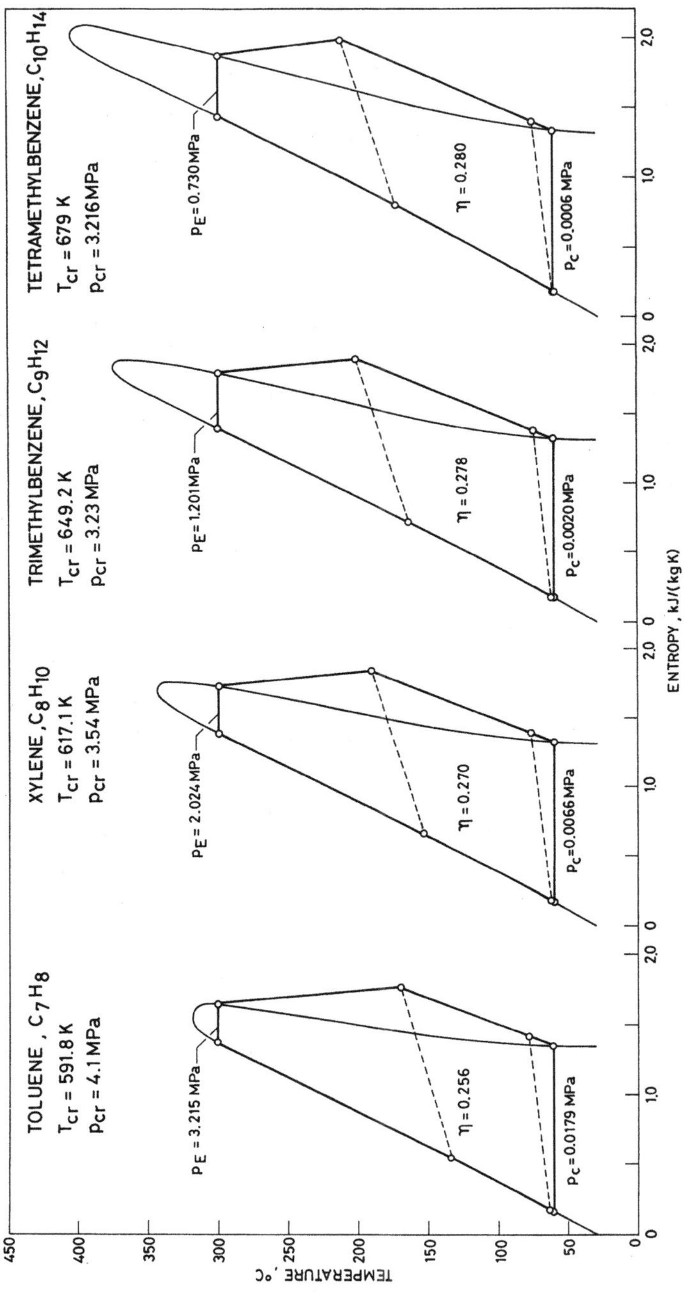

FIGURE 16.8. Temperature-entropy diagrams for methyl-substituted benzenes.

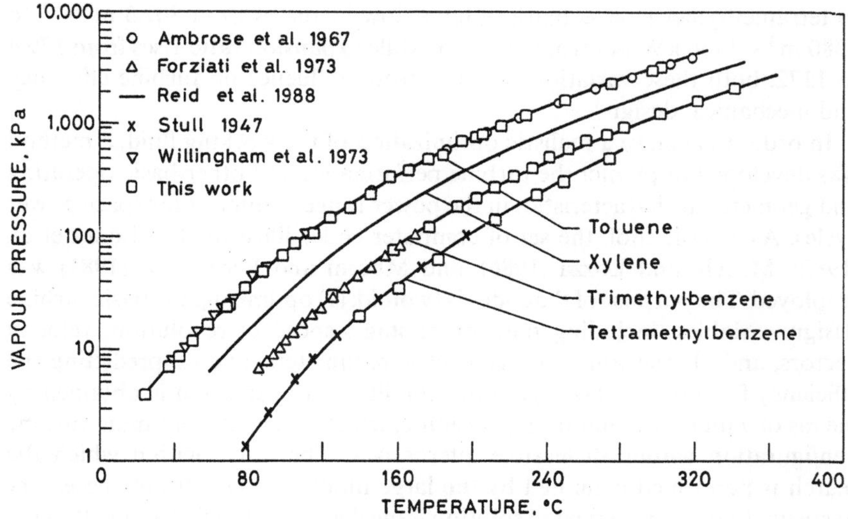

FIGURE 16.9. Experimental vapor pressure curves for candidate working fluids.

methyl group increases the molecular complexity 6 and related properties and raises the critical temperature, thus reducing the saturation pressure for a given condensation temperature. The comparative shape of the saturation curves and the configuration of saturated cycles at $t_C = 60°C$, $t_E = 300°C$ are represented in Fig. 16.8. Operating pressures and conversion efficiencies are also indicated. The best fluid, tetramethyl-benzene, exhibits a 0.283 efficiency at $p_C = 0.623$ kPa, $p_E = 730$ kPa, while the corresponding values for toluene are 0.256, 17.92 kPa, and 321.5 kPa, respectively.

Although the saturation pressure versus temperature curves were available from the literature (Ambrose et al. 1967; Forziati et al. 1973; Reid et al. 1988; Stull 1947; Willingham et al. 1973), some direct measurements were performed by means of an experimental apparatus primarily built for evaluating the fluid's thermal stability (as illustrated later). As shown in Fig. 16.9, measured vapor pressures are in good agreement with literature values. Other thermodynamic data, sufficient to generate complete state diagrams by means of an appropriate computer program (Invernizzi 1984) were excerpted from the literature (Reid et al. 1988). As a result, a thorough and reliable knowledge of the fluid's thermodynamic behavior became available.

Turbine Performance Evaluation

The use of different fluids implies such a wide variation of turbine operating parameters that turbine efficiency is likely to vary strongly in different cases. with reference to the data of Fig. 16.8, for instance, by changing from toluene to tetramethylbenzene, exhaust volume flow increases from 307.5 $m^3\ s^{-1}$ to 6380 $m^3\ s^{-1}$ (30 kW isentropic power), while expansion ratio rises from 179.4 to 1172: both these variations have a strong influence on turbine efficiency and mechanical design.

In order to achieve a realistic optimization of the working fluid, a method was developed to predict the turbine performance and other basic operating and geometrical characteristics in the power range of interest for space power cycles. As a basic tool, the set of computer codes illustrated in Lozza et al. (1982), Macchi and Lozza (1986), and Macchi and Perdichizzi (1981) was employed. The purpose of the codes is twofold: (1) optimizing the basic turbine design variables (including number of stage, speed of revolution, velocity vectors, and all relevant blade geometric parameters) and (2) predicting the efficiency for any specified operating condition. The solution is obtained by means of a numerical iteration procedure, which selects the optimum turbine configuration within all feasible alternatives. The range within which the search is performed is limited by the large number of constraints necessary to ensure that the investigated solution is mechanically feasible and within the applicability range of correlations used for predicting flow losses and deviations occurring in blade passages. Since no limits are specified for the speed of revolution and the investigated fluids have relatively small enthalpy drops (and require therefore peripherical speeds well below mechanical capabilities of turbine structural materials), the selected optimized solutions operate *either* at optimum specific speed N_s or specific diameter D_s. It is believed that the codes, which have been extensively used for preliminary design of several axial flow turbines operating in ORC (Angelino et al. 1984) and are regularly tested against test cases made available in the technical literature, provide a fairly accurate method for predicting the efficiency achievable by a turbine designed according to state-of-the-art rules.

Since it would be obviously impractical to incorporate this rather complex turbine design method in the thermodynamic analysis of power cycles, an attempt was made to correlate the turbine efficiency to a few quantities related to fluid thermodynamic properties and power cycle characteristics. As discussed in Lozza et al. (1982), the two most significant parameters for predicting the efficiency penalties related to flow compressibility and/or small blade dimensions are the volumetric flow ratio (VFR) and the so-called size parameter (SP), defined as follows:

$$\mathrm{VFR} = \frac{V_{\mathrm{out}}}{V_{\mathrm{in}}} \tag{4}$$

$$\mathrm{SP} = \frac{V_{\mathrm{out}}^{0.5}}{\Delta h_{\mathrm{is}}^{0.25}} = \frac{D}{D_s} \tag{5}$$

where V_{out} and V_{in} are the outlet/inlet volume flow rates, evaluated for multistage machines at the exit of the last stage and at the inlet of the first stage, respectively.

Generally speaking, the efficiency of a turbine decreases by increasing VFR because of the losses related to the occurrence of high Mach numbers and, for reaction turbines, to the negative influence of wide flow area variations across the rotor blades. Moreover, for multistage single-shaft turbines, large VFR prevent operating the various stages at optimum specific speed. As shown in Eq. 5, the parameter SP, which is a function solely of the thermodynamic cycle and power output, is proportional, for a given (optimized) value of D_s (Balje 1981) to actual turbine dimensions: low SP values penalize the turbine efficiency because of large losses caused by the increase of relative blade thickness, clearances, roughness, etc. Since SP is defined with reference to outlet volume flow rate, the contemporary occurrence of high VFR and low SP yields very small flow areas in the first portion of the machine (for instance, the quoted 30-kW toluene and tetramethyl-benzene turbines would have a total throat area in the first stator blades as low as 20.3 and 90.1 mm^2, respectively), with further efficiency penalties.

In attempting to derive a practical working instrument, the turbine design procedure was applied to a large number of turbines. A total of 170 one-stage and 90 two-stage turbines, all using methyl-substituted benzenes, were optimized. Turbine configurations featuring more than two stages were judged impractically complex for the modest power level considered, as a rule, in the present study. As expected, all the resulting data on optimum turbine efficiency could be organized as a function of the two parameters VFR and SP with very limited dispersion in the performance diagrams of Fig. 16.10(a) and (b) (Davoli 1988), relating to one- and two-stage turbines, respectively.

Volume flow ratios up to 400 for the single stage and 2000 for the two-stage configuration are embraced by diagrams. Larger expansion ratios, leading to efficiencies below 66 percent were not considered. While the computing method underlying the two diagrams warrants that the indicated performance is achievable through a well-designed expander, it should be pointed out that better solutions could be found in particular cases, mainly by relaxing some of the conservative assumptions that were adopted at high Mach numbers. In particular, recently published data (Kurzrock 1989) seem to indicate that efficiencies above 75 percent can be reached by highly supersonic single-stage impulse turbines at relatively high VFR and low SP.

In the same diagrams, the operating points of a number of saturated cycles using various fluids for two power levels (30 and 200 kW) and a fixed condensation temperature of 60°C are indicated. It can be seen that a maximum vaporization temperature of 250°C could be managed with a one-stage turbine at an adiabatic efficiency higher than 66 percent. On the contrary, two-stage turbines give an acceptable performance also at $t_E = 300°C$ (superheated cycles at optimum turbine inlet pressure, for inlet temperatures up to 400°C give operating points very close to those of saturated cycle at $t_E = 300°C$ and were not reported in the diagrams). The following main trends emerge from the inspection of Fig. 16.10: (1) vaporization temperature is the main parameter affecting turbine performance (as a typical case, the efficiency of a trimethyl-

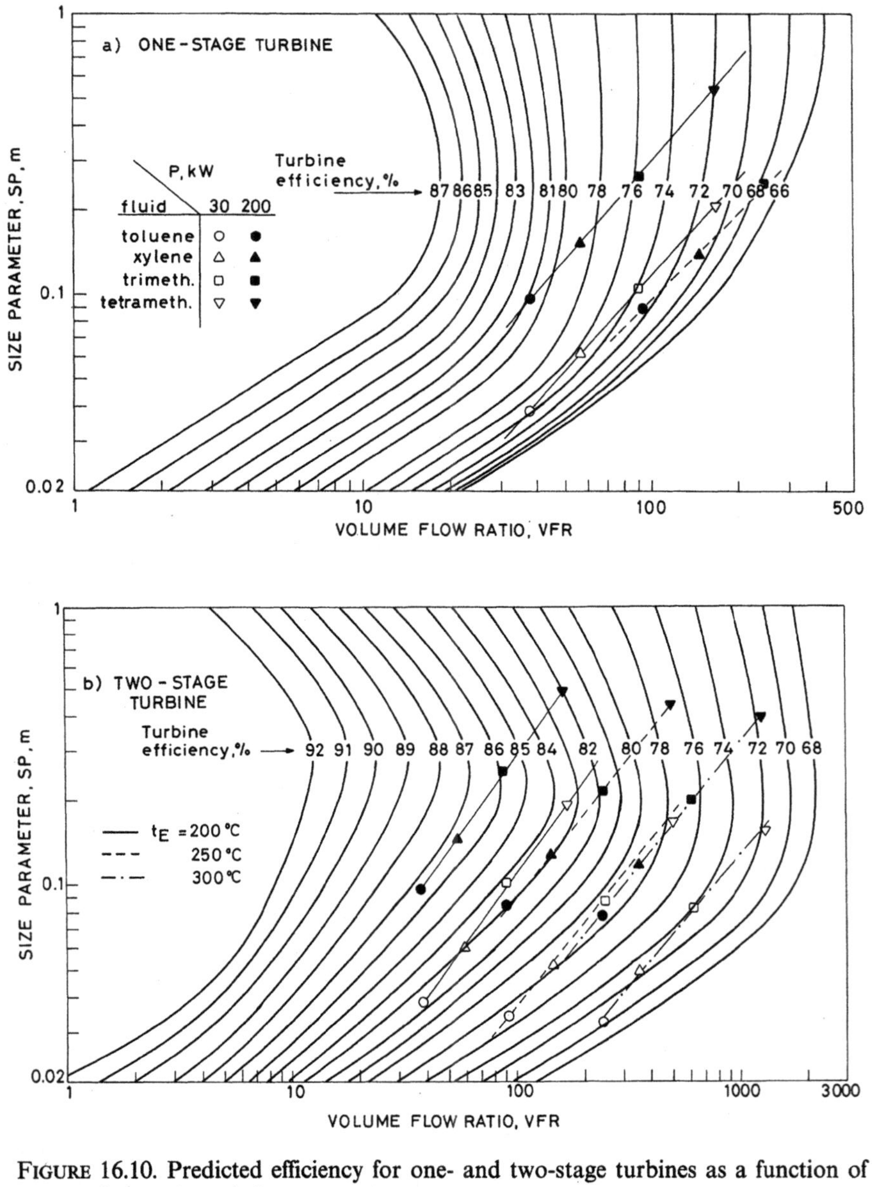

FIGURE 16.10. Predicted efficiency for one- and two-stage turbines as a function of volume flow ratio and size parameter.

benzene, 30-kW, two-stage turbine drops from 84 percent to 74 percent by changing t_E from 200° to 300°C); (2) trimethylbenzene, on the average, offers a good turbine efficiency, at least in the case of a two-stage turbine, notwithstanding the large expansion ratio associated with its use; and (3) the very low values of size parameter SP for toluene at $P = 30$ kW implies miniature, high speed of rotation turbine wheels, while other lower pressure fluids yield more practical dimensions and rotating speeds.

Thermo-Fluid Dynamic Cycle Optimization

The availability of a reliable tool for the prediction of the expander efficiency allows the determination of optimum power cycles based on either thermo- or fluid dynamic criteria. The optimization performed here does not imply the system optimization for a given mission, which represents a further and by far more ambitious goal. Two basic configurations will be considered: saturated and superheated cycles.

Saturated Cycles

As a general rule, owing to the high molecular complexity of the organic fluids considered, saturated cycles are preferable; superheating should not be recommended since, while it is not requested as a means of preventing moisture formation, it overloads the regenerator and, for a given top temperature, reduces the thermodynamic quality of the cycle.

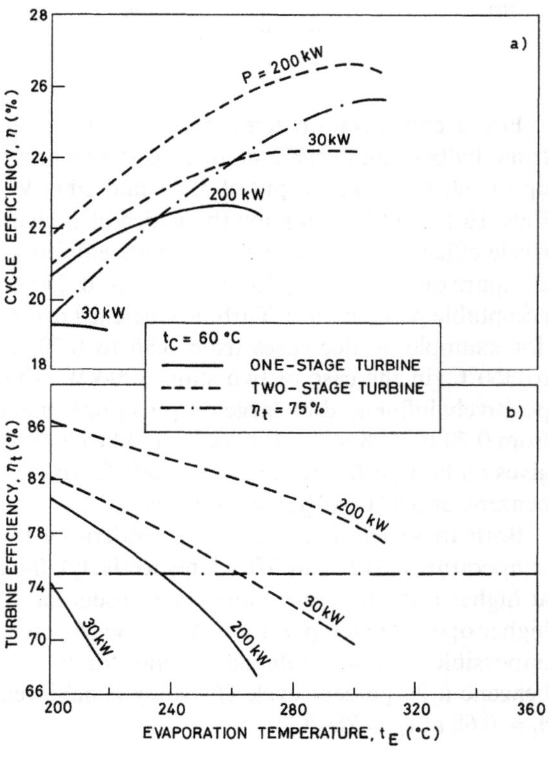

FIGURE 16.11. Saturated cycle and turbine efficiency for toluene working fluid.

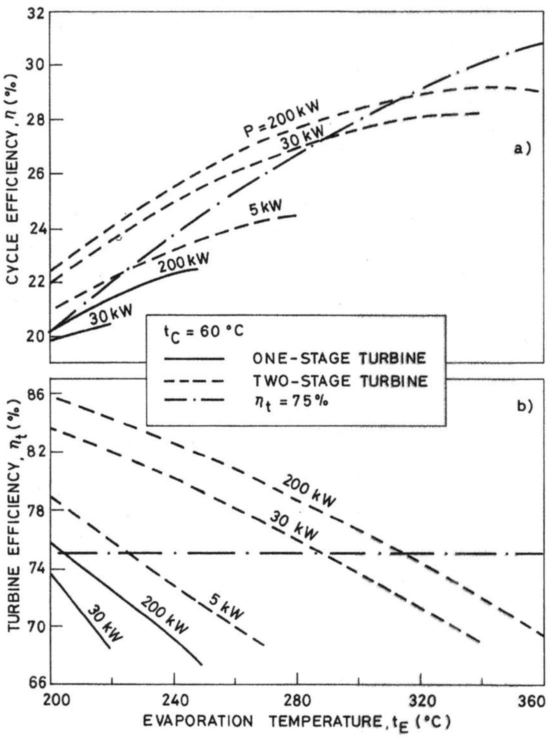

FIGURE 16.12. Saturated cycle and turbine efficiency for trimethylbenzene working fluid.

For a condensation temperature of 60°C, the efficiency of toluene and trimethylbenzene cycles is computed at increasing vaporization temperatures, up to 360°C, for an output of 5, 30, and 200 kW. The results are reported in Figs. 16.11 and 16.12(a) and (b), which also give cycle and turbine efficiency. Cycle efficiency for a constant turbine efficiency of 0.75 is also reported for comparison. Only for t_E lower than about 250°C one-stage turbines offer an acceptable performance. Turbine efficiency decreases steadily at increasing t_E (for example, η_t decreases from 0.86 to 0.70 when t_E increases from 200° to 360°C, in the case of two-stage 200-kW trimethylbenzene cycles) and is positively influenced by large outputs (for toluene at $t_E = 300$°C, η_t increases from 0.70 to 0.78 when P rises from 30 to 200 kW). For $P = 30$ kW toluene gives its best performance at $t_E = 280$°C with $\eta = 24$ percent and trimethylbenzene at 330°C with $\eta = 28$ percent.

Both these temperatures are considerably lower than the fluid's critical temperatures (319° and 376°C, respectively); the loss in turbine performance at higher inlet pressures more than offsets the thermodynamic gains due to higher operating temperatures. With the 66 percent limiting value for η_t it was impossible to design toluene turbines for $P = 5$ kW, while using trimethylbenzene a 24 percent cycle efficiency is achieved for the same output, with $\eta_t = 0.68$ at $t_E = 280$°C.

Super-Heated Cycles

If the thermal stability of the fluid allows operation at supercritical temperatures or if the saturated cycle expansion ratio cannot be handled efficiently by the turbine, super-heated cycles represent a preferable solution. A summary of superheated cycles performance for toluene and trimethylbenzene for an output of 30 kW is given in Fig. 16.13, by which optimum vaporization temperatures (or the equivalent turbine inlet pressure) for various turbine inlet temperatures (up to 400°C) can be inferred.

At $t_{max} = 360°C$ toluene cycles reach an optimum efficiency of slightly less than 29 percent at $t_E = 270°C$, $p_E = 2149$ kPa while trimethylbenzene's best performance is slightly above 30 percent at $t_E = 290°C$, $p_E = 1038$ kPa. Obviously such optima are directly dependent on assumed turbine characteristics as summarized, in our case, by the graphs of Fig. 16.10. Trimethylbenzene at 360°C yields the same efficiency as toluene at 400°C. Super-heating implies that external heat is introduced in the working fluid at an average temperature considerably lower than the top temperature. This influences negatively the cycle thermodynamics, as shown in Fig. 16.14(a) and (b) giving the quality factor QF as a function of t_E for $P = 30$ kW; trimethylbenzene gives the best performance with an optimum QF around 65 percent. Saturated cycles at

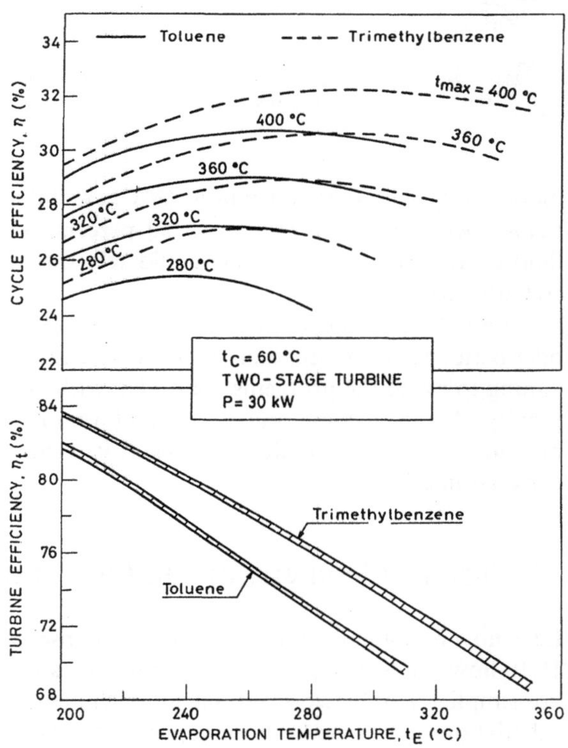

FIGURE 16.13. Super-heated cycle and turbine efficiency for toluene and trimethylbenzene.

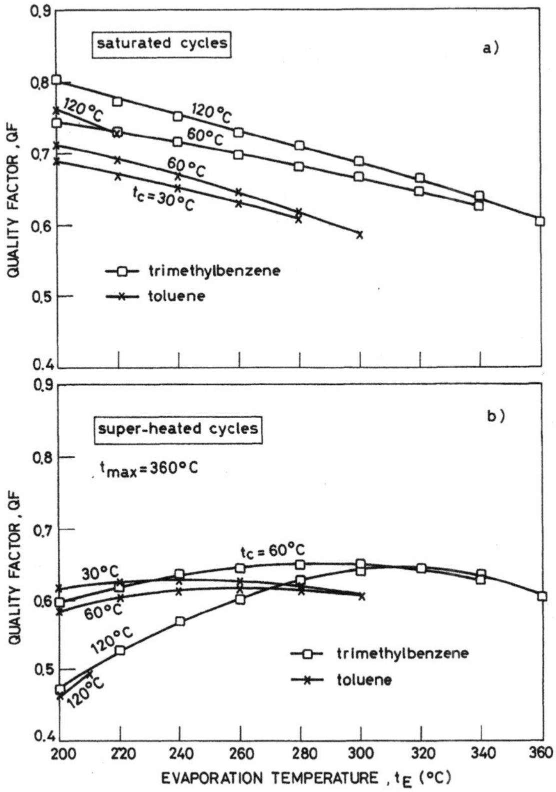

FIGURE 16.14. Achievable thermodynamic performance, in terms of quality factor QF, for toluene and trimethylbenzene cycles.

moderate t_E (t_c is kept constant at 60°C) exhibit a far better thermodynamic quality with values of QF up to 80 percent. This suggests that multiple-fluid cycles (binary or ternary) could rely on very high efficient elemental constituents.

Although according to our analysis, cycles at supercritical pressures do not offer better performance than that of subcritical cycles, they could be adopted aiming to avoid the problems connected with vaporization in the absence of gravity. Actual turbine and pump performances, both penalized by high-pressure operation, should be checked in view of achieving an acceptable conversion efficiency.

Turbine Geometrical and Operating Characteristics

Let's now investigate in more detail than consented by the inspection of Fig. 16.10 how the working fluid nature influences the characteristics of turbines (for simplicity, only two-stage machines will be considered). For all fluids, the calculations are carried out for super-heated cycles having t_{max} = 360°C and

FIGURE 16.15. Cycle efficiency, overall and first-stage turbine efficiency as a function of shaft power output for the considered working fluids.

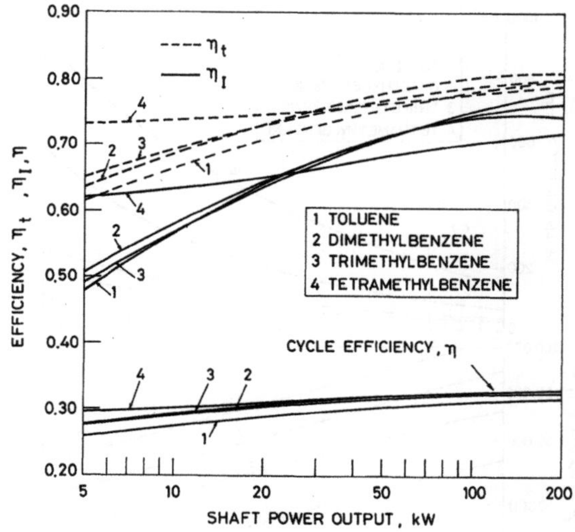

$t_C = 60°C$, with an evaporating pressure optimized according to the method illustrated in the previous section. The results of calculations are illustrated in Figs. 16.15 and 16.16 as a function of power output, in the range from 5 to 200 kW. The inspection of the curves in Fig. 16.15 suggests the following comments:

1. The power output has a strong influence on turbine efficiency (dotted lines in Fig. 16.15); in the considered power range, efficiency increases from about 60 to 65 percent (5 kW) to over 80 percent (200 kW).
2. The influence of working fluid thermodynamic properties on the turbine efficiency depends on the considered power level; at low power, the best performance is achieved by tetramethylbenzene, while di- and trimethylbenzene turbines become more efficient over 30 kW. Toluene turbines reach efficiencies close to other fluids only over 100 kW.
3. As shown by the curves giving the turbine first-stage efficiency (Fig. 16.15), this behavior is mostly related to the poor performance achievable by small output turbines in the first part of the expansion, where the volume flow rate is very small; below 30 kW, the blade height at the inlet of the first rotor is lower than 2 mm for all fluids (Fig. 16.16(a)). Larger blade heights could of course be obtained by designing partial admission impulse stages rather than full admission, small degree of reaction turbines like the ones selected for this analysis, but, at least according to the adopted loss prediction method, the overall turbine efficiency would not significantly benefit from this change, owing to losses related to higher Mach numbers, ventilation, blade scavenging, etc.
4. The working fluid nature plays a fundamental role in setting the practical feasibility of the turbines; high condensation pressure fluids, like toluene (and partly dimethylbenzene), require unpractically high speeds of revolu-

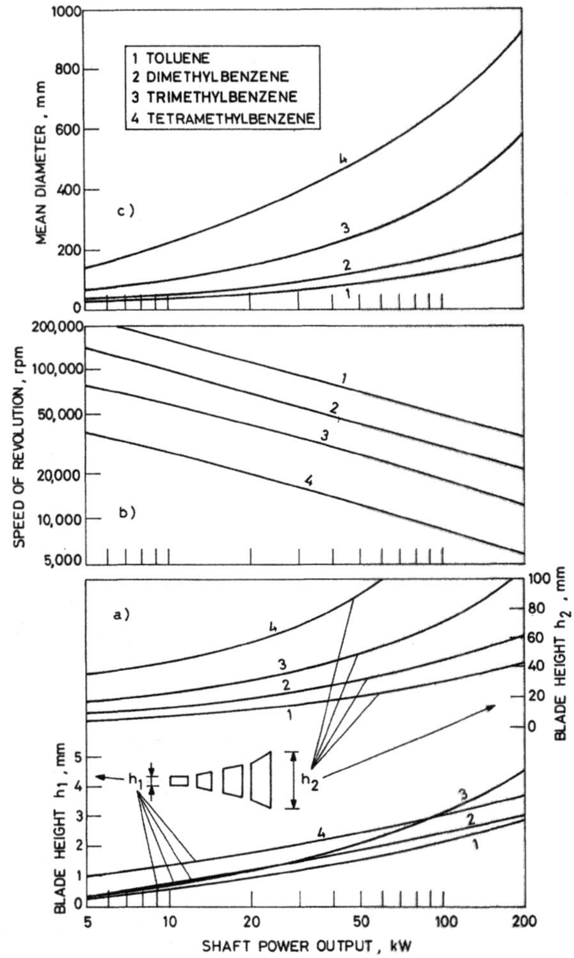

FIGURE 16.16. Characteristics of optimized two-stage axial flow turbines as a function of shaft power output for the considered working fluids: (a) blade height h_1 and h_2; (b) speed of revolution; and (c) mean turbine diameter.

tion at low power output (over 100,000 rpm for power levels below 20 kW, see Fig. 16.16(b), while low condensation pressure fluids, like tetramethylbenzene (and partly trimethylbenzene), require exceedingly high dimensions at turbine discharge for large power output (mean diameter over 500 mm for power levels over 50 kW, see Fig. 16.16(c)).

5. If the turbine efficiency predictions are combined with thermodynamic cycle calculations, the overall cycle efficiency curves drawn in the lower part of Fig. 16.15 are obtained. In the low power output range, the adoption of tetramethylbenzene yields much higher cycle efficiency than toluene, while bi- and trimethylbenzenes are preferable in the higher power range.

Maximum predicted values of cycle efficiency range between 29.5 percent (5 kW) and 32.5 percent (200 kW) and confirm the potential interest in ORC for space power generation. It should be pointed out that even better cycle efficiencies could be achieved by ORC, whenever condensing temperatures lower than the assumed 60°C are acceptable. As an example, for trimethylbenzene, assuming a condensation temperature of 45°C, a three-stage turbine in the 100-kW power range with a 360°C inlet temperature, 1.2-MPa inlet pressure would have an efficiency above 80 percent. Turbine dimensions and rotating speed (528 mm mean diameter of last stage and 9500 rpm, respectively) are well within the conventional turbomachinery practice. The resulting cycle efficiency would be over 35 percent.

Thermal Stability

Although ORC, if properly designed, exhibit a very high thermodynamic quality, their ultimate absolute performance is strictly dependent on maximum operating temperature. Tests of various natures are usually performed to establish its maximum safe value. However, the concept of thermal stability itself is somewhat ambiguous.

Sometimes an effort is made to use containing materials as inert as possible, in view of determining a hopefully intrinsic molecular stability of the fluid (Johns et al. 1962). In other experiments, materials are selected a priori on the ground of the plant engineering requirements and a mutual compatibility, rather than the fluid absolute stability, is actually investigated (Basiulis and Prager 1976). Other times a single physical property is assumed, more or less arbitrarily, as a univocal symptom of thermal degradation (isothermal pressure rise (Fabuss et al. 1963), formation of deposits (Shayeson 1969), etc.). Catalysis or corrosion sometimes plays a determining role in the test outcome (Moroni et al. 1974; Scholten 1980).

For these reasons, most of the results available in the literature give only an approximate and relative indication of the fluid thermal stability (Blake et al. 1961; Johns et al. 1962). Strictly speaking, only accurate tests performed through experimental facilities simulating the actual engine behavior (see Niggeman and Siberl (1969) for diphenyl, Cole et al. (1987) and Havens et al. (1987) for toluene), with reference to both materials and thermal processes, supply data which can be directly used to define the limiting useful temperature of an ORC fluid.

It was beyond the scope of this work to perform such demanding tests; a simpler, yet sufficiently accurate, method was devised to investigate the thermal stability of organic fluids and was applied to two trimethylbenzene isomers, which represent the most promising candidate for the proposed application. Also the stability of toluene, selected as the best known reference fluid, was extensively investigated by the same method.

The test apparatus (Invernizzi 1990) consists of a stainless-steel vessel of about 200 cc internal volume, maintained at the controlled temperature in a thermostatic oven. Fluid internal pressure and temperature are continuously monitored and recorded. Both subcritical and supercritical fluid states can be investigated. Information about the fluid thermal stability is obtained by: (1) the isothermal variation of pressure with time, which reflects a change in the fluid average molecular mass at supercritical temperatures or a variation in vapor pressure due to changes in chemical composition, at subcritical temperatures; (2) the modification of the low-temperature portion of the saturation pressure curve, after the fluid is kept at the test temperature for a given time, mainly owing to the formation of noncondensible light molecular fragments; and (3) the chemical analysis of the fluid at the end of the test.

Method (1) is widely employed as a screening tool for thermally stable fluids (Blake et al. 1961; Fabuss et al. 1963; Johns et al. 1962a, 1962b), but cannot supply truly quantitative information on fluid degradation, with the nature and partition of decomposition unknown products. However it can be used as an empirical test to determine a conventional decomposition threshold. In our apparatus it was found that a degradation below 1 percent in 100 hours was hardly detectable. In Fig. 16.17 pressure versus time curves at supercritical temperatures for toluene and 1,2,4-trimethylbenzene are reported, both exhibiting an incipient decomposition at about 400°C, and a very clear molecular breakdown at about 420°C. The pressure rise during the first 20 hours for toluene is probably due to a passivation process of the vessel material, after which decomposition proceeds at a very slow rate.

Not even method (2) can give quantitative information about fluid degradation, since it is selectively sensitive to the formation of noncondensible products. Even at temperatures at which the fluid is certainly stable, a departure of the saturation pressure curve from that of the untreated fluid is often evidenced, possibly due to the degradation of impurities, which are usually less stable than the basic fluid, or to a reaction between the fluid and the active metal surface, before the latter undergoes the usual passivation process. The method gives a direct indication of the difficulty of maintaining a very low condensation pressure in the presence of an even negligible degradation, in a truly sealed, unvented system. The results obtained by this method on toluene and trimetilbenzene are reported in Fig. 16.18. No indication about a decomposition threshold can be inferred by the pressure curves reported.

Method (3) requests an accurate chemical analysis of the fluid before and after testing. Such analyses were performed on the whole fluid gathered in the liquid phase at the end of the test by a specialized institution. The potential danger that sizable amounts of decomposition products remaining in the gas phase at ambient temperature are lost and ignored in the chemical analysis was checked through method (2) and found always negligible. At 400°C both toluene and trimethylbenzene exhibited less than 1 percent decomposition in 100 h; at 423°C after 100 h about 20 percent of the original toluene changed its chemical structure, having been converted mostly to trimethylbenzene. A similar decomposition rate was found for trimethylbenzene at 423°C. The

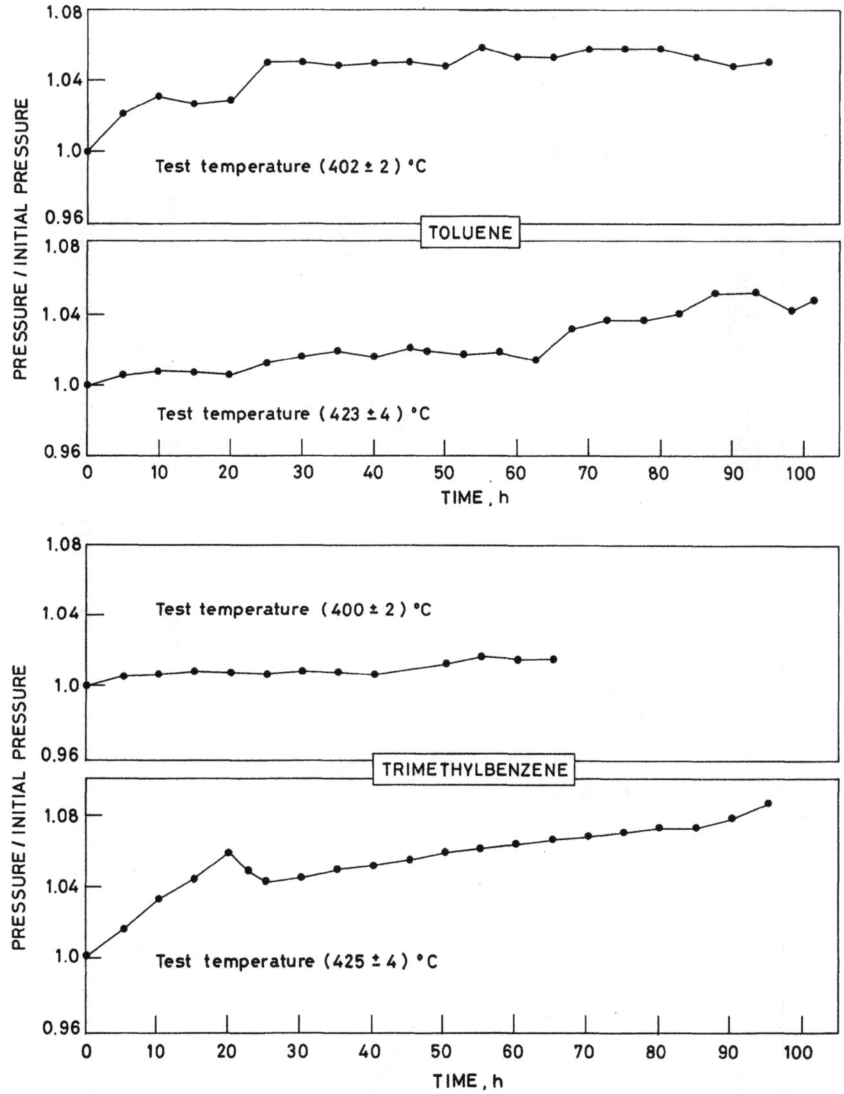

FIGURE 16.17. Isothermal variation of pressure with time at about 400° and 425°C for toluene and trimethylbenzene, showing the effect of thermal degradation.

thermal stability of 1,3,5-trimethylbenzene was found to be similar to that of 1,2,4-trimethylbenzene.

On the basis of all the tests performed toluene and trimethylbenzene could be tentatively credited a similar thermal stability, allowing operation at top temperatures of 350°–380°C.

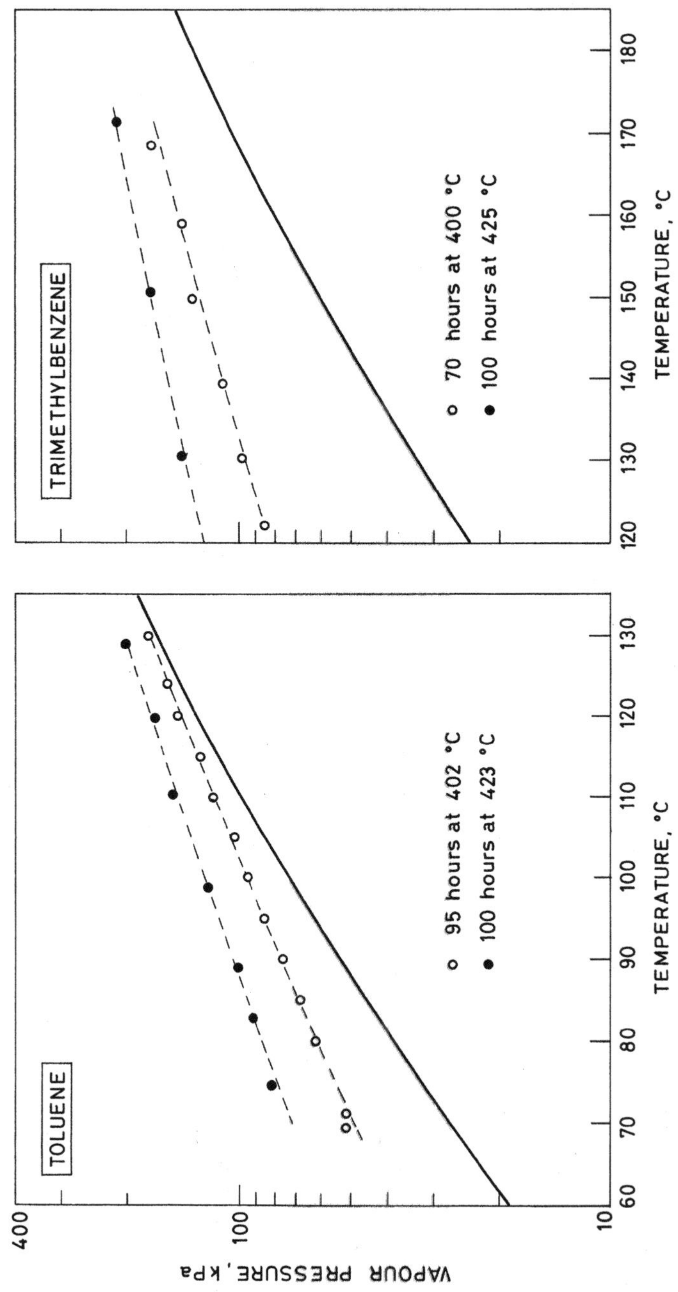

FIGURE 16.18. Effect of gaseous degradation products on vapor pressure for toluene and trimethylbenzene.

Conclusions

On the ground of the results obtained in the present work, the following conclusions can be derived:

1. In organic fluid power cycles the best overall performance is achieved in the regenerative configuration using fluids with a moderately complex molecule condensing at the minimum practical reduced temperature.
2. A detailed turbine analysis is requested to perform a reliable fluid and cycle optimization; for instance, super-heated cycles are shown to be more efficient than saturated cycles at the same top temperature only when the actual turbine performance is taken into account.
3. The selection of the optimum fluid is directly dependent on the power output; for given cycle-limiting temperatures, within a homogeneous class of fluids a decreasing critical temperature, causing increasing working pressures, is advisable at increasing power. With reference to the methyl-substituted benzenes class, at $P = 30$ kW, for example, trimethylbenzene offers at the same time the best overall performance ($\eta = 31.3$ percent at $t_{max} = 360°C$ $t_{min} = 60°C$) and the least unconventional turbine design (35,000 rpm rotating speed, 180 mm mean exhaust diameter, in comparison with a 100,000-rpm rotating speed, 70-mm mean diameter for a 29.3 percent efficient toluene cycle). The large volume flow leaving the turbine, however, is likely to cause increasing plumbing and heat exchangers shell sizes.
4. Optimized organic power cycles represent, even in the low power range, high-quality energy-conversion devices for space and terrestrial applications both in terms of relative efficiency (QF as high as 80 percent) and absolute performance (35 percent conversion efficiency for a 100-kW engine at $t_{max} = 360°C$, $t_{min} = 45°C$).

Acknowledgment. The authors wish to thank Mr. R. Biscuola, Mr. S. Sensolo, and Mr. G. Zucchetti, who carried out all the drawings.

Nomenclature

D = turbine diameter, m.
D_s = turbine-specific diameter (Eq. 5).
M = molecular mass, kg kmole^{-1}.
n = turbine wheels rotating speed, s^{-1}.
N_s = turbine-specific speed ($= n(V_{out})^{1/2}(\Delta h_{is})^{-3/4}$).
P = turbine effective power, kW.
p_C = condensation pressure, Pa.
p_{cr} = critical pressure, Pa.
p_E = evaporation pressure, Pa.
p_{in} = turbine inlet pressure, Pa.

p_{out} = turbine outlet pressure, Pa.
p_{vpr} = reduced vapor pressure ($= p_{vpr}/p_{cr}$).
QF = cycle quality factor (Eq. 2).
R = universal gas constant, J mole^{-1} K^{-1}.
SP = turbine size parameter, m (Eq. 5).
S = molar entropy, J mole^{-1} K^{-1}.
S^* = parameter proportional to regenerator heat transfer surface, K^{-1} (Eq. 3).
T, t = temperature, K, °C.
T_C, t_C = condensation temperature, K, °C.
T_{cr}, t_{cr} = critical temperature, K, °C.
T_E, t_E = evaporation temperature, K, °C.
T_{max}, t_{max} = maximum temperature, K, °C.
T_{min}, t_{min} = minimum temperature, K, °C.
T_r = reduced temperature ($= T/T_{cr}$).
T_{rC} = reduced condensation temperature ($= T_c/T_{cr}$).
V_{in} = inlet turbine volume flow rate, m^3 s^{-1}.
V_{out} = outlet turbine volume flow rate, m^3 s^{-1}.
VFR = volumetric turbine flow ratio (Eq. 4).
W = specific net work, kJ kg^{-1}.
Δh_{is} = isentropic turbine entalphy drop, kJ kg^{-1}.
Δh_r = regenerated heat, kJ kg^{-1}.
ΔT = cycle temperature extension ($T_{max} - T_{min}$), K.

Greek Letters

β_t = turbine expansion ratio (p_{in}/p_{out}).
η = effective cycle efficiency.
η_{id} = ideal cycle efficiency ($= 1 - T_{min}/T_{max}$).
η_t = turbine efficiency.
$\wedge$ = regenerator log-mean temperature difference, °C.
σ = parameter of the working fluid molecular complexity, K (Eq. 1).
τ = maximum-to-minimum cycle temperature ratio.
ω = acentric factor ($= -\log p_{vpr}$ (at $T_r = 0.7$) $- 1.0$).

References

Ambrose, D., Broderick, B.E., and Townsend, R., 1967, "The Vapour Pressures Above the Normal Boiling Point and the Critical Pressures of Some Aromatic Hydrocarbons," *J. of the Chemical Society* (*A*), 633–641.

Angelino, G., Gaia, M., and Macchi, E., 1984, "A Review of Italian Activity in the Field of Organic Rankine Cycle," *Proc. of International VDI Seminar on "ORC-HP Technology Working Fluids Problems,"* Zurich, VBD-Berichte 539, 465–482.

Angelino, G., and Invernizzi, C., 1988, "General Method for the Thermodynamic Evaluation of Heat Pump Working Fluids," *Int. J. of Refrigeration*, *11*, Jan., 17–25.

Balje, O.E., 1981, *Turbo-machines. A Guide to Design, Selection and Theory,* John Wiley & Sons.

Basiulis, A., and Prager, R.C., 1976, "Compatibility and Reliability of Heat-Pipe Materials," *Progress in Astronautical and Aeronautical, 49,* 515–529.

Blake, E.S., Hammann, W.C., Edwards, J.W., Reichard, T.E., and Ort, M.R., 1961, "Thermal Stability as a Function of Chemical Structure," *J. of Chemical and Engineering Data, 6,* 1, 87–98.

Boyle, R.V., Coombs, M.G., and Kudija, C.T., 1988, "Solar Dynamic Power Option for the Space Station," *Proc. of the 23d IECEC,* paper 889163, Denver.

Casci, C., and Angelino, G., 1969, "The Dependence of Power Cycles' Performance on Their Location Relative to the Andrews Curve," *ASME Paper 69-GT-65,* Cleveland.

Chandoir, D.W., Niggemann, R.E., and Bland, T.J., 1985, "A Solar Dynamic ORC Power System for Space Station Application," *Proc. of the 20th IECEC,* paper 859085, Miami Beach.

Cole, R.L., Demirgian, J.C., and Allen, J.W., 1987, "Predicting Toluene Degradation in Organic Rankine-Cycle Engines," *Proc. of the 22d IECEC,* paper 879075, Philadelphia.

Davoli, M., 1988, High Temperature Organic Fluid Cycles for Space Application" (in Italian), *graduation thesis,* Politecnico di Milano.

Downing, R.S., and Parekh, M.B., 1985, "Thermal Energy Storage for an Organic Rankine Cycle Solar Dynamic Powered Space Station," *Proc. of the 20th IECEC,* paper 859061, Miami Beach.

Fabuss, M.A., Borsanyi, A.S., Fabuss, B.M., and Smith, J.O., 1963, "Thermal Stability Studies of Pure Hydrocarbons in a High Pressure Isoteniscope," *J. of Chemical and Engineering Data, 8,* 1, 64–69.

Faget, N.M., Fraser, W.M., and Simon, W.E., 1985, "Thermal Energy Storage for a Space Solar Dynamic Power System," *Proc. of the 20th IECEC,* paper 859057, Miami Beach.

Forziati, A.F., Norris, W.R., and Rossini, F.D., "Chapter," T. Boublik, V. Fried, and E. Hala (Eds.), "*The Vapour Pressures of Pure Substances,* Elsevier, 439.

Havens, V.N., Ragaller, D.R., Sibert, L., and Miller, D., 1987, "Toluene Stability Space Station Rankine Power System," *Proc. of the 22d IECEC,* paper 879161, Philadelphia.

Heidenreich, G., Bland, T., and Niggemann, R., 1985, "Receiver for Solar Dynamic Organic Rankine Cycle Powered Space Station," *Proc. of the 20th IECEC,* paper 859220, Miami Beach.

Invernizzi, C., 1984, "Calculation of Thermodynamic Properties for Some Halo-Substitute Aromatic Hydrocarbons," (in Italian), La Termotecnica, April, 47–54.

Invernizzi, C., 1990, "Thermal Stability Investigation of Organic Working Fluids: An Experimental Apparatus and Some Calibration Results" (in Italian), *La Termotecnica,* April 69–76.

Jin Song, S., and Louis, J.F., 1988, "Liquid Lithium Thermal Energy Storage for Solar Dynamic Power Systems," *Proc. of the 23d IECEC,* paper 889510, Denver.

Johns, I. B., McElhill, E.A., and Smith, J.O., 1962a, "Thermal Stability of Organic Compounds," *Industrial and Engineering Chemistry Product Research and Development, 1,* 1, March, 2–6.

Johns, I.B., McElhill, E.A., and Smith, J.O., 1962b, "Thermal Stability of Some Organic Compounds," *J. of Chemical and Engineering Data,* 7, 2, 277–281.

Kurzrock, J.W., 1989, "Experimental Investigation of Supersonic Turbine Performance," *ASME Paper 89-GT-238*, Toronto.

Lozza, G., Macchi, E., and Perdichizzi, A., 1982, "On the Influence of the Number of Stages on the Efficiency of Axial-Flow Turbines," *ASME Paper 82-GT-43*, London.

Macchi, E., and Lozza, G., 1986, "Comparison of Partial vs Full Admission for Small Gas Turbines at Low Specific Speeds," *Turbo & Jet-Engines, 3*, 4, 307–317.

Macchi, E., and Perdichizzi, A., 1981, "Efficiency Prediction for Axial-Flow Turbines Operating with Nonconventional Fluids," *J. of Engineering for Power, 103*, Oct., 718–724.

Moroni, V., Macchi, E., and Giglioli, G., 1974, "Investigation on Thermal Stability and Corrosion Effects of Dichloro-Difluoro-Methane in View of Its Possible Application as Working Fluid in a Power Plant," *La Termotecnica, 28*, 4, 209–221.

Niggemann, R.E., and Sibert, L.A., 1969, "Organic Working Fluid Thermal Stability Investigation," *Report No. SAN-651-101.*

Nored, D.L., and Bernatowicz, D.T., 1986, "Electric Power System Design for the U.S. Space Station," *Proc. of the 21st IECEC*, paper 869321, San Diego.

Phillips, W.M., and Stearns, J.W., 1985, "Advanced latent Heat of Fusion Thermal Energy Storage for Solar Power Systems," *Proc. of the 20th IECEC*, paper 859058, Miami Beach.

Pietsch, A., and Trimble, S., 1985, "Space Station Brayton Power System," *Proc. of the 20th IECEC*, paper 859154, Miami Beach.

Portinari, F., 1988, "Generalized Method for the Prediction of Power Cycles' Performance" (in Italian), *graduation thesis*, Politecnico di Milano.

Reid, R.C., Prausnitz, J.M., and Poling, B.E., 1988, *The Properties of Gases and Liquids*, 4th ed., McGraw-Hill.

Scholten, W., 1980, *Working Fluids* (in German), VDI-Berichte No. 377, 5–11.

Shayeson, M.W., 1969, "Thermal Stability Measurement of Fuels for the U.S. Supersonic Transport Engine," *ASME paper*, 14th Annual International Gas Turbine Conference and Products Show, Cleveland.

Stull, D.R., 1947, "Vapor Pressure of Pure Substances: Organic Compounds," *Industrial and Engineering Chemistry, 39*, 4, 517–540.

Teren, F., 1987, "Space Station Electric Power System Requirements and Design," *Proc. of the 22d IECEC*, paper 879003, Philadelphia.

Trudell, J.J., Dalsania, V., Baumeister, J.F., and Jefferies, K.S., 1988, "Thermal Distortion Analysis of the Space Station Dynamic Concentrator," *Proc. of the 23d IECEC*, paper 889166, Denver.

Valade, F.H., 1988, "Solar Concentrator Advanced Development Program Update," *Proc. of the 23d IECEC*, paper 88167, Denver.

Van Landingham, E., 1988, "Space Power Technology to Meet Civil Mission Requirements," *Proc. of the 23d Intersociety Energy Conversion Engineering Conference (IECEC)*, paper 889025, Denver.

West, C.D., 1988, "A Historical Perspective on Stirling Engine Performance," *Proc. of the 23d IECEC*, paper 889004, Denver.

Willingham, C.J., Taylor, W.J., Pignocco, J.M., and Rossini, F.D., 1973, "Chapter," T. Boublik, V. Fried, and E. Hala, (Eds.), *The Vapour Pressures of Pure Substances*, Elsevier, 324.

V
Flight Dynamics

17
Highly Loaded Turbines for Space Applications: Rotor Flow Analysis and Performance Evaluation

F. BASSI, C. OSNAGHI AND A. PERDICHIZZI

ABSTRACT: The flow field in a supersonic rotor cascade has been analyzed for different operating conditions. Cascade performance, i.e., losses, blade pressure distribution, and inlet and outlet flow angles, are obtained by testing the cascade in a supersonic wind tunnel. A numerical investigation is performed by means of a Navier-Stokes code and a shock-fitting inviscid code. The computed results are discussed and compared to the experimental data so as to determine the degree of accuracy now achieveable in the prediction of flows characterized by high Mach numbers and large viscous effects. An analysis of the unique incidence angle for different inlet Mach numbers is carried out by numerical codes and simplified methods. The predicted incidence angles are shown to be in good agreement with the experimental data.

Introduction

In liquid fuel engine rockets, fuel and oxidizer, i.e., hydrogen and oxygen, are pressurized by turbopumps driven by supersonic turbines. Although the available pressure ratio is very high ($\beta = 20$–30), the requirement of limiting the overall mass leads the optimal solution toward single or two-stage turbines. Therefore these turbines are generally highly loaded and characterized by quite low specific speeds (NASA, 1974). The constraints on the maximum values of peripheral velocity and rotational speed, due to stress limitations and pump cavitation problems, imply the adoption of quite unusual velocity triangles (Fig. 17.1), if compared to aeronautical or industrial applications. As a consequence, both absolute and relative Mach numbers larger then 2.0 often result.

The design of supersonic nozzles does not represent a difficult problem and good efficiencies can be commonly achieved, especially if advanced numerical techniques are employed. On the other hand, the rotor design is a hard task for the following reasons:

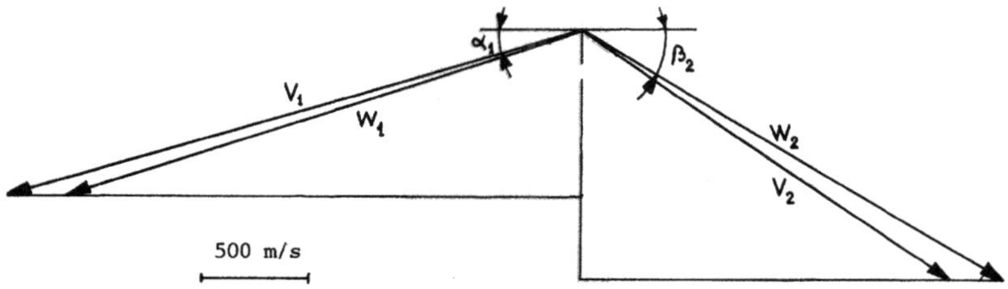

1: STATOR EXIT

Absolute Mach Number M_1 = 2.39
Absolute Angle α_1 = 16.00 deg
Relative Mach Number M_{r1} = 2.19
elative Angle β_1 = 17.49 deg

2: ROTOR EXIT

Absolute Mach Number M_2 = 1.71
Absolute Angle α_2 = 33.25 deg
Relative Mach Number M_{r2} = 1.87
Relative Angle β_2 = 30.00 deg

FIGURE 17.1. Supersonic turbine velocity triangles.

- The supersonic entry requires the determination of the unique incidence angle related to the inlet Mach number and to the profile geometry in the first part of the suction side.
- The large deflection of the supersonic flow implies high blade loading; moreover, shock-wave boundary layer interactions make the boundary layer on most of the suction side prone to separate.

In this field, information is available on complete stage performance (Kurzrock 1989; Ohlsson 1964; Verdonk and Dufurnet 1987), but only few experimental data (Colclough 1966a, 1966b, Liccini, 1949) can be found in the open literature on rotor cascade performance and on the details of the flow behavior throughout the blade. Therefore experimental investigations should be performed to overcome this lack of data. Furthermore additional information on cascade performance can be obtained by advanced computing techniques that now range from inviscid calculations to Navier-Stokes codes with possibly very sophisticated turbulence models.

The present work aims at a better understanding of the behavior of supersonic flow in rotor cascades so as to provide useful information for future design. This is primarily achieved by carrying out a detailed experimental investigation of the flow field in a typical rotor cascade tested in a transonic/supersonic wind tunnel and secondly by running two different numerical codes to get additional information on the internal flow. Another goal of this investigation is to find out which features of these flows can be predicted by numerical techniques and which accuracy level can be achieved.

An introduction to the nozzle and rotor design criteria is presented in the next section, including a description of simplified approaches for unique incidence prediction. The experimental investigation on a typical rotor cascade and the codes used in the numerical investigation are described in the following sections. A description of experimental and numerical results is given, together with a critical analysis of the codes performances. Finally some conclusions are given.

Design Criteria

Supersonic turbines for space applications are requested to give the maximum specific work per stage in order to limit mass and size; as a consequence, the stage is designed to operate normally at limit loading.

Nozzle Design

Due to the supersonic flow throughout the rotor the flow conditions at nozzle outlet cannot change even if the downstream pressure changes due to off-design conditions. Hence the nozzles are designed to give the design Mach number and flow direction with minimum losses. This is in contrast with the design criteria used for supersonic turbines with transonic flow in the rotor; in this case the nozzle outlet area is generally choosen to produce at the design operating point a certain amount of afterexpansion downstream of the blade with a flow rotation toward the axial direction. This avoids large additional losses at high-backpressure off-design conditions.

Many semiempirical methods are available for nozzle design (Deich et al. 1964), but the inverse method of characteristics is more suitable to draw the divergent part of the nozzle. Referring to Fig. 17.2(A), which represents a typical supersonic turbine nozzle, the pressure side of the channel after the throat can be assumed a straight line; the suction side AB can be choosen as an arbitrary convex arc, with the condition that the right running characteristic starting at point B reaches the opposite wall in point the F that is exactly at the design Mach number. The arc BC is calculated starting from the line BF by means of simple wave theory, in order to obtain uniform flow on the straight characteristic FC. The suction side is completed by the straight line CD, which reduces to zero if the axial Mach number is one, i.e., if the limit loading condition is attained.

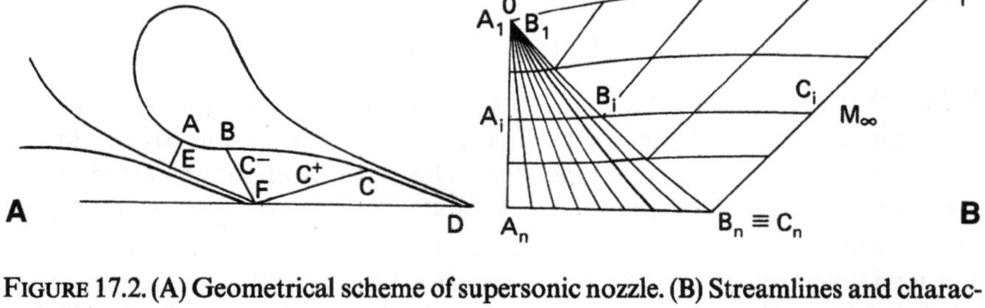

FIGURE 17.2. (A) Geometrical scheme of supersonic nozzle. (B) Streamlines and characteristics maps for a nozzle.

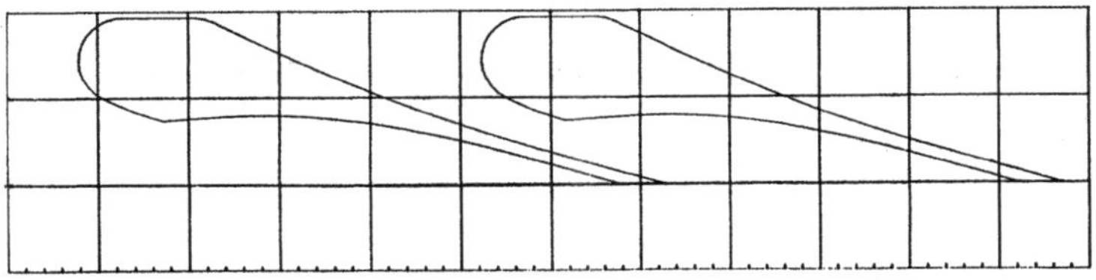

FIGURE 17.3. Typical supersonic nozzle profile.

A rational scheme for the choice of the suction side profile consists of solving, by the classical method of characteristics, an expansion region over a corner, starting from a uniform flow just over Mach one, to obtain the design Mach number in F. Many streamlines can be traced (e.g., A_iB_i in Fig. 17.2(b)) and, starting from the line BF it is easy to construct a set of streamlines B_iC_i in the simple wave region BFC. The streamlines A_iC_i represent possible suction side walls for the nozzle, determining different ratios between the divergent length and the throat width, i.e., different blade solidities. Boundary layer considerations suggest the choice of the best streamline as solid wall. Often the optimization leads to the largest solidity, as is the case for the nozzle shown in Fig. 17.3.

Due to favourable pressure gradients, no strong interaction occurs between boundary layer and inviscid flow; hence the displacement thickness of the boundary layer, calculated from the inviscid velocity distribution, is sufficient to take into account viscous phenomena. This leads to a geometry correction corresponding to an outlet Mach number reduction of about 0.1 to 0.2, depending on the Mach number level.

High supersonic nozzle blades require large trailing edge thickness due to the extreme mechanical stresses; as a consequence, it is recommended to limit the solidity to reduce the blockage effect at the outlet section.

Rotor Design

Two major problems in the design of highly supersonic impulse rotors are the unique incidence and the presence of high-compression gradients related to the design criterion for the flow inside the channel and to the inlet shock wave.

Unique Incidence

When the flow upstream of an annular cascade is supersonic, but axially subsonic, and the blade geometry is given, the flow angle and Mach number at upstream infinity are not independent. This is true both for unstarted regime (i.e., sonic line inside the channel) and for started regime (flow entirely supersonic inside the channel). The incidence depends only upon the geometry of the semivaned region at inlet and is generated by the shock system upstream of the blades. The correct evaluation of the incidence is a matter of great importance, since very small errors in the flow direction can lead to large

Mach number differences, with respect to the design conditions, in the region between nozzle and rotor; therefore significant errors in the pressure level and rotor loading can occur.

The simplest scheme for the unique incidence prediction is based on the assumption of isentropic flow rotation from the angle $\beta_{1\infty}$ at upstream infinity to the angle β_1 at the inlet section (o), Fig. 17.6. For given Mach number $M_{1\infty}$ and angle β_1 assuming the flow in (o) one-dimensional it results:

$$\beta_1 - \beta_{1\infty} = \nu(\mathrm{M}_1) - \nu(\mathrm{M}_{1\infty})$$

where the Prandtl-Mayer function ν has been introduced. On the other hand, the continuity equation between the upstream normal section ($s \sin \beta_1$) and the internal section (o) gives a relation between angle and Mach numbers determining $\beta_{1\infty}$ and M_1.

Many theories try to correct the isentropic model, using different assumptions for the shock system in the entrance region of the blade (Chauvin et al. 1970), if the leading edge is thick and blunt a correlation is given (Starken et al. 1984) based on the losses arising from a detached shock wave This scheme is not satisfactory for application in highly loaded supersonic rotors, where the leading edge is thick, but the nose is in general sharp and the section (o) is not well defined. To overcome the uncertainties related to these methods it was decided to get more information from the experimental and numerical investigations presented in the following.

Channel Design

The most useful idea in the design of high-deflection supersonic blades is based on the plane vortex flow (Boxer et al. 1952; Osnaghi 1971; Vanco and Goldman 1968). A special geometry of the channel (inlet transition region) converts the inlet flow into a free vortex, which can be maintained easily by two circular walls until the desired flow deflection has been obtained. Then the vortex flow is converted back to a uniform one at the outlet (outlet transition region).

The free vortex flow minimizes the pressure gradients along the walls inside the channel especially if a wall correction is applied to compensate the boundary layer growth; however, strong adverse gradients are present in the inlet transition region near the pressure side. To avoid separation it is recommended to limit the minimum Mach number, i.e., the pressure-side Mach number. Goldman and Sculling (1968) suggest that $(p_{\max} - p_1)/(1/2\rho V^2)_1 \leq 1/2$.

To generalize the design criteria we characterize the free vortex streamlines by the nondimensional radius $r = R/R_s$ and the critical Mach number $M^* = V/V_s$, where R_s and V_s are, respectively, the radius of the streamline with Mach number equal to one and the sonic speed; in a free vortex flow $M^* = 1/r$. A practical scheme for drawing the convex and concave transition region profiles is based on two general maps of the streamlines (Fig. 17.4(a) and (b)) originated from a vortex flow at different nondimensional radii r. These maps can be

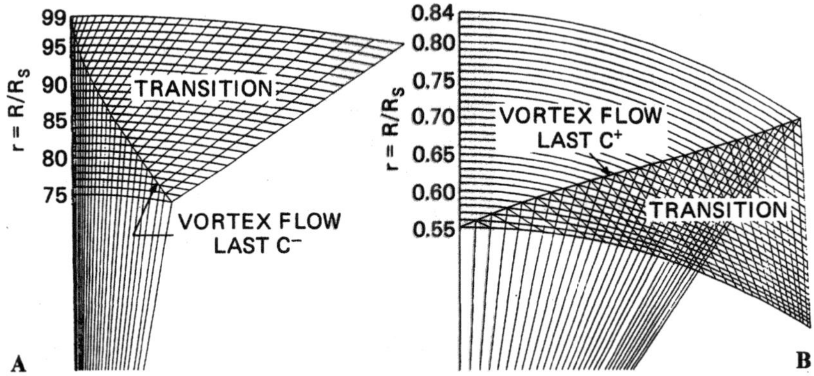

FIGURE 17.4. (A) Pressure side transition region. (B) Suction side transition region.

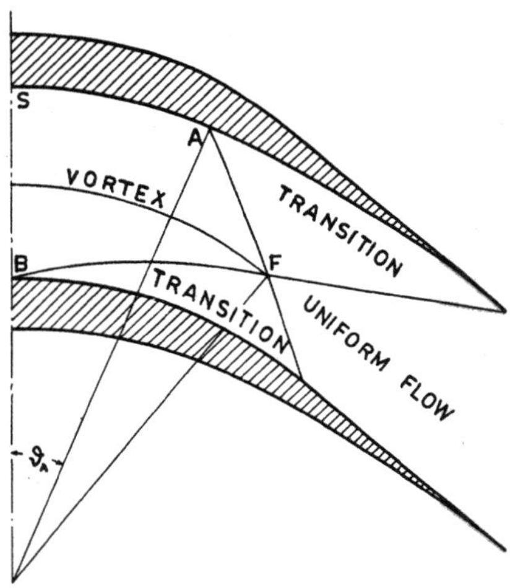

FIGURE 17.5. Geometrical scheme of the outlet region of a free vortex blade.

superimposed with the same center to represent different flow at infinity, on suction wall, and pressure wall. As an example (Fig. 17.5), given the pressure-side and suction-side critical Mach numbers M_A^* and M_B^* for the vortex flow and the critical Mach number for the uniform outlet flow M_F^*, we can draw the outlet transition region using map (a) between r_F and r_A and map (b) between r_F and r_B.

Experimental Investigation

The blading considered in this investigation is the rotor cascade of the turbine that drives the oxygen pump of the European rocket Ariane 5; the blade profile, the cascade geometry, and the operating design conditions are presented in

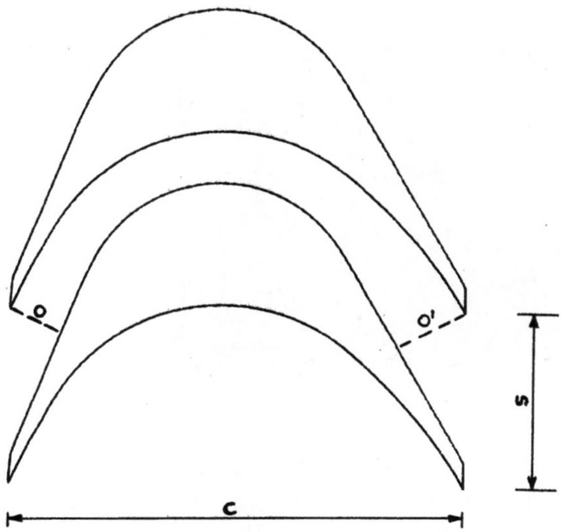

Solidity	$s/c = 0.385$	**Inlet geometric angle**	$\sin^{-1}(o/s) = 25.3$
Aspect ratio	$h/c = 0.96$	**Outlet geometric angle**	$\sin^{-1}(o'/s) = 30.0$
Inlet Mach number	$M_1 = 2.19$	**Inlet flow angle**	$\beta_1 = 17.5$ **deg**
Outlet Mach number	$M_2 = 1.71$	**Outlet flow angle**	$\beta_2 = 30.0$ **deg**

FIGURE 17.6. Rotor blade geometry and operating conditions.

Fig. 17.6. The blade is characterized by a sharp leading edge so as to avoid the formation of a strong detached bow wave in front of the blade; the suction and pressure profiles were built up by a free vortex method design.

The cascade was tested at C.N.P.M. (Istituto per Ricerche sulla Propulsione e sull'Energetica, Milano) in the transonic blow-down wind tunnel for turbine cascades, under a research grant sponsored by Fiat Avio s.p.a.. Due to the high supersonic inlet Mach number the usual tunnel setup was modified as shown in Fig. 17.7. To ensure the desired inlet Mach number, a calibrated nozzle, designed for a uniform exit flow, was installed just ahead of the cascade inlet. The nozzle geometry was defined by using the characteristic method and taking into account the displacement thickness of the boundary layer on the side walls. The nozzle design Mach number was assumed to be larger than the cascade inlet one to compensate the Mach number decrease due to the first shock generated at the leading edge of the first blade. The actual nozzle Mach number could be slightly adjusted, to get the test conditions, by translating the upper wall with respect to the lower one, i.e., through a small variation of the area ratio.

In testing cascades with supersonic entry, an important problem to be solved is getting periodic flow conditions, not only at outlet, but also at inlet; indeed, according to the unique incidence theory, for a defined inlet Mach number, only one direction of the incoming flow produces periodic flow

FIGURE 17.7. Wind tunnel configuration and test section.

conditions in front of the blades. If the cascade is not set with the right angle with respect to the inlet flow, i.e., if the angle of attack is not the right one, an oblique shock (or expansion) wave starting from the first blade deflects the flow to make it assume the direction of the unique incidence. In this case a periodic flow might be obtained too, but for a Mach number lower (or higher) than the design one, due to the presence of the shock (or expansion) wave.

In these tests the cascade was fitted in the wind tunnel on a rotating disk, and several runs of the tunnel were done at different inlet angles in order to determine the one producing satisfactory periodicity conditions at the design Mach number.

The variation of the static-to-static pressure ratio across the cascade was obtained by changing the pressure level at cascade outlet; this was achieved by an adjustable tailboard attached to the trailing edge of the last blade (Fig. 17.7); the angular position of the tailboard determined the degree of after-expansion at the blade exit and thereby the outlet flow angle. Moreover the use of the tailboard contributed to improve the flow periodicity downstream of the cascade.

For the pressure ratios considered the following measurements were carried out:

- Upstream flow traverse at $x/c = 0.37$ ahead of the leading edge, to evaluate the Mach number, the inlet flow angle, and the inlet total pressure.
- Downstream flow traverse at $x/c = 0.37$ downstream of the trailing edge, to determine the exit flow conditions and the cascade losses.
- Blade static pressures to obtain the pressure coefficient distribution on the profile; the blade pressure coefficient is defined as:

$$C_p = \frac{1 - p/p_{\text{exit}}}{1/2\gamma M_{\text{exit}}^2}$$

Upstream and downstream flow measurements were done by means of a miniaturized probe especially designed for transonic and supersonic flows with a head diameter of only 1.5 mm. Tle probe was calibrated for Mach numbers ranging from 0.5 to 2.9. The calibration was performed by using calibrated nozzles at a reference Mach number. Yaw and pitch angle were varied by ± 24 deg and ± 16 deg, respectively.

Schlieren pictures were taken to check flow periodicity and to visualize the shock wave pattern. A conventional Schlieren system using two concave mirrors and a 5-mW He-Ne laser was employed. Perspex windows were selected instead of optical glass windows in order to minimize blade-fitting problems; unfortunately this impaired the picture quality.

The probe measurements in highly supersonic cascade flows are affected by typical sources of inaccuracy. One is related to the use of a probe with finite head in flows characterized by very high gradients induced by shock waves; referring to previous experience gained in transonic and supersonic cascade testing at C.N.P.M., the error in the Mach number is estimated to be about 0.03 to 0.05. Another source of possible inaccuracy, which affects mainly the loss evaluation, is due to the fact that it is practically impossible to get periodic flow in a linear cascade; this error is estimated to be about 0.03.

Numerical Investigation

The supersonic flow field through the rotor was investigated by means of an inviscid method of characteristics with shock-fitting and a time-marching Navier-Stokes solver.

Inviscid Code

The inviscid calculation is based on the method of characteristics for steady supersonic two-dimensional flows (Bassi 1978a; Tome 1972). The flow is adiabatic, but not necessarily isentropic and irrotational; jumps in entropy and vorticity are introduced explicitly by the Rankine-Hugoniot relations across shock waves. The equations of motion for two-dimensional inviscid flows are written in a rotating frame, on an axisymmetric blade-to-blade surface in a transformed intrinsic coordinate system (X, Y) formed by streamlines and normals. The relations between physical and transformed coordinates are

$$dm = \alpha(X, Y)\cos\theta\, dX - \beta(X, Y)\sin\theta\, dY$$

$$r d\phi = \alpha(X, Y)\sin\theta\, dX + \beta(X, Y)\cos\theta\, dY$$

where m and ϕ represent meridional and angular coordinates and θ is the angle between vector velocity and the coordinate m. The transformation functions α and β must satisfy the integrability conditions

$$\frac{\partial \alpha}{\partial Y} = -\beta \frac{\partial \theta}{\partial X}$$

$$\frac{\partial \beta}{\partial X} = \alpha \frac{\partial \theta}{\partial Y}.$$

The characteristic lines are defined by the directions

$$\Lambda_0 = 0$$

$$\Lambda_{1,2} = \pm a \tan^{-1}\left(\frac{\alpha}{\beta}\tan\lambda\right)$$

where $\lambda = a\sin^{-1}(1/M)$ is the Mach angle. The compatibility equations along the characteristic directions Λ_1 and Λ_2 are:

$$\frac{\sqrt{M^2-1}}{\rho W^2}\,dp \pm d\theta = (C_1 \pm C_2)\alpha\, dX$$

where W is the relative velocity, C_1 and C_2 depend on the blade-to-blade surface geometry in the meridional plane and on the angular velocity about the symmetry axis; C_1 and C_2 are zero for plane flows. The compatibility equations along the characteristic direction Λ_0 simplify to

$$s = \text{const}$$

$$H_R = h + \frac{W^2}{2} - \frac{U^2}{2} = \text{const}$$

where s and h are entropy and enthalpy, U is the peripheral velocity, and H_R is the total enthalpy in the rotating frame. The entropy jump across shock waves is explicitly given by the Rankine-Hugoniot relations.

Calculation Method

In the computational plane (X,Y), X and Y constant lines represent, respectively, streamlines and normals; hence side boundaries (walls and jets) are represented by Y constant lines. Due to the hyperbolic nature of the flow equations in the X direction, a space-marching technique is used to advance explicitly the solution, starting from an initial normal line, where the flow properties are given. Each point on a new normal line is calculated taking information from the adjacent points on the previous normal; the integration step ΔX is thus limited by the stability requirement that the physical domain of dependence lies inside the numerical one. Shock waves are explicitly taken into account and followed across the computational grid; shock points are calculated by matching the compatibility equations in the upstream and downstream regions by using the Rankine-Hugoniot relations. Slip-line points are calculated by imposing equal pressure and flow direction across the discontinuity. It is worth mentioning that the code has virtually no restriction in the computation of complicated shock-wave and slip-line configurations.

An interesting feature of the code is that it can handle the supersonic inlet and outlet regions of cascades where the periodic boundaries allow information to propagate upstream if the flow is axially subsonic. The space-marching procedure does not allow direct imposition of the periodicitv condition: this condition results from the calculation of the flow in a number of the adjacent blade channels (10–20), i.e., from the numerical simulation of the flow in a semi-infinite cascade in a wind tunnel. Different flow solutions are obtained in the entrance and outlet regions by changing, respectively, the inlet boundary conditions at upstream infinity and the lateral boundary condition downstream of the cascade. From the computed solutions at half a chord ahead of the leading edges and at half a chord downstream of the trailing edges, integral conservation equations allow to obtain, respectively, the inlet reference uniform flow and the mixed out-flow conditions .

Navier-Stokes Code

The compressible form of the unsteady Navier-Stokes equations is used to compute the viscous flow; the Navier-Stokes equations in integral conservation form in a cartesian coordinate system are

$$\frac{\partial}{\partial t}\int_V W\,dV + \oint_{\partial V} (Fn_x + Gn_y)\,dS = 0 \tag{1}$$

where

$$W = \begin{bmatrix} \rho \\ \rho u \\ \rho v \\ \rho E \end{bmatrix}, F = \begin{bmatrix} \rho u \\ \rho u^2 + p + \sigma_x \\ \rho uv + \tau_{xy} \\ \rho uH + u\sigma_x + v\tau_{xy} + q_x \end{bmatrix}, G = \begin{bmatrix} \rho v \\ \rho vu + \tau_{yx} \\ \rho v^2 + p + \sigma_y \\ \rho vH + u\tau_{yx} + v\sigma_y + q_y \end{bmatrix},$$

and

$$p = (\gamma - 1)\rho\left[E - \frac{1}{2}(u^2 + v^2)\right], \qquad H = E + \frac{p}{\rho},$$

and n_x, n_y are the components of the unit vector normal to the surface that encloses the volume V, fixed in space. The working fluid is air and it is assumed to be thermally and calorically perfect, with a ratio γ of constant specific heats equal to 1.4. The linear relation between stress and rate of strain tensors (for isotropic fluids) and between heat flux vector and temperature gradient give the following relations:

$$\sigma_x = -\lambda\left(\frac{\partial u}{\partial x} + \frac{\partial v}{\partial y}\right) - 2\mu\frac{\partial u}{\partial x}$$

$$\tau_{xy} = \tau_{yx} = -\mu\left(\frac{\partial u}{\partial y} + \frac{\partial v}{\partial x}\right)$$

$$\sigma_y = -\lambda\left(\frac{\partial u}{\partial x} + \frac{\partial v}{\partial y}\right) - 2\mu\frac{\partial v}{\partial y}$$

$$q_x = -k\frac{\partial T}{\partial x} \qquad q_y = -k\frac{\partial T}{\partial y}$$

$$\lambda = -\frac{2}{3}\mu \qquad Pr = \frac{\mu c_p}{k}.$$

The Stokes hypothesis is used to relate λ and the molecular viscosity μ. The Prandtl number is assumed constant and equal to 0.72 and is used to obtain the molecular conductivity k. A simple power law describes the dependence of molecular viscosity from temperature.

For turbulent flows the Navier-Stokes equations are mass-weighted averaged and the effect of turbulence is taken into account by means of the eddy viscosity hypothesis, i.e., the viscosity and thermal conductivity coefficients are defined as

$$\mu = \mu_l + \mu_t \qquad k = \frac{c_p\mu_l}{Pr_l} + \frac{c_p\mu_t}{Pr_t}$$

where the subscripts l and t stand for laminar and turbulent. The simple zero equation turbulence model of Baldwin and Lomax is used to compute μ_t: this model proved to be sufficiently accurate in the computation of a number of flows, both external and internal (Colantuoni et al 1989; Holst 1987), and has the obvious advantage that it does not require the solution of additional differential equations for turbulent variables; the counterpart of these advantages is that one has to expect some deficiencies in the modelization of turbulent phenomena, especially for complicated shock boundary layer interactions and large regions of separated flow. In this paper the Baldwin-Lomax model has been used with the values of the model constants reported in (Baldwin and Lomax 1978), some computations reported in the literature (Visbal and Knight 1984), with minor modifications to the original model do not seem to give substantial improvements in the predictive capabilities of the model. The turbulent Prandtl number is assumed to be constant and equal to 0.9.

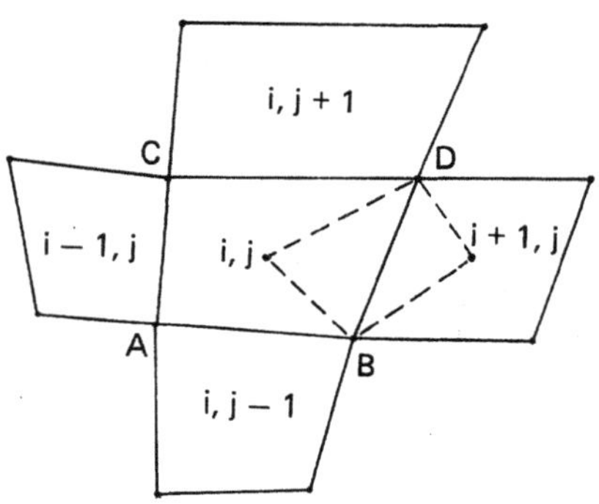

FIGURE 17.8. Discretization for viscous fluxes.

Numerical Solution

The numerical solution is based on the algorithm introduced by Jameson et al. (1981) for inviscid flows and extended to Navier-Stokes equations (Bassi et al. 1988). The finite-volume approach is used to discretize the system of Eq. 1; the computational domain is discretized into quadrilateral cells and the governing equations are applied to each cell. The inviscid fluxes on cell faces, needed for the computation of surface integrals, are taken as mean values between neighboring cells; this corresponds to a centered approximation on a smooth grid. The velocity and temperature derivatives on cell faces, needed to evaluate the viscous fluxes, are computed by applying Green's theorem to the auxiliary control volumes shown in Fig. 17.8; note that the variables at grid points B and D are evaluated as arithmetic averages in the surrounding cells.

The centered discretization of convective fluxes does not contain any dissipation and the physical dissipation effects are usually confined in small shear layer regions; hence, artificial dissipation terms need to be included in the scheme in order to avoid even-odd decoupling and oscillations near shock waves. Jameson et al. (1981) proposed an effective dissipation model that contains a blend of second- and fourth-order dissipation terms.

The solution is advanced in time with a three-stage Runge-Kutta scheme as follows

$$W^{(0)} = W^n$$
$$W^{(k)} = W^{(0)} - \alpha_k \Delta t R(W^{(k-1)}) \qquad k = 1, \ldots, 3$$
$$W^{n+1} = W^{(3)}$$

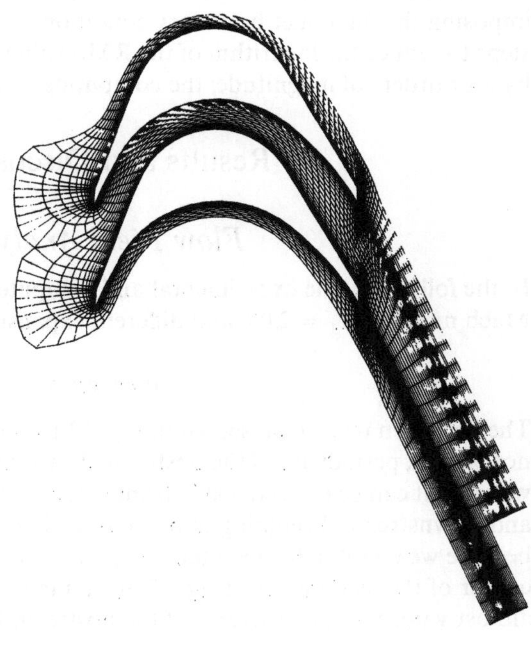

FIGURE 17.9. Grid for viscous computations (readuced to 145 × 21 for clarity).

where the residual R is given by

$$R(W^{(k)}) = \frac{1}{V}[Q(W^{(k)}) - D(W^{(0)})]$$

where Q is the discrete operator for the convective and physical diffusive terms and D is the discrete operator for the artificial dissipation terms. For efficiency purposes, the artificial dissipation terms are evaluated only at the first stage of the Runge-Kutta algorithm. Good damping of high-frequency modes is obtained by using $\alpha_1 = \alpha_2 = 0.6$, $\alpha_3 = 1$. Convergence toward steady state can be enhanced by using local time steps and a FAS multigrid algorithm (Jameson 1983).

Navier-Stokes computations have been carried out on the C-type grid, shown in Fig. 17.9, with 288 × 40 cells. An elliptic mesh generator was used to obtain a quite smooth grid. A fine discretization was used in the high-gradient regions near the leading and trailing edges; the mesh is highly refined near the blade so as to have the centers of the cells adjacent to the blade inside the laminar sublayer.

Computational Data

The inviscid and viscous codes were run on an HP-835 computer with 32-MB RAM memory and 3 Mflops CPU speed.

The inviscid computation of 20 blade channels with 20 streamlines in each passage took approximately 15 minutes.

The viscous computations, starting from zero velocity field and suddenly imposing the final inlet boundary conditions, took approximately 8000 time steps to reduce the logarithm of the RMS value of the density time derivative by four orders of magnitude; the computing time was 14 hours.

Results and Discussion

Flow Field Analysis

In the following, the experimental and computed flow properties for an inlet Mach number $M_1 = 2.09$, and different expansion ratios are described.

Inlet Flow

The Schlieren picture presented in Fig. 17.10 shows that an acceptable, albeit not perfect, periodicity of the upstream flow was obtained. The detached bow waves that can be noticed just in front of the wedges propagate both upstream and downstream, becoming immediately oblique shocks. The upstream legs become weaker due to the intense expansion fans generated from the second corner of the leading edge (Fig. 17.6), and in front of the next blade they are almost vanished. The traverse of the upstream flow (Fig. 17.11) confirms the

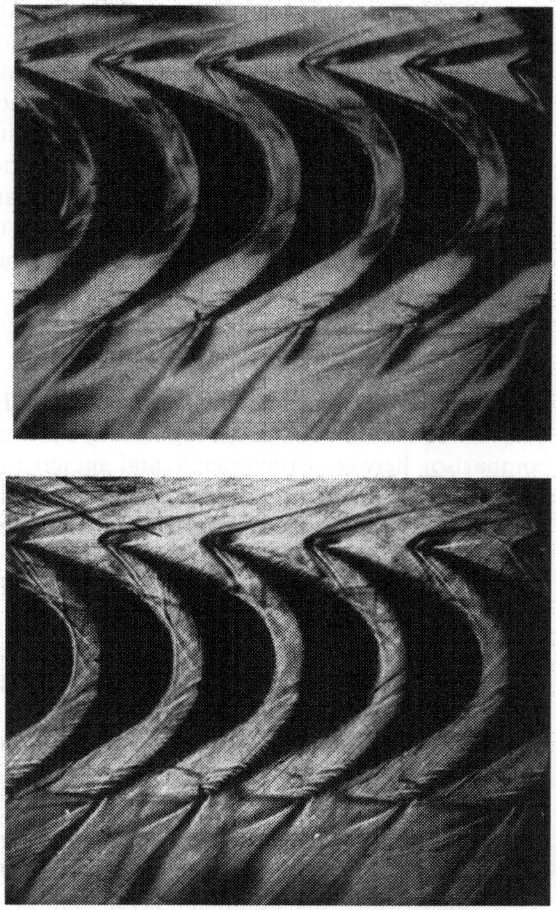

FIGURE 17.10. Schlieren pictures for $M_1 = 2.09$ and $M_{2is} = 2.35$.

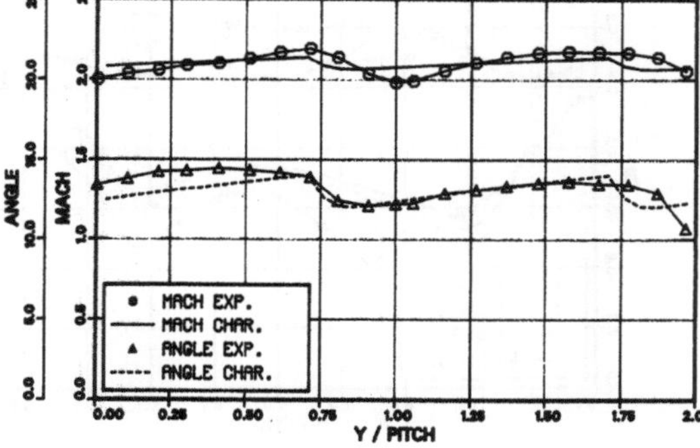

FIGURE 17.11. Upstream traverse results.

presence of alternate shocks, denoted by the Mach number decrease, and expansion fans where the flow angle rises from about 12 degrees up to 14 degrees and the Mach number undergoes a slight increase.

The numerical results of the method of characteristics are in good agreement with the experimental results both for the inlet angle and for the Mach number. The Mach number distributions from the Navier-Stokes (Fig. 17.14) and the characteristics (Fig. 17.15) codes show that the leading edge shock and the expansion fan present patterns quite similar to those shown by the Schlieren picture; however, Navier-Stokes results suffer from a loss of definition due to mesh coarsening.

Internal Flow

The pressure distributions are represented in Fig. 17.12 by means of the pressure coefficient; due to the high Mach number level it is thought that the comparison between experimental and numerical results is more significant if carried out in terms of pressure rather than in terms of isentropic Mach number. The experimental results show that the blade pressure coefficient distribution is consistent with the theoretical one corresponding to a free vortex design. However, significant differences with respect to the theoretical flow solution deserve to be discussed. In the first part of the suction side the pressure is almost constant. As the profile curvature takes place there is a significant expansion of the flow up to $x/c = 0.22$; here the downstream leg of the leading edge shock wave interacts with the suction-side bondary layer. Downstream of this station a very complex flow configuration takes place. Indeed the shock is not cancelled by the profile curvature and therefore a reflected shock takes place on the suction side; moreover, the compression waves generated by the pressure side curvature behind the leading edge tend to focus just in this region, producing an additional adverse pressure gradient. As a consequence the boundary layer separates at about $x/c = 0.31$. The

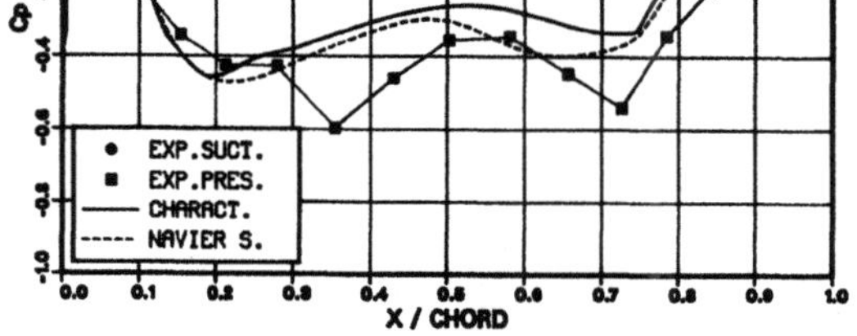

FIGURE 17.12. Blade pressure coefficient.

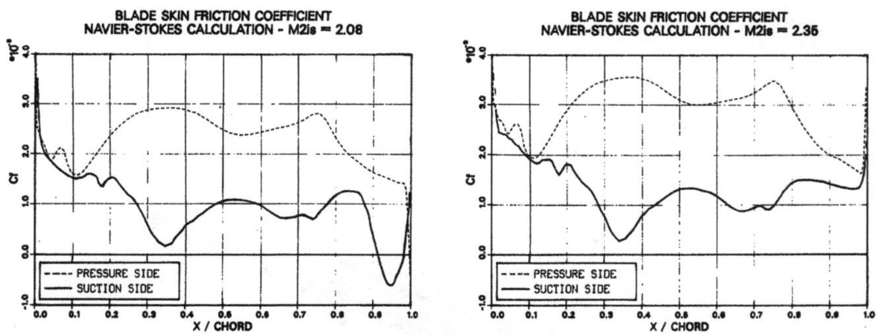

(a) $M_1 = 2.09$ and $M_{2is} = 2.08$ (b) $M_1 = 2.09$ and $M_{2is} = 2.35$

FIGURE 17.13. Blade skin friction coefficient from viscous calculation.

separation persists for a large part of the suction side producing an important blockage effect on the main flow, due to the large boundary layer displacement thickness. This flow behavior is evinced by the Schlieren visualization.

Beyond $x/c = 0.35$ the flow near the pressure side undergoes a continuous expansion (probably enforced by the reduction of the separated region, i.e., by a minor displacement thickness) that stops due to a weak shock occuring at $x/c = 0.56$ (see Fig. 17.10). Then there is a compression region that is probably related to an extension of the separation occurring on the suction side beyond $x/c = 0.5$.

The blade pressure coefficients predicted by the two numerical codes are in good agreement with the experiments only up to the separation, then higher values are found both on the pressure and the suction sides. This should be due to an incorrect estimation of the blockage effect caused by the separation. In supersonic flow the displacement thickness increase produces an appreciable pressurization of the whole flow, that is consistent with the lower values of the experimental blade pressure coefficient. This feature is not predicted, of course, by the method of characteristics. The viscous calculation underestimates the effects of the shock boundary layer interaction; in fact the skin friction coefficient (Fig. 17.13) approaches zero at about $x/c = 0.35$, but actually the flow does not separate. This could be due both to an insufficient mesh refinement and to the intrinsic limitations of the adopted turbulence model. The failure of the turbulence model is probably amplified by the small width of the channel. In wide channels or in external flows an inaccuracy in the estimation of the extent of the separation or of its development does not induce so significant an error in the main flow.

Outlet Flow

The flow configuration in the exit region is typical for blades with thick trailing edges. The Schlieren picture taken at $M_{2is} = 2.35$ shows an intense expansion

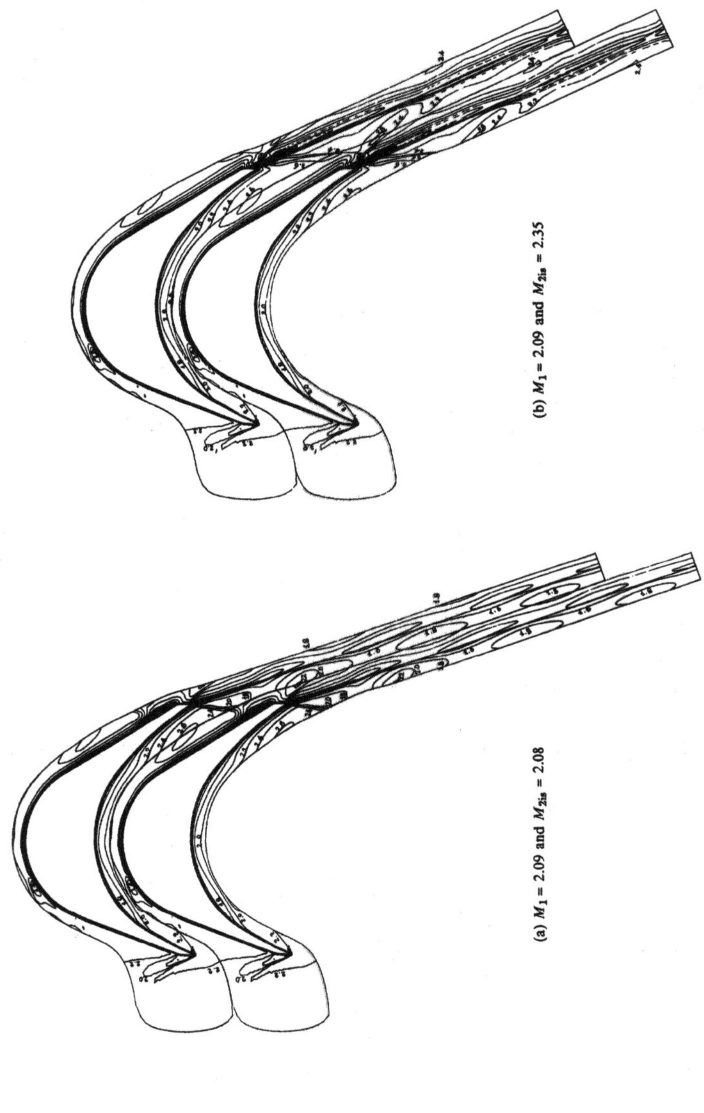

(a) $M_1 = 2.09$ and $M_{2is} = 2.08$

(b) $M_1 = 2.09$ and $M_{2is} = 2.35$

FIGURE 17.14. Mach number distribution from viscous calculation.

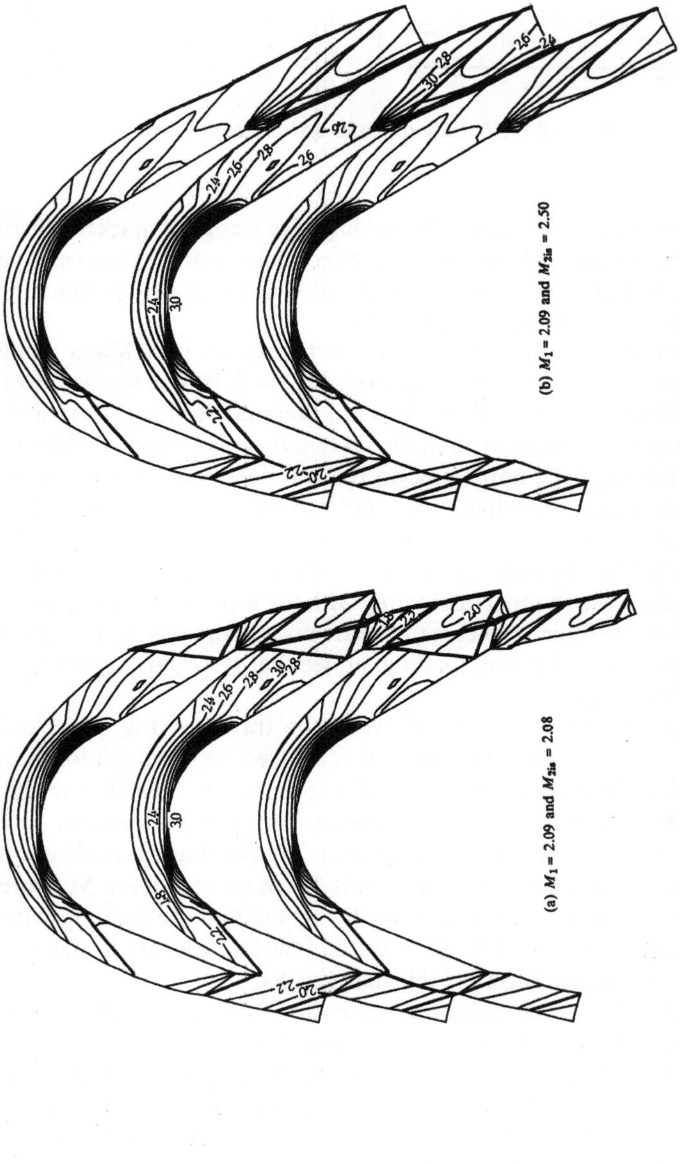

FIGURE 17.15. Mach number distribution from inviscid calculation.

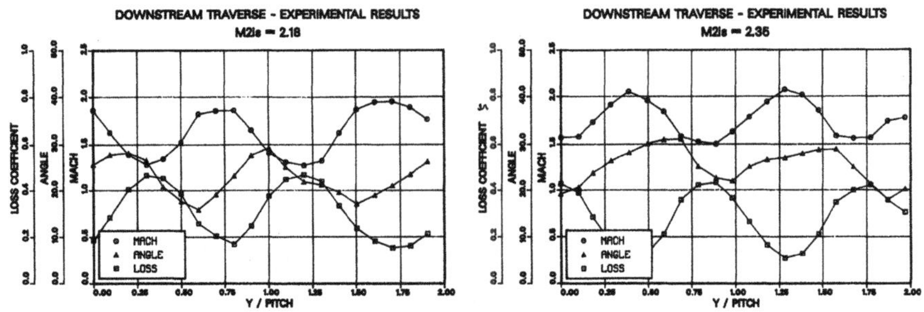

FIGURE 17.16. Experimental downstream traverses.

fan produced by the suction-side trailing edge wedge, a marked left running oblique shock starting from the point of confluence of pressure- and suction-side flow, and the wake; for this expansion ratio the right running shock towards the suction side is quite weak.

The computations were carried out at the design inlet Mach number for different expansion ratios, ranging from $M_{2is} = 2.08$ to 2.50; the flow field in terms of Mach number distributions are presented in Figs. 17.14 and 17.15. For expansion ratios lower than the design, there is a marked oblique shock from the trailing edge impinging on the last part of the suction side; in the viscous calculation it produces a separation of the thick suction-side boundary layer, which persists up to the trailing edge, as can be seen also from the blade skin friction distribution (Fig. 17.13). For the design expansion ratio this shock goes outside of the channel interacting with the wake, while the other shock, if compared to the Shlieren picture, appears underestimated. In the characteristic computation at the limit loading, notice that the right running shock has completely disappeared.

The experimental data of the downstream traverses (Fig. 17.16) at design conditions show that most of the flow is affected by very high loss values up to $\zeta = 0.4$, indicating the presence of a quite wide wake that is the direct consequence of the important separation taking place on most of the suction side. The loss level is furthermore increased by the shock wave losses that, as is well known, become important at this Mach number level. Measurements from traverses at different blade heights showed that the secondary flows are confined approximately within one third of the blade height from the endwall and that they do not affect the midspan results.

The Navier-Stokes computation traverses at design conditions (Fig 17.17) show a deeper but much less wide wake that denotes the previously mentioned failure of the turbulence model and of its capability of predicting the separated regions; the flow angle distribution is much more flat than the experimental one and this is probably an effect of the downstream boundary condition which sets a constant pressure along the boundary.

At $M_{2is} = 2.08$ more loss is found because of the greater importance of the

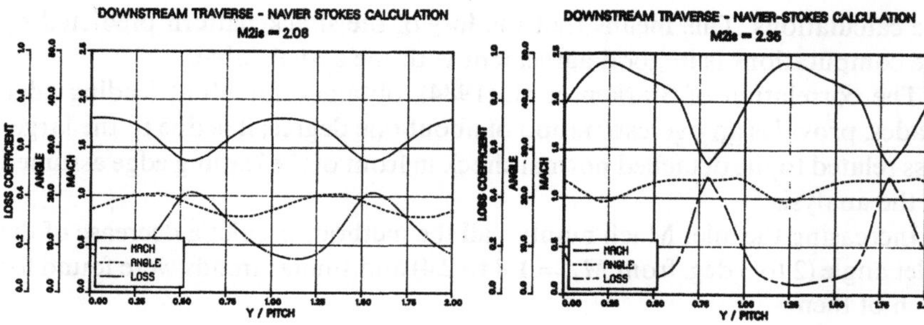

FIGURE 17.17. Viscous computation downstream traverses.

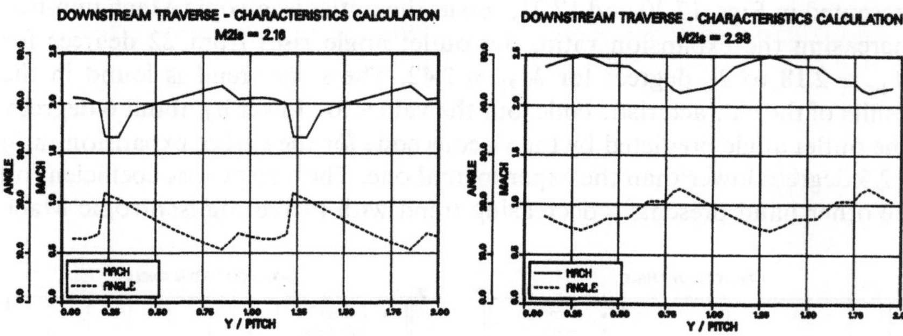

FIGURE 17.18. Inviscid computation downstream traverses.

right-running trailing edge shock and of its interaction with the suction-side boundary layer, which causes a noticeable separation. In this case the computed results compare more favourably with the experiment.

The traverses from the characteristic computations (Fig. 17.18) present higher Mach number and lower flow angle; this is clearly due to the lack of the viscous wake.

Inlet and Outlet Mean Flow Properties

Many computations were carried out for different inlet and outlet Mach numbers and the corresponding mean flow angles were evaluated, in order to verify the ability of the 2-D computations to improve the classical correlations.

Unique Incidence Analysis

In Fig. 17.19 the results of the analysis of the unique incidence carried out for different inlet Mach numbers are presented. It can be seen that the simplest method, based on the isentropic flow assumption, provides an underestimation of the inlet angle of about one degree with respect to the experimental data, while better results (with a 0.4-deg error) are obtained by both the calculations. This means that the loss of the shock pattern predicted by the computations is in good agreement with the actual value.

The correlation of Starken et al. (1984), obtained for blunt leading edge blades, provides an overestimation of about one degree; it is due to the larger loss related to the detached normal shock in front of the leading edge assumed in the analysis.

Increasing the inlet Mach number all the methods predict a decrease of the inlet angle (2 to 3 deg. from $M_1 = 1.8$ to 2.4) and similar trends were found for each of them.

Mixed-Out Flow

The mixed-out flow properties for the experiments and computations are presented in Figs. 17.20 and 17.21, versus the outlet isentropic Mach number. Increasing the expansion ratio, the outlet angle rises from 22 degrees for $M_{2is} = 2.18$ to 26 degrees for $M_{2is} = 2.42$. The same trend is found in the results of the characteristic code, but the values are lower by about 6 degrees. The outlet angle predicted by the viscous code for the design expansion ratio is 2.5 degrees lower than the experimental one. The energy loss coefficient on the other hand presents a decreasing trend with increasing isentropic Mach

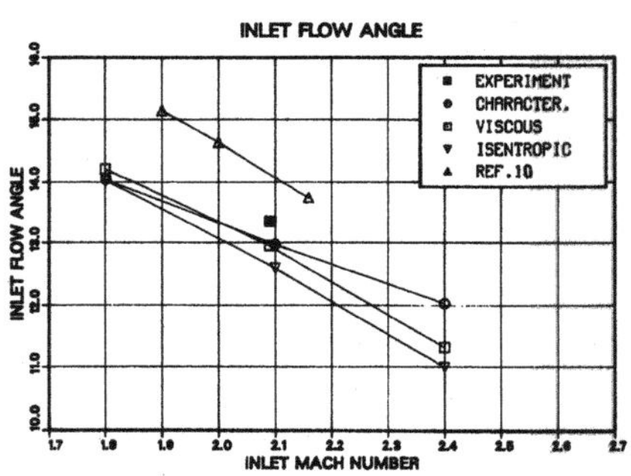

FIGURE 17.19. Inlet angles vs. inlet Mach number.

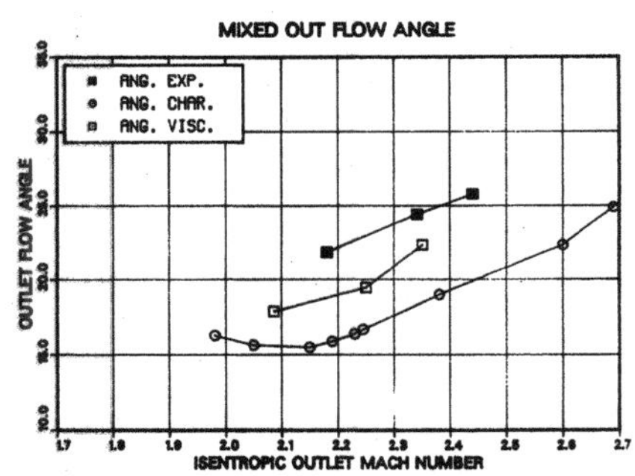

FIGURE 17.20. Outlet flow angle.

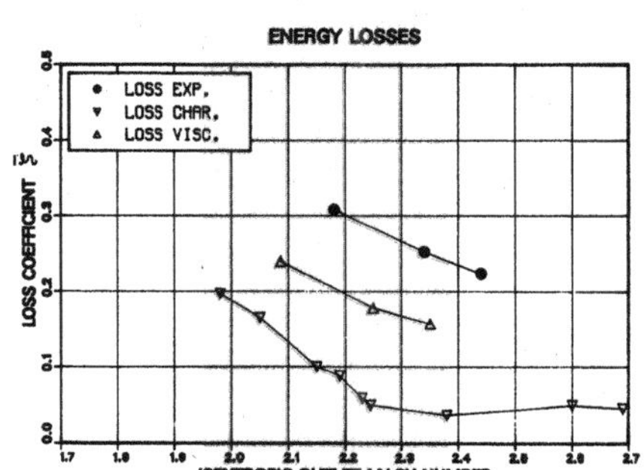

FIGURE 17.21. Losses.

number; this loss reduction is partly due to the minor strength of the right running trailing edge shock and partly to the reheat effect related to larger enthalpy drop. The loss level predicted by the inviscid code, which accounts only for shock losses, is of course much lower, however, it has to be pointed out that the trend is well-enough captured. For isentropic Mach numbers larger than 2.35 the loss coefficient remains constant. The viscous computation, which should account for both the loss sources (i.e., shock losses and viscous dissipations), provides actually an underestimation by about 30 percent of the loss coefficient. The turbulence model is the principal responsible for the unsatisfactory code performance; in fact it is plausible that the well-known limitations of the model (i.e., the unphysical breakdown of turbulence at separation and the absence of history effects) are crucial in this flow characterized by a large blockage effect of the boundary layer.

Conclusions

The present investigation has shown the details of the flow field and the performance of a typical rotor cascade for supersonic turbines designed through the free vortex method. The loss level is found to be quite high because of the simultaneous presence of important shock wave losses and of large viscous losses due to the boundary layer development in long and narrow channels, with the occurrence of adverse shock-induced pressure gradients. In a new design carried out on the basis of the information provided by this investigation, some margin of improvement of the efficiency may be expected if the entry transition region is redesigned in the attempt of avoiding the suction-side boundary layer separation; this could be achieved both reducing the strength of the shock wave that enters the channel (i.e., making the pressure-side angle equal to that of the flow just ahead of the leading edge) and designing the inlet transition region taking into account accurately the boundary layer development. Anyway the loss level of the present cascade may be considered acceptable for space applications where the main goal is to obtain high specific work with the minimum mass.

The numerical analysis has shown that many features of the flow configuration like the unique incidence angle and the shock wave pattern have been properly captured by the calculations. Nevertheless significant differences have been found in the blade pressure distribution, in the outlet angle, and in the loss level. For the Navier-Stokes code the turbulence model, which is proved to provide good performance both in external and usual turbomachinery flows, is supposed to be responsible for this inaccuracy. This shows that for supersonic turbomachinery flows there is a need of experimental data to be used for validation of numerical codes and assessing turbulence models

The unique incidence angle was found fairly well predicted also by the isentropic solution; this is due to the weakness of the shock configuration in the entry region produced by the sharp leading edge of the blade

Acknowledgments. The experimental investigation was carried out as part of a research program subsidized by Fiat Avio s.p.a., Torino, Italy. This support and the authorization to publish the results are kindly acknowledged.

Nomenclature

c = blade chord.
c_p = constant pressure-specific heat.
Cf = skin friction coefficient.
Cp = pressure coefficient.
E = total internal energy.
F, G = x and y flux vectors.
h = enthalpy; also blade height.
H = total enthalpy.
H_R = rothalpy.
k = molecular conductivity.
m = meridional coordinate.
M = Mach number.
n = unit vector.
p = pressure.
Pr = Prandtl number.
q = heat flux.
r = radius.
s = spacing: also entropy.
t = time.
T = temperature.
u, v = x and y velocity components.
U = peripheral velocity.
V = velocity; also volume.
W = vector of conserved variables; also relative velocity.
x, y = axial and tangential coordinates.
X, Y = transformed intrinsic coordinates.

Greek Symbols

α, β = transformation functions; also angles between absolute or relative velocity and tangential direction.
γ = specific heats ratio.
ζ = local energy loss coefficient defined as $\zeta = \dfrac{v_{2is}^2(y) - v_2^2(y)}{\bar{v}_{2is}}$.
θ = angle between velocity and meridional coordinate.
λ = second coefficient of molecular viscosity; also Mach angle.
μ = molecular viscosity.
ν = Prandtl-Mayer function.
ρ = density.
σ = normal stress.

τ = shear stress.
ϕ = angular coordinate.

Subscripts

is = isentropic.
r = relative.
s = sonic.
x, y = x and y component.
1 = upstream.
2 = downstream.
∞ = infinity.
$\bar{}$ = mixed out.

References

Baldwin, B., and Lomax, H., 1978, "Thin Layer Approximation and Algebraic Model for Separated Turbulent Flows," *AIAA paper 78-257.*

Bassi, F., 1978, "Calcolo non Isentropico di Flussi Supersonici per to Studio di Schiere di Pale di Turbomacchine," *Congresso Nazionale ATI*, Ancona.

Bassi, F., Grasso, F., and Savini, M., 1988, "Numerical Solution of Compressible Navier-Stokes Flows," *AGARD CP-437, 1.*

Boxer, E., Sterret, J.R., and Woldarskie, J., 1952, "Application of Supersonic Vortex Flow Theory to the Design of Supersonic Impulse Compressor or Turbine Blade Sections," *NACA RM L52B06.*

Chauvin, J., Sieverding, C., and Griepentrog, H., 1970, "Flow in Cascades with a Transonic Regime," *Proc. of the Symposium on Flow Research on Blading*, Elsevier, Amsterdam.

Colantuoni, S., Terlizzi, A., and Grasso, F., 1989, "A Validation of a Navier-Stokes 2D Solver for Transonic Turbine Cascade Flows," *AIAA paper 89-2451.*

Colclough, C.D., 1966a, "Design of Turbine Blades Suitable for Supersonic Relative Inlet Velocities and the Investigation of their Performance in Cascades: Part I—Theory and Design," *J. of Mechanical Science, 8*, 1.

Colclough, C.D., 1966b, "Design of Turbine Blades Suitable for Supersonic Relative Inlet Velocities and the Investigation of their Performance in Cascades: Part II—Experiments, Results and Discussion," *J. of Mechanical Science, 8*, 2.

Deich, M.E., et al., 1964, "Investigation and Calculation of Axial Turbine Stages," Wright Patterson Air Force Base, Ohio.

Eriksson, L., 1984, "Development of a Supersonic Turbine Stage for the HM60 Engine," *AIAA-84-1464.*

Goldman, L., and Sculling, J., 1968, "Analytical Investigation of Supersonic Turbomachinery Blading," *NASA TN-D-4421.*

Holst, T.L., 1987, "Viscous Transonic Airfoil Workshop Compendium of Results," *AIAA paper 87-1460.*

Jameson, A., 1983, "Transonic Flow Calculations," *MAE Report no. 1651*, Princeton University.

Jameson, A., Schmidt, W., and Turkel, E., 1981, "Numerical Solutions of the Euler Equations by Finite Volume Methods Using Runge-Kutta Time-Stepping Schemes," *AIAA paper n. 81-1259.*

Kurzrock, J.W., 1989, "Experimental Investigation of Supersonic Turbine Performance," *ASME paper 89-GT-238.*

Liccini, L.L., 1949, "Analytical and Experimental Investigation of 90° Supersonic Turning Passages, Suitables for Supersonic Compressors or Turbines," *NACA RM L9G07.*

Meauzé, G., and Fourmaux, A., 1987, "Numerical Simulation of Flows in Axial and Radial Turbomachines Using Euler Solvers," *Small High Pressure Ratio Turbines*, VKI LS 07.

NASA, 1974, "Liquid Rocket Engine Turbines," *NASA SP-81110.*

Ohlsson, G.O., 1964, "Supersonic Turbines," *J. of Engineering for Power*, 86, 1, Jan

Osnaghi, C., 1971, "Progetto di Palette ad Azione per Turbine Supersoniche," *La Termotecnica*, 4.

Starken, H., Yongxing, Z., and Schreiber, H.A., 1984, "Mass Flow Limitation of Supersonic Blade Rows due to Leading Edge Blockage," *ASME paper 84-GT-233.*

Tomé, C. 1972, "Une Application de la Methöde des Caracteristiques en Coordonnées Intrinsiques," *La Recherche Aerospatiale*, 6.

Vanco, M.R., and Goldman, L.J., 1968, "Computer Program for Design of Two-dimensional Supersonic Nozzle with Sharp-Edged Throat," *NASA TM X-1502.*

Verdonk, G., and Dufurnet, T., 1987, Development of a Supersonic Steam Turbine with a Single Stage Pressure Ratio of 200 for Generator and Mechanical Drive," *Small High Pressure Ratio Turbines*, VKI LS 07.

Verneau, A., 1987, "Supersonic Turbines for Organic Fluid Rankine Cycles from 3 to 1300 KW," *Small High Pressure Ratio Turbines.* VKI LS 07.

Visbal, M. and Knight, D, 1984, "The Baldwin-Lomax Turbaulence Model for Two-Dimensional Shock-Wave/Boundary-Layers Interactions," *AIAA J. 22*, 7.

18
Perspectives on Wind Shear Flight

A. MIELE, T. WANG AND G.D. WU

ABSTRACT: Wind shears originating from downbursts have been the cause of many aircraft accidents in the past two decades. In turn, this has led to considerable research on wind shear avoidance systems and wind shear recovery systems. This paper reviews recent advances in wind shear recovery systems. It summarizes the work done at Rice University on trajectory optimization and trajectory guidance for two basic flight conditions: takeoff and abort landing. Future research directions are discussed with particular reference to detection and recovery. It appears that, in the relatively near future, an advanced wind shear control system can be developed, that is, capable of functioning in different wind models and covering the spectrum of flight conditions having interest in a wind shear encounter.

Introduction

Low-altitude wind shear is a threat to the safety of aircraft in takeoff and landing. Over the past 20 years, some 30 aircraft accidents have been attributed to wind shear (National Academy Press 1983). The most notorious ones are the crash of Eastern Airlines Flight 066 at JFK International Airport (1975), the crash of Pan Am Flight 759 at New Orleans International Airport (1982), and the crash of Delta Airlines Flight 191 at Dallas–Fort Worth International Airport (1985). These crashes involved the loss of some 400 people and an insurance settlement in excess of $500 million (Fujita 1985, 1986; Gorney 1987; National Transportation Safety Board 1983, 1986; Wingrove and Bach 1987).

To offset the wind shear threat, there are two basic systems: wind shear avoidance systems and wind shear recovery systems. A wind shear avoidance system is designed to alert the pilot to the fact that a wind shear encounter might take place; here, the intent is avoidance of a microburst. A wind shear recovery system is designed to guide the pilot in the course of a wind shear encounter; here, the intent is to fly smartly through a microburst, if an inadvertent encounter takes place. Obviously, wind shear avoidance systems and wind shear recovery systems are not mutually exclusive, but complementary to one other.

Wind Shear Avoidance Systems

Wind shear avoidance systems include ground-based mechanical systems, ground-based radar systems, and airborne systems (Bowles and Targ 1988; Wilson et al. 1984).

Ground-based mechanical systems involve the installation of anemometers at different points of an airport. These anemometers provide the wind direction and intensity. If a wind velocity divergence is detected, and if the divergence exceeds a certain threshold value, then an alert takes place. Depending on the situation, the director of an airport can choose among several options: he can alert the pilots of planes flying nearby; he can shut off a particular runway; and he can shut off the entire airport. An example of such a system is the low-level wind shear alert system (LLWAS); this system is simple, has low cost, but does not have great accuracy.

Ground-based radar systems involve the installation of Doppler radars on the ground, capable of measuring the wind distribution around the airport area. The difficulty of this technique is the presence of ground clutter, which can have a significant effect on the signal-to-noise ratio. Compared with anemometer systems, ground-based radar systems are more accurate, but their cost is high.

Airborne systems involve the installation of either a radar or a lidar on an airplane, capable of seeing ahead in the direction along which the airplane is flying. Compared with ground-based radar systems, airborne systems have a lower signal-to-noise ratio because of limitations to transmitter power and antenna or telescope size. Airborne systems have less accuracy than ground-based radar systems.

Wind Shear Recovery Systems

Wind shear recovery systems include various guidance schemes, such as maximum angle of attack guidance, constant pitch guidance, and variable pitch guidance or advanced guidance (Boeing Airliners 1985; Bray 1986; Bowles and Targ 1988; Chu and Bryson 1987; Federal Aviation Administration 1987; Frost 1983; Frost and Bowles 1984; Miele 1988; Miele et al. 1986a, 1986, 1987, 1988, 1989; Psiaki and Stengel 1986).

In the maximum-angle-of-attack guidance, the aircraft is rotated at the maximum permissible rate (3 deg/sec) until the stick-shaker angle of attack is reached; afterward, the aircraft is held at the stick-shaker angle of attack. This system, which has been in use for some 20 years, does not require any new instrumentation. However, as shown by computer simulations, its performance is poor in both takeoff and abort landing.

In the constant-pitch guidance (Boeing Airliners 1985; Bray 1986; Federal

Aviation Administration 1987), the aircraft is rotated at the maximum permissible rate (3 deg/sec) until a certain target pitch is obtained (15 deg of fuselage pitch); afterward, the aircraft is held at the angle of attack corresponding to the target pitch, subject to the stick-shaker limitation. This system, which was adopted in the FAA Windshear Training Aid (1987), does not require any new instrumentation. Its performance is better than that of the maximum-angle-of-attack guidance, but worse than that of the advanced guidance schemes discussed later.

Advanced guidance schemes are based on the properties of the optimal trajectories and are constructed in such a way that they approximate the behavior of the optimal trajectories while using only local information on the wind shear and the downdraft (Chu and Bryson 1987; Miele 1988; Miele et al. 1986, 1987, 1988, 1989; Psiaki and Stengel 1986). There are two possible operational modes: the totally automated mode and the semiautomated or display mode; the latter is preferred for flight operations.

In the semiautomated mode, an advanced guidance scheme includes three basic parts: sensors, software, and display. The sensors supply the local measurements of the wind shear, downdraft, and certain components of the state of the aircraft. The software consists of a feedback-control algorithm, which processes the information supplied by the sensors and arrives at the guidance values for the pitch angle or angle of attack. These values lead to the satisfaction of a guidance law, obtained from the analysis of the optimal trajectories (Miele 1988; Miele et al. 1986, 1987, 1988). The display shows either the actual pitch and guidance pitch or the actual angle of attack and guidance angle of attack. Then, in the semiautomated mode, the pilot must act on the controls in such a way that the actual pitch approaches the guidance pitch or the actual angle of attack approaches the suidance angle of attack.

Examples of advanced guidance schemes are the acceleration guidance and gamma guidance, developed by the Aero-Astronautics Group of Rice University (Miele et al. 1986, 1987, 1988, 1989).

An advanced guidance scheme requires the installation of instrumentation measuring the wind acceleration and downdraft. These quantities can be computed indirectly from measurements already available in aircraft equipped with inertial instrumentation, such as the Boeing B-747 and the Lockheed L-1011. On the other hand, for older-type aircraft (for instance, the Boeing B-727), the installation of additional sensors is necessary. The implementation of this system is not too costly.

Comment

Wind shear avoidance systems and wind shear recovery systems must be viewed as complementary to one another, in the following sense: the former are essential to the practice of avoidance of microbursts; at the same time, if avoidance is not possible, the latter must take over.

Among the wind shear avoidance systems, ground-based radar systems and airborne radar or lidar systems appear to be the best. Among the wind shear recovery systems, constant-pitch guidance and advanced guidance (variable-pitch guidance) appear to be the best. In the writers' opinion, further research is necessary before making large commitments of funds to any particular system.

Rice University Research on Wind Shear

This research was started in 1984 at the suggestion of Captain W.W. Melvin of Delta Airlines and ALPA (Air Line Pilots Association). The motivation for the research was that, while there had been previous efforts on the meteorological, aerodynamic, instrumentation, and stability aspects of the wind shear problems, relatively little had been done on the flight mechanics aspects. It was felt that a fundamental study was needed in order to better understand the dynamic behavior of an aircraft in a wind shear; that the determination of good strategies for coping with wind shear situations was essentially an optimal control problem; that the methods of optimal control theory were needed; and that, only after having found optimal control solutions, one could properly address the guidance problem.

An overview of the research performed during the years 1984 to 1989 is presented in the following pages with reference to various topics of optimization and guidance arising in connection with two basic flight conditions: takeoff and abort landing (Miele 1988; Miele et al. 1986, 1987, 1988, 1989). Then, some current and future studies are reviewed. Finally, the conclusions are given.

System Description

In this paper, we make use of the relative wind-axes system in connection with the following assumptions: (1) the aircraft is a particle of constant mass; (2) flight takes place in a vertical plane; (3) Newton's law is valid in an Earth-fixed system; and (4) the wind flow field is steady.

With these premises, the equations of motion include the kinematic equations:

$$\dot{x} = V\cos\gamma + W_x, \tag{1a}$$

$$\dot{h} = V\sin\gamma + W_h, \tag{1b}$$

and the dynamic equations

$$\dot{V} = (T/m)\cos(\alpha + \delta) - D/m - g\sin\gamma - (\dot{W}_x\cos\gamma + \dot{W}_h\sin\gamma), \tag{2a}$$

$$\dot{\gamma} = (T/mV)\sin(\alpha + \delta) + L/mV - (g/V)\cos\gamma + (1/V)(\dot{W}_x\sin\gamma - \dot{W}_h\cos\gamma). \tag{2b}$$

Because of assumption (4), the total derivatives of the wind-velocity components and the corresponding partial derivatives satisfy the relations

$$\dot{W}_x = (\partial W_x/\partial x)(V\cos\gamma + W_x) + (\partial W_x/\partial h)(V\sin\gamma + W_h), \tag{3a}$$

$$\dot{W}_h = (\partial W_h/\partial x)(V\cos\gamma + W_x) + (\partial W_h/\partial h)(V\sin\gamma + W_h). \tag{3b}$$

These equations must be supplemented by the functional relations

$$T = T(h, V, \beta), \tag{4a}$$

$$D = D(h, V, \alpha), \qquad L = L(h, V, \alpha), \tag{4b}$$

$$W_x = W_x(x, h), \qquad W_h = W_h(x, h), \tag{4c}$$

and by the analytical relations

$$\theta = \alpha + \gamma, \tag{5a}$$

$$\gamma_e = \arctan[(V\sin\gamma + W_h)/(V\cos\gamma + W_x)]. \tag{5b}$$

The differential system (Eqs. 1–4) involves four state variables $[x(t), h(t), V(t), \gamma(t)]$ and two control variables $[\alpha(t), \beta(t)]$. However, the number of control variables reduces to one (the angle of attack), if the power setting is specified in advance. The quantities θ, γ_e can be computed a posteriori, once the values of the state and control are known.

Inequality Constraints

The angle of attack α and its time derivative $\dot{\alpha}$ are subject to the inequalities

$$\alpha \le \alpha_*, \tag{6a}$$

$$-\dot{\alpha}_* \le \dot{\alpha} \le \dot{\alpha}_*, \tag{6b}$$

where α_* is a prescribed upper bound and $\dot{\alpha}_*$ is a prescribed positive constant.

The power setting β and its time derivative $\dot{\beta}$ are subject to the inequalities

$$\beta_* \le \beta \le 1, \tag{7a}$$

$$-\dot{\beta}_* \le \dot{\beta} \le \dot{\beta}_*, \tag{7b}$$

where β_* is a prescribed lower bound and $\dot{\beta}_*$ is a prescribed positive constant.

Wind Model

The wind model employed in this chapter involves the combination of shear (transition from head wind to tail wind) and downdraft (Ivan 1986; Zhu and Etkin 1985). Analytically, it is represented by the relations

$$W_x = \lambda A(x), \tag{8a}$$

$$W_h = \lambda(h/h_*)B(x), \tag{8b}$$

with

$$\Delta W_x = \lambda \Delta W_{x*}, \tag{8c}$$

$$\Delta W_h = \lambda(\Delta W_{x*}/2)h/h_*. \tag{8d}$$

Here, the parameter λ characterizes the intensity of the wind shear/downdraft combination; the function $A(x)$ represents the profile of the horizontal wind versus the horizontal distance; and the function $B(x)$ represents the profile of the vertical wind versus the horizontal distance. Also, ΔW_x is the horizontal wind velocity difference (maximum tail wind minus maximum head wind); ΔW_h is the vertical wind velocity difference (maximum updraft minus maximum downdraft); $\Delta W_{x*} = 100$ fps is a reference value for the horizontal wind velocity difference; and $h_* = 1000$ feet is a reference value for the altitude.

Decreasing values of λ (hence, decreasing values of ΔW_x) correspond to milder wind shears; conversely, increasing values of λ (hence, increasing values of ΔW_x) correspond to more severe wind shears. Therefore, by changing the value of λ, one can generate shear/downdraft combinations ranging from extremely mild to extremely severe.

To sum up, the wind shear model (Eqs. 8) has the following properties: (1) it represents the transition from a uniform head wind to a uniform tail wind, with nearly constant shear in the core of the downburst; (2) the downdraft achieves maximum negative value at the center of the downburst; (3) the

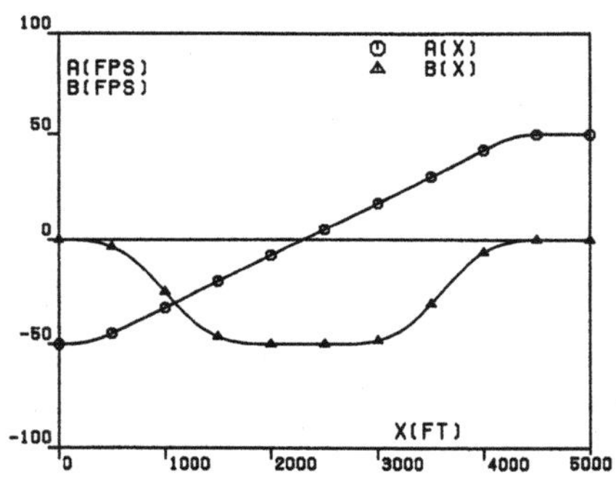

FIGURE 18.1. The functions $A(x)$ and $B(x)$.

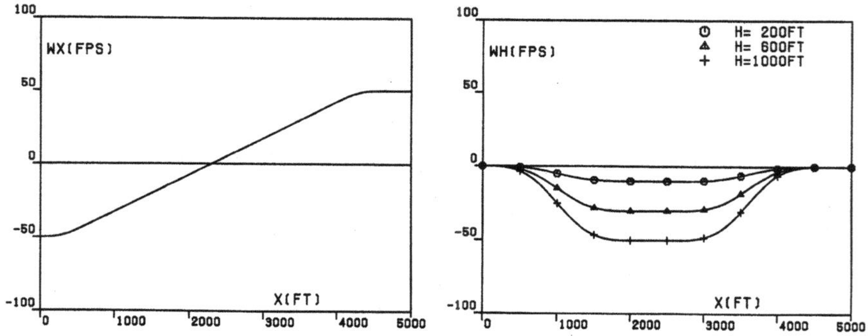

(a) (b)

FIGURE 18.2. (a) The function $W_x(x)$ for $\Delta W_x = 100$ fps; (b) the function $W_h(x, h)$ for $\Delta W_x = 100$ fps.

downdraft vanishes at $h = 0$; and (4) the wind velocity components nearly satisfy the continuity equation and the irrotationality condition in the core of the downburst.

Figure 18.1 shows the basic functions $A(x)$, $B(x)$; and Fig. 18.2 shows the resulting functions $W_x(x)$ and $W_h(x, h)$ for a particular value of the horizontal wind velocity difference, namely, $\Delta W_x = 100$ fps.

Takeoff

Here, we discuss three aspects of the takeoff problem: optimal trajectories, guidance trajectories, and survival capability. We assume that maximum power setting is employed; hence, in Eq. 4a, we set

$$\beta = 1, \qquad 0 \leq t \leq \tau, \tag{9}$$

where τ is the final time.

Optimal Trajectories

Optimal takeoff trajectories can be determined by minimizing the peak value of the deviation of the absolute path inclination from a reference value. The resulting performance index is given by

$$I = \max_t |\gamma_e - \gamma_{eR}|, \qquad 0 \leq t \leq \tau, \tag{10a}$$

$$\gamma_e = \arctan[(V \sin\gamma + W_h)/(V \cos\gamma + W_x)], \tag{10b}$$

$$\gamma_{eR} = \gamma_{e0}, \tag{10c}$$

and is to be minimized subject to the constraints in Eqs. (1)–(9). This is a minimax or Chebyshev problem of optimal control, which can be converted into a Bolza problem via suitable transformations, omitted here for the sake of brevity. After the transformation, the Bolza problem can be solved numerically by means of the sequential gradient-restoration algorithm for optimal control problems (Miele 1975, 1980; Miele and Wang 1986; Miele et al. 1974, 1970).

The most unfavorable takeoff conditions occur if the wind shear encounter takes place at maximum or near-maximum takeoff weight in a hot summer day. For these conditions, several hundred optimal trajectories were computed for three Boeing aircraft (B-727, B-737, and B-747) and different wind shear intensities. From the computer runs, certain general conclusions became apparent:

1. The optimal trajectories achieve minimum velocity at the end of the shear;
2. The optimal trajectories require an initial decrease in the angle of attack, followed by a gradual increase; the maximum permissible angle of attack, the stick-shaker angle of attack, is achieved near the end of the shear.

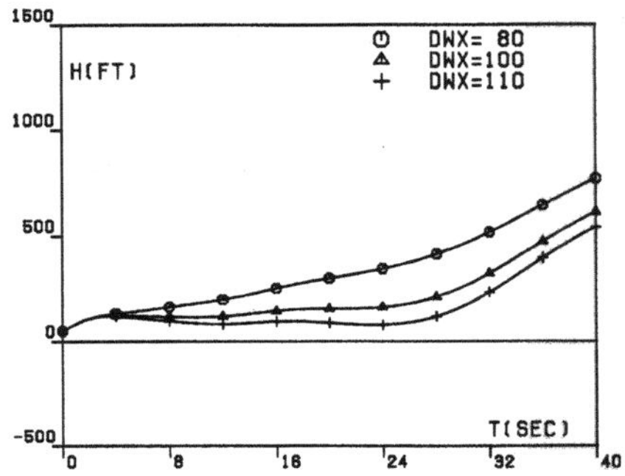

FIGURE 18.3. Optimal takeoff trajectories: altitude h versus time t for several values of ΔW_x.

3. For weak-to-moderate wind shears, the optimal trajectories are characterized by a continuous climb; the average value of the path inclination decreases as the intensity of the shear increases.
4. For relatively severe wind shears, the optimal trajectories are characterized by an initial climb, followed by nearly horizontal flight, followed by renewed climbing after the aircraft has passed through the shear region.
5. Weak-to-moderate wind shears and relatively severe wind shears are survivable employing an optimized flight strategy; however, extremely severe wind shears are not survivable, even employing an optimized flight strategy.

For a particular case, that of the Boeing B-727-200 aircraft (takeoff weight $W = 180{,}000$ lb, flap deflection $\delta_F = 15$ deg, ambient temperature 100°F, initial altitude $h_0 = 50$ ft), Fig. 18.3 shows the altitude profile $h(t)$ of the optimal trajectory for several values of the horizontal wind velocity difference, namely, $\Delta W_x = 80$, 100, 110 fps. Note that, as ΔW_x increases, the minimum altitude of the optimal trajectory decreases, but the aircraft survives the wind shear encounter. However, should the horizontal wind velocity difference be further increased to $\Delta W_x = 120$ fps, the B-727 would crash, even flying an optimal trajectory.

Guidance Trajectories

The computation of the optimal trajectories requires global information on the wind flow field; that is, it requires the knowledge of the wind components at every point of the region of space in which the aircraft is flying. In practice, global information is not available; even if it were available, there would not be enough computing capability onboard or enough time to process it adequately. As a consequence, the optimal trajectories are merely benchmark trajectories that it is desirable to approach in actual flight.

Since global information is not available, one is forced to employ local information on the wind flow field, in particular, on the wind acceleration and the downdraft. Therefore, the guidance problem must be addressed in these terms: assuming that local information is available on the wind acceleration, the downdraft, and certain components of the state of the aircraft, we wish to guide an aircraft automatically or semiautomatically in such a way that the key properties of the optimal trajectories are preserved.

Based on the idea of preserving the properties of the optimal trajectories, two guidance schemes were developed at Rice University: acceleration guidance (AG), based on the relative acceleration; and gamma guidance (GG), based on the absolute path inclination.

Acceleration Guidance

Basic to the acceleration guidance are the laws

$$\text{(R1)} \qquad \dot{V}/g + C_1 F = 0, \tag{11a}$$

$$\text{(R2)} \qquad V - C_2 V_0 = 0, \tag{11b}$$

which are derived from the study of the optimal trajectories. In Eqs. 11a and b, (R1) denotes the shear region and (R2) the aftershear region; C_1 and C_2 are constants; and F is the shear/downdraft factor (Miele et al. 1987):

$$F = \dot{W}_x/g - W_h/V, \tag{12}$$

which combines the effect of the shear and the downdraft into a single scalar quantity.

The guidance laws (Eqs. 11) are implemented via the feedback control forms

$$\text{(R1)} \qquad \alpha - \tilde{\alpha}(V) = K_1(\dot{V}/g + C_1 F), \tag{13a}$$

$$\text{(R2)} \qquad \alpha - \tilde{\alpha}(V) = K_2(V - C_2 V_0). \tag{13b}$$

In Eqs. 13a and b, K_1 and K_2 are the gain coefficients; α is the guidance angle of attack; and $\tilde{\alpha}(V)$ is the nominal angle of attack (this is the angle of attack associated with static equilibrium in the direction normal to the flight path in the absence of shear and downdraft). For a recent textbook on feedback control, see Franklin et al. (1986).

Gamma Guidance

Basic to the gamma guidance are the laws

$$\text{(R1)} \qquad \gamma_e - \gamma_{e0}(1 - C_1 F) = 0, \tag{14a}$$

$$\text{(R2)} \qquad \gamma_e - C_2 \gamma_{e0} = 0, \tag{14b}$$

which are derived from the study of the optimal trajectories. In Eqs. 14a and b, C_1 and C_2 are constants and F is the shear/downdraft factor (Eq. 12).

The guidance laws (Eqs. 14) are implemented via the feedback control forms

$$\text{(R1)} \qquad \alpha - \tilde{\alpha}(V) = -K_1[\gamma_e - \gamma_{e0}(1 - C_1 F)], \tag{15a}$$

$$\text{(R2)} \qquad \alpha - \tilde{\alpha}(V) = -K_2(\gamma_e - C_2 \gamma_{e0}). \tag{15b}$$

In Eqs. 14a and b, K_1 and K_2 are the gain coefficients; α is the guidance angle of attack; and $\tilde{\alpha}(V)$ is the nominal angle of attack.

Switch Condition

In both the acceleration guidance and the gamma guidance, there is a switch condition, based on the velocity change, which regulates the transition from region (R1) to region (R2).

Trajectory Comparison

For a particular case, that of the Boeing B-727-200 aircraft (takeoff weight $W = 180{,}000$ lb, flap deflection $\delta_F = 15$ deg, ambient temperature 100°F, initial altitude $h_0 = 50$ ft, horizontal wind velocity difference $\Delta W_x = 100$ fps), Fig. 18.4 shows the altitude profile $h(t)$ of the optimal trajectory (OT) as well as the altitude profiles of the trajectories resulting from the acceleration guidance (AG), the constant-pitch guidance (CPG), and the maximum angle of attack guidance (MAAG). While the MAAG trajectory crashes, the CPG trajectory scrapes the ground; on the other hand, the AG trajectory is close to the OT. An analogous remark holds for the GG trajectory, which is not shown, since it is nearly identical with the AG trajectory.

Survival Capability

Here, we analyze the survival capability of an aircraft in a severe wind shear. Indicative of this survival capability is the wind shear/downdraft combination that results in the minimum altitude being equal to the ground altitude.

To analyze this important problem, we recall the one-parameter family of wind shear models (Eqs. 8), in which the parameter λ characterizes the intensity of the wind shear/downdraft combination. By increasing the value of λ, more intense wind shear/downdraft combinations are generated until a critical value λ_c (hence, a critical value ΔW_{xc}) is found, such that $h_{\min} = 0$ for a given trajectory type.

The results are shown in Table 18.1, which supplies the critical wind velocity difference ΔW_{xc} for the optimal trajectory and various guidance trajectories. Table 18.1 also shows the wind shear efficiency ratio WER, defined to be

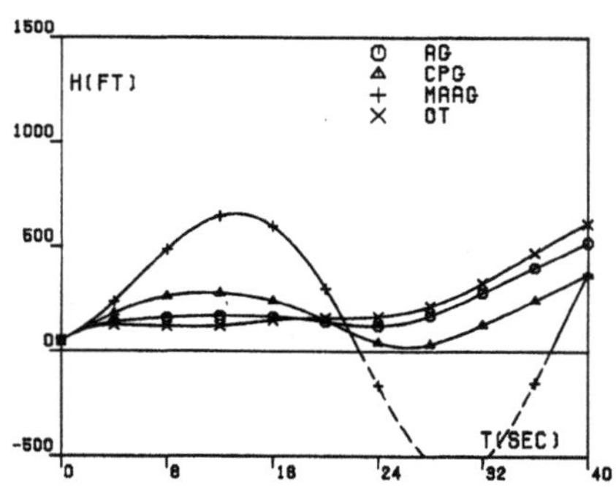

FIGURE 18.4. Comparison of takeoff trajectories: altitude h versus time t for $\Delta W_x = 100$ fps.

TABLE 18.1. Takeoff survival capability ΔW_{xc} and wind shear efficiency ratio WER (B-727).

Trajectory	h_0 (ft)	ΔW_{xc} (fps)	WER
OT	50	119.5	1.000
AG	50	113.0	0.946
GG	50	115.3	0.965
CPG	50	101.8	0.852
MAAG	50	57.7	0.483

Takeoff weight = 180,000 lb; transition distance = 4000 ft.

$$\mathrm{WER} = (\lambda_c)_{\mathrm{PT}}/(\lambda_c)_{\mathrm{OT}} = (\Delta W_{xc})_{\mathrm{PT}}/(\Delta W_{xc})_{\mathrm{OT}}. \tag{16}$$

Here, the subscript PT denotes a particular trajectory and the subscript OT denotes the optimal trajectory. Clearly, if the wind shear efficiency of the OT is defined to be 100 percent, that of the AG trajectory is 95 percent, that of the GG trajectory is 96 percent, that of the CPG trajectory is 85 percent, and that of the MAAG trajectory is 48 percent.

Abort Landing

Here, we discuss three aspects of the abort landing problem: optimal trajectories, guidance trajectories, and survival capability. We assume that, prior to the wind shear onset, the aircraft is in quasi-steady flight along an approach path with absolute path inclination $\gamma_e = -3.0$ deg. We also assume that, at the wind shear onset, the pilot increases the power setting β from the initial value $\beta = \beta_0$ to the maximum value $\beta = 1$ at the constant time rate $\dot{\beta}_0$. Afterward, the maximum value is held. Hence, in Eq. 4a, we set

$$\beta = \beta_0 + \dot{\beta}_0 t, \qquad 0 \le t \le \sigma, \tag{17a}$$

$$\beta = 1, \qquad \sigma \le t \le \tau. \tag{17b}$$

Here, $\sigma = (1 - \beta_0)/\dot{\beta}_0$ is the time at which maximum power setting is achieved and τ is the final time.

Optimal Trajectories

Optimal abort landing trajectories can be determined by minimizing the peak value of the altitude drop. The resulting performance index is given by

$$I = \max_t\, (h_R - h), \qquad 0 \le t \le \tau, \tag{18}$$

and is to be minimized subject to the constraints in Eqs. (1)–(8) and (17). This is a minimax or Chebyshev problem of optimal control, which can be converted into a Bolza problem via suitable transformations, omitted for the sake of brevity. After the transformation, the Bolza problem can be solved numerically by means of the sequential gradient-restoration algorithm for optimal control problems (Miele 1975, 1980; Miele and Wang 1986a, 1986b; Miele et al. 1974, 1970).

The most unfavorable abort landing condition occurs if the wind shear encounter takes place at maximum or near-maximum landing weight on a hot summer day. For these conditions, several hundred optimal trajectories were computed for three Boeing aircraft (B-727, B-737, and B-747) and various combinations of wind shear intensity and initial altitude. From the computer runs, certain general conclusions became apparent:

1. The optimal trajectory includes three branches: a descending flight branch, followed by a nearly horizontal flight branch, followed by an ascending flight branch after the aircraft has passed through the shear region.
2. Along an optimal trajectory, the point of minimum velocity is reached at the end of the shear.
3. The peak altitude drop depends on the wind shear intensity and initial altitude; it increases as the wind shear intensity and initial altitude increase.
4. Weak-to-moderate wind shears and relatively severe windshears are survivable employing an optimized flight strategy; however, extremely severe wind shears are not survivable, even employing an optimized flight strategy.

For a particular case, that of the Boeing B-727-200 aircraft (landing weight $W = 150{,}000$ lb, flap deflection $\delta_F = 30$ deg, ambient temperature 100°F, initial altitude $h_0 = 600$ ft), Fig. 18.5 shows the altitude profile $h(t)$ of the optimal trajectory for several values of the horizontal wind velocity difference, namely, $\Delta W_x = 100$, 120, and 140 fps. Note that, as ΔW_x increases, the minimum altitude of the optimal trajectory decreases, but the aircraft survives the wind shear encounter. However, should the horizontal wind velocity difference be further increased to $\Delta W_x = 190$ fps, the B-727 would crash, even flying an optimal trajectory.

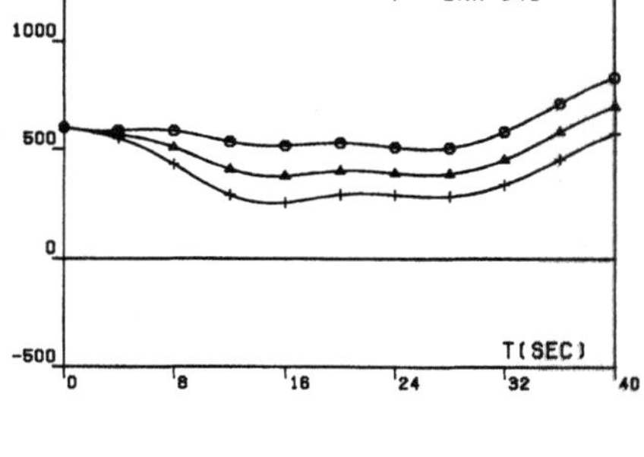

FIGURE 18.5. Optimal abort landing trajectories: altitude h versus time t for several values of ΔW_x.

Guidance Trajectories

Based on the idea of preserving the properties of the optimal trajectories, two guidance schemes were developed at Rice University: acceleration guidance (AG), based on the relative acceleration, and gamma guidance (GG), based on the absolute path inclination.

Acceleration Guidance

Basic to acceleration guidance are the laws

$$\text{(R1)} \qquad \dot{V}/g = 0, \tag{19a}$$

$$\text{(R2)} \qquad \dot{V}/g + C_2 F = 0, \tag{19b}$$

$$\text{(R3)} \qquad V - C_3 V_0 = 0, \tag{19c}$$

which are derived from the study of the optimal trajectories. In Eqs. 19a, b, and c, (R1) denotes the portion of the shear region where descending flight occurs; (R2) denotes the portion of the shear region where level flight occurs; and (R3) denotes the aftershear region; C_2 and C_3 are constants; and F is the shear/downdraft factor, defined by Eq. 12.

The guidance laws (Eqs. 19) are implemented via the feedback control forms

$$\text{(R1)} \qquad \alpha - \tilde{\alpha}(V) = K_1(\dot{V}/g), \tag{20a}$$

$$\text{(R2)} \qquad \alpha - \tilde{\alpha}(V) = K_2(\dot{V}/g + C_2 F), \tag{20b}$$

$$\text{(R3)} \qquad \alpha - \tilde{\alpha}(V) = K_3(V - C_3 V_0). \tag{20c}$$

In Eqs. 20a, b, and c, K_1, K_2, K_3 are the gain coefficients; α is the guidance angle of attack; and $\tilde{\alpha}(V)$ is the nominal angle of attack (this is the angle of attack associated with static equilibrium in the direction normal to the flight path in the absence of shear and downdraft).

Gamma Guidance

Basic to the gamma guidance are the laws

$$\text{(R1)} \qquad \gamma_e - (A_1 + B_1 h_0/h_* + C_1 F) = 0, \tag{21a}$$

$$\text{(R2)} \qquad \gamma_e - \gamma_{e\tau}(1 - C_2 F) = 0, \tag{21b}$$

$$\text{(R3)} \qquad \gamma_e - C_3 \gamma_{e\tau} = 0, \tag{21c}$$

which are derived from the study of the optimal trajectories. In Eqs. 21a, b, and c, A_1, B_1, C_1, C_2, C_3 are constants and F is the shear/downdraft factor.

The guidance laws (Eqs. 21) are implemented via the feedback control forms

$$\text{(R1)} \qquad \alpha - \tilde{\alpha}(V) = -K_1[\gamma_e - (A_1 + B_1 h_0/h_* + C_1 F)] \tag{22a}$$

$$\text{(R2)} \qquad \alpha - \tilde{\alpha}(V) = -K_2[\gamma_e - \gamma_{e\tau}(1 - C_1 F)], \tag{22b}$$

$$\text{(R3)} \qquad \alpha - \tilde{\alpha}(V) = -K_3[\gamma_e - C_3 \gamma_{e\tau}]. \tag{22c}$$

In Eqs. 22a, b, and c, K_1, K_2, K_3 are the gain coefficients; α is the guidance angle of attack; and $\tilde{\alpha}(V)$ is the nominal angle of attack.

Switch Conditions

In both the acceleration guidance and the gamma guidance, there are two switch conditions. The first switch condition is based on h_0, F and regulates the transition from region (R1) to region (R2). The second switch condition is based on the velocity change and regulates the transition from region (R2) to region (R3).

Trajectory Comparison

For a particular case, that of the Boeing B-727-200 aircraft (landing weight $W = 150{,}000$ lb, flap deflection $\delta_F = 30$ deg, ambient temperature 100°F, initial altitude $h_0 = 600$ ft, horizontal wind velocity difference $\Delta W_x = 120$ fps), Fig. 18.6 shows the altitude profile $h(t)$ of the optimal trajectory (OT) as well as the altitude profiles of the trajectories resulting from the acceleration guidance (AG), the constant-pitch guidance (CPG), and the maximum-angle-of-attack guidance (MAAG). While the MAAG trajectory crashes, the CPG trajectory survives, albeit with a minimum altitude about half that of the optimal trajectory. On the other hand, the AG trajectory is close to the OT. An analogous remark holds for the GG trajectory, which is not shown, since it is nearly identical with the AG trajectory.

Survival Capability

Here, we analyze the survival capability of an aircraft in a severe wind shear. Indicative of this survival capability is the wind shear/downdraft combination that results in the minimum altitude being equal to the ground altitude.

To analyze this important problem, we recall the one-parameter family of wind shear models (Eqs. 8), in which the parameter λ characterizes the intensity of the wind shear/downdraft combination. By increasing the value of λ, more intense wind shear/downdraft combinations are generated until a critical

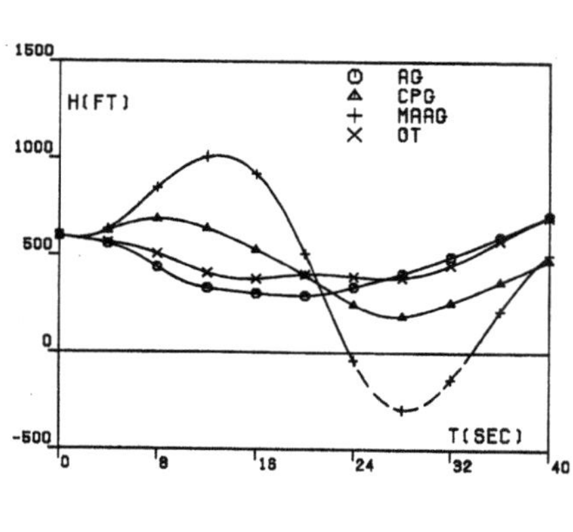

FIGURE 18.6. Comparison of abort landing trajectories: altitude h versus time t for $\Delta W_x = 120$ fps.

TABLE 18.2. Abort landing, survival capability ΔW_{xc} and wind shear efficiency ratio WER (B-727).

Trajectory	h_0 (ft)	ΔW_{xc} (fps)	WER
OT	600	187.1	1.000
AG	600	179.1	0.957
GG	600	184.2	0.985
CPG	600	139.4	0.745
MAAG	600	81.7	0.437

Landing weight = 150,000 lb; transition distance = 4000 ft.

value λ_c (hence, a critical value ΔW_{xc}) is found, such that $h_{min} = 0$ for a given trajectory type.

The results are shown in Table 18.2, which supplies the critical wind velocity difference ΔW_{xc} for the optimal trajectory and various guidance trajectories. Table 18.2 also shows the wind shear efficiency ratio WER, defined by Eq. 16. Clearly, if the wind shear efficiency of the OT is defined to be 100 percent, that of the AG trajectory is 96 percent, that of the GG trajectory is 98 percent, that of the CPG trajectory is 75 percent, and that of the MAAG trajectory is 44 percent.

Current and Future Studies

Several new projects are currently underway at Rice University with reference to the following problem areas: identification of the wind along a given trajectory; real-time wind identification; early detection of a wind shear encounter; advanced wind shear recovery systems; and advanced wind shear control systems.

Wind Identification

These studies have been motivated chiefly by the 1985 crash of Delta Flight 191 at Dallas–Fort Worth International Airport. For previous studies, see Bach and Parks (1987), Bach and Wingrove (1989), Fujita (1986), Gorney (1987), NTSB (1985), and Wingrove and Bach (1987).

Suppose that a particular flight trajectory is known, in the sense that some of the components of the state and/or the control are known from flight measurements. The wind identification problem consists of determining the wind components along the given flight trajectory in such a way that an appropriate performance index is minimized. The performance index measures the deviation of the experimentally determined flight trajectory from the computed flight trajectory.

There are two basic approaches to the wind identification problem: the kinematic approach and the dynamic approach. Both approaches yield consistent results.

In the kinematic approach, one uses only available information about the position, velocity, and acceleration of the aircraft. Clearly, this is a direct method. The research is focusing on ways and means for improving the precision of the identification procedure, above all in cases where a smaller number of sensors is available.

In the dynamic approach, one uses available information on the state and control components, and hence on the forces acting on the aircraft (thrust, drag, lift, and weight). Clearly, this is an indirect method, in which both the kinematic equations and the dynamic equations are employed as differential constraints. The research is focusing on ways and means for improving the precision of the identification procedure, above all in cases where a smaller number of sensors is available.

Real-Time Wind Identification

The approach of the previous section is suitable to the analysis of aircraft accidents, but is not suitable for real-time wind shear detection. This is because the analysis of aircraft accidents exploits the total time history of a flight trajectory, while real-time wind shear detection exploits only the previous time history. For real-time wind identification, the research is focusing on techniques of the predictor-corrector type, in the following sense: (1) previous information is employed in order to predict the wind components at a particular time instant, and (2) present information is used to correct the errors associated with (1).

Advanced Wind Shear Warning System

There is a direct correlation between early wind shear detection and capability of the aircraft to survive a wind shear encounter. Therefore, the development of an advanced wind shear warning system requires the use of real-time identification techniques; such system must be characterized by small computational time, coupled with limited memory requirements.

In the real world, an element of complication is the presence of turbulence, which confuses the clarity of the instantaneous shear detection process and creates the possibility of false wind shear warnings. To prevent this danger, research is focusing on the use of previous wind shear measurements in addition to instantaneous wind shear measurements; see the previous section.

A characteristic of microburst-associated wind shears is that unfavorable shears are both preceded and followed by favorable shears: transition from maximum head wind to maximum tail wind is preceded by transition from nearly-zero wind to maximum head wind and is followed bv transition back from maximum tail wind to nearly-zero wind. It is felt that, by memorizing previous wind shear measurement signals, an increasing head wind can be detected and the pilot can be warned in time.

Advanced Wind Shear Recovery System

Based on the optimization, guidance, and identification studies already performed, it appears that the development of an advanced wind shear recovery system is not only possible, but feasible in the relatively near future. The research is focusing on the incorporation of four basic properties: completeness, continuation, near-optimality, and simplicity. These properties are now explained:

1. The system should be able to function in takeoff, abort landing, and penetration landing.
2. The system should cover a variety of situations, ranging from zero wind shear to moderate wind shear to strong-to-severe wind shear; the switch from no–wind shear operation to wind shear operation should be smooth.
3. The system should be constructed so as to supply a good approximation to the properties of the optimal trajectories.
4. The system should be as simple as possible and should emphasize the use of existing instrumentation, whenever possible.

With reference to Property 2, note that any adverse wind gradient (inner core of a downburst) is both preceded and followed by a favorable wind gradient. The advanced wind shear recovery system must not only react in a near-optimal way to adverse wind gradients, but must exploit to the best advantage of the aircraft favorable wind gradients. This means that, in an increasing head wind scenario, kinetic energy must be increased; conversely, in a decreasing tail wind scenario, potential energy must be increased. Clearly, this requires that not only the current wind shear signals be measured, but that previous wind shear signals be recorded and memorized, such that favorable wind gradients can be detected and used.

To sum up, it is felt that an advanced wind shear recovery system, endowed with the properties described herein, should improve considerably the survival capability of aircraft in a severe wind shear.

Advanced Wind Shear Control System

This system results from the combination of the warning function and the recovery function of the previous sections into a single system.

Conclusions

Over the past five years, considerable research has been performed at Rice University on two aspects of the wind shear problem: determination of optimal trajectories and development of near-optimal guidance schemes. It now appears that, in the relatively near future, an advanced wind shear control system can be developed, capable of functioning in different wind models and of covering the entire spectrum of flight conditions, including takeoff, abort landing, and penetration landing.

The advanced wind shear control system must have the properties of completeness, continuation, near-optimality, and simplicity. It must be such that the warning function and the recovery function are incorporated into a single system. Concerning the warning function, the system must be characterized by noise resistance and early detection. Concerning the recovery function, the system must not only be capable of reacting in a near-optimal way to adverse wind gradients, but must exploit to the best advantage of the aircraft favorable wind gradients.

Acknowledgment

This research was supported by NASA Langley Research Center, by Boeing Commercial Airplane Company, by Air Line Pilots Association, and by Texas Advanced Technology Program.

Nomenclature

D = drag force, lb.
g = acceleration of gravity, ft sec^{-2}.
h = altitude, ft.
L = lift force, lb.
m = mass, lb ft^{-1} sec^2.
S = reference surface area, ft^2.
t = running time, sec.
T = thrust force, lb.
V = relative velocity, ft sec^{-1}.
$W = mg$ = weight, lb.
W_h = h-component of wind velocity, ft sec^{-1}.
W_x = x-component of wind velocity, ft sec^{-1}.
x = horizontal distance, ft.
α = angle of attack (wing), rad.
β = engine power setting.
γ = relative path inclination, rad.
γ_e = absolute path inclination, rad.
δ = thrust inclination, rad.
δ_F = flap deflection, rad.
θ = pitch attitude angle (wing), rad.
λ = wind intensity parameter.
τ = final time, sec.

References

Bach, R.E. Jr., and Parks, E.K., 1987, "Angle of Attack Estimation, for Analysis of Windshear Encounters," *J. of Aircraft, 24*, 11, 789–792.

Bach, R.E. Jr., and Wingrove, R.C., 1989, "Analysis of Windshear from Airline Flight Data," *J. of Aircraft, 26*, 2, 103–109.

Boeing Airliners, 1985, "Flight Path Control in Windshear," *Boeing Airliners*, Jan.–March, 1–12.

Bowles, R.L., and Targ, R., 1988, "Windshear Detection and Avoidance: Airborne Systems Perspective," *Paper No. ICAS-88-112*, 16th Congress of the International Council of the Aeronautical Sciences, Jerusalem, Israel.

Bray, R.S., 1986, "Aircraft Performance and Control in Downburst Windshear," *Paper No. SAE-86-1698*, SAE Aerospace Technology Conference and Exposition, Long Beach, Calif.

Chu, P.Y., and Bryson, A.E. Jr., 1987, "Control of Aircraft Landing Approach in Windshear," *Paper No. AIAA-87-0632*, AIAA 25th Aerospace Sciences Meeting, Reno, Nev.

Federal Aviation Administration, 1987, *Windshear Training Aid*, 1, 2, Federal Aviation Administration, Washington, D.C.

Franklin, G.F., Powell, J.D., and Emani-Naeini, A., 1986, *Feedback Control of Dynamic Systems*, Addison-Wesley Publishing Co., Reading, Mass.

Frost, W., 1983, "Flight in a Low Level Windshear," *Contractor Report No. 3678*, NASA.

Frost, W., and Bowles, R.L., 1984, "Windshear Terms in the Equations of Aircraft Motion," *J. of Aircraft, 21*, 11, 866–872.

Fujita, T.T., 1985, *The Downburst*, Department of Geophysical Sciences, University of Chicago, Chicago, Ill.

Fujita, T.T., 1986, *DFW Microburst*, Department of Geophysical Sciences, University of Chicago, Chicago, Ill.

Gorney, J.L., 1987, "An Analysis of the Delta 191 Windshear Accident," *Paper No. AIAA-87-0626*, AIAA 25th Aerospace Sciences Meeting, Reno, Nev.

Ivan, M., 1986, "A Ring-Vortex Downburst Model for Flight Simulation," *J. of Aircraft, 23*, 3, 232–236.

Miele, A., 1975, "Recent Advances in Gradient Algorithms for Optimal Control Problems," *J. of Optimization Theory and Applications, 17*, 5/6, 361–430.

Miele, A., 1980, "Gradient Algorithms for the Optimization of Dynamic Systems," *Control and Dynamic Systems, Advances in Theory and Application*, C.T. Leondes (Ed.), Academic Press, New York, 16, 1–52.

Miele, A., 1990, "Optimal Trajectories of Aircraft and Spacecraft," *Aircraft Trajectories: Computation, Prediction, and Control*, A. Benoit (Ed.), AGARDograph No. AG-301, AGARD/NATO, Paris, France, 201–256.

Miele, A., and Wang, T., 1986, "Primal-Dual Properties of Sequential Gradient-Restoration Algorithms for Optimal Control Problems, Part 1, Basic Problem," *Integral Methods in Science and Engineering*, F.R. Payne et al. (Eds.), Hemisphere Publishing Corp., Washington, D.C., 577–607.

Miele, A., and Wang, T., 1986, "Primal-Dual Properties of Sequential Gradient-Restoration Algorithms for Optimal Control Problems, Part 2, General Problem," *J. of Mathematical Analysis and Applications, 119*, 1/2, 21–54.

Miele, A., Damoulakis, J.N., Cloutier, J.R., and Tietze, J.L., 1974, "Sequential Gradient-Restoration Algorithm for Optimal Control Problems with Nondifferential Constraints," *J. of Optimization Theory and Applications, 13*, 2, 218–255.

Miele, A., Pritchard, R.E., and Damoulakis, J.N., 1970, "Sequential Gradient-Restoration Algorithm for Optimal Conrol Problems," *J. of Optimization Theory and Applications, 5*, 4, 235–282.

Miele, A., Wang, T., and Melvin, W.W., 1986, "Guidance Strategies for Near-Optimum Take-Off Performance in a Windshear," *J. of Optimization Theory and Applications, 50*, 1, 1–47.

Miele, A., Wang, T., and Melvin, W.W., 1986, "Optimal Take-Off Trajectories in the Presence of Windshear," *J. of Optimization Theory and Applications, 49*, 1, 1–45.

Miele, A., Wang, T., and Melvin, W.W., 1987, "Optimization and Acceleration Guidance of Flight Trajectories in a Windshear," *J. of Guidance, Control, and Dynamics, 10*, 4, 368–377.

Miele, A., Wang, T., and Melvin, W.W., 1987, "Quasi-Steady Flight to Quasi-Steady Flight Transition in a Windshear: Trajectory Optimization and Guidance," *J. of Optimization Theory and Applications, 54*, 2, 203–240.

Miele, A., Wang, T., and Melvin, W.W., 1988, "Quasi-Steady Flight to Quasi-Steady Flight Transition for Abort Landing in a Windshear: Trajectory Optimization and Guidance," *J. of Optimization Theory and Applications, 58*, 2, 165–207.

Miele, A., Wang, T., and Melvin, W.W., 1989, "Penetration Landing Guidance Trajectories in the Presence of Windshear," *J. of Guidance, Control, and Dynamics, 12*, 6, 806–814.

Miele, A., Wang, T., Melvin, W.W., and Bowles, R.L., 1987, "Maximum Survival Capability of an Aircraft in a Severe Windshear," *J. of Optimization Theory and Applications, 53*, 2, 181–217.

Miele, A., Wang, T., Melvin, W.W., and Bowles, R.L., 1988, "Gamma Guidance Schemes for Flight in a Windshear," *J. of Guidance, Control, and Dynamics, 11*, 4, 320–327.

Miele, A., Wang, T. Melvin, W.W., and Bowles, R.L., 1989, "Acceleration, Gamma, and Theta Guidance for Abort Landing in a Windshear," *J. of Guidance, Control, and Dynamics, 12*, 6, 815–821.

Miele, A., Wang, T., Tzeng, C.Y., and Melvin, W.W., 1987, "Optimal Abort Landing Trajectories in the Presence of Windshear," *J. of Optimization Theory and Applications, 55*, 2, 165–202.

Miele, A., Wang, T., Tzeng, C.Y., and Melvin, W.W., 1989, "Abort Landing Guidance Trajectories in the Presence of Windshear," *J. of the Franklin Institute, 326*, 2, 185–220.

Miele, A., Wang, T., Wang, H., and Melvin, W.W., 1988, "Optimal Penetration Landing Trajectories in the Presence of Windshear," *J. of Optimization Theory and Applications, 57*, 1, 1–40.

National Academy Press, 1983, "Low Altitude Windshear and Its Hazard to Aviation," National Academy Press, Washington, D.C.

National Transportation Safety Board, 1983, "Aircraft Accident Report: Pan American World Airways, Clipper 759, Boeing 727-235, N4737, New Orleans International Airport, Kenner, Louisiana, July 9, 1982," *Report No. NTSB-AAR-8302*, National Transportation Safety Board, Washington, D.C.

National Transportation Safety Board, 1986, "Aircraft Accident Report: Delta Air

Lines, Lockheed L-1011-3851, N726DA, Dallas–Fort Worth International Airport, Texas, August 2, 1985," *Report No. NTSB-AAR-8605*, National Transportation Safety Board, Washington, D.C.

Psiaki, M.L., and Stengel, R.F., 1986, "Optimal Flight Paths through Microburst Wind Profiles," *J. of Aircraft*, *23*, 8, 629–635.

Wilson, J.W., et al., 1984, "Microburst Wind Structure and Evaluation of Doppler Radar for Airport Windshear Detection," *NCAR Report*, National Center for Atmospheric Research, Boulder, Colo.

Wingrove, R.C., and Bach, R.E. Jr., 1987, "Severe Winds in the DFW Microburst

MIX
Papier aus verantwortungsvollen Quellen
Paper from responsible sources
FSC® C105338

If you have any concerns about our products,
you can contact us on
ProductSafety@springernature.com

In case Publisher is established outside the EU,
the EU authorized representative is:
Springer Nature Customer Service Center GmbH
Europaplatz 3, 69115 Heidelberg, Germany

Printed by Libri Plureos GmbH
in Hamburg, Germany